TUNNELS AND UNDERGROUND CITIES: ENGINEERING AND
INNOVATION MEET ARCHAEOLOGY, ARCHITECTURE AND ART

PROCEEDINGS OF THE WTC2019 ITA-AITES WORLD TUNNEL CONGRESS, NAPLES, ITALY, 3-9 MAY, 2019

Tunnels and Underground Cities: Engineering and Innovation meet Archaeology, Architecture and Art

Volume 4: Ground improvement in underground constructions

Editors

Daniele Peila
Politecnico di Torino, Italy

Giulia Viggiani
University of Cambridge, UK
Università di Roma "Tor Vergata", Italy

Tarcisio Celestino
University of Sao Paulo, Brasil

CRC Press
Taylor & Francis Group
Boca Raton London New York

CRC Press is an imprint of the
Taylor & Francis Group, an **informa** business

A BALKEMA BOOK

Cover illustration:

View of Naples gulf

CRC Press/Balkema is an imprint of the Taylor & Francis Group, an informa business

© 2020 Taylor & Francis Group, London, UK

Typeset by Integra Software Services Pvt. Ltd., Pondicherry, India

Published by: CRC Press/Balkema
      Schipholweg 107C, 2316XC Leiden, The Netherlands
      e-mail: Pub.NL@taylorandfrancis.com
      www.crcpress.com – www.taylorandfrancis.com

ISBN: 978-0-367-46868-2 (Hbk)
ISBN: 978-1-003-03162-8 (eBook)

Table of contents

Tunnels and Underground Cities: Engineering and Innovation meet Archaeology, Architecture and Art, Volume 4: Ground improvement in underground constructions – Peila, Viggiani & Celestino (Eds)
© 2020 Taylor & Francis Group, London, ISBN 978-0-367-46868-2

Preface

The World Tunnel Congress 2019 and the 45th General Assembly of the International Tunnelling and Underground Space Association (ITA), will be held in Naples, Italy next May.

The Italian Tunnelling Society is honored and proud to host this outstanding event of the international tunnelling community.

Hopefully hundreds of experts, engineers, architects, geologists, consultants, contractors, designers, clients, suppliers, manufacturers will come and meet together in Naples to share knowledge, experience and business, enjoying the atmosphere of culture, technology and good living of this historic city, full of marvelous natural, artistic and historical treasures together with new innovative and high standard underground infrastructures.

The city of Naples was the inspirational venue of this conference, starting from the title Tunnels and Underground cities: engineering and innovation meet Archaeology, Architecture and Art.

Naples is a cradle of underground works with an extended network of Greek and Roman tunnels and underground cavities dated to the fourth century BC, but also a vibrant and innovative city boasting a modern and efficient underground transit system, whose stations represent one of the most interesting Italian experiments on the permanent insertion of contemporary artwork in the urban context.

All this has inspired and deeply enriched the scientific contributions received from authors coming from over 50 different countries.

We have entrusted the WTC2019 proceedings to an editorial board of 3 professors skilled in the field of tunneling, engineering, geotechnics and geomechanics of soil and rocks, well known at international level. They have relied on a Scientific Committee made up of 11 Topic Coordinators and more than 100 national and international experts: they have reviewed more than 1.000 abstracts and 750 papers, to end up with the publication of about 670 papers, inserted in this WTC2019 proceedings.

According to the Scientific Board statement we believe these proceedings can be a valuable text in the development of the art and science of engineering and construction of underground works even with reference to the subject matters "Archaeology, Architecture and Art" proposed by the innovative title of the congress, which have "contaminated" and enriched many proceedings' papers.

Andrea Pigorini
SIG President

Renato Casale
Chairman of the Organizing Committee WTC2019

Tunnels and Underground Cities: Engineering and Innovation meet Archaeology,
Architecture and Art, Volume 4: Ground improvement in
underground constructions – Peila, Viggiani & Celestino (Eds)
© 2020 Taylor & Francis Group, London, ISBN 978-0-367-46868-2

Acknowledgements

REVIEWERS

The Editors wish to express their gratitude to the eleven Topic Coordinators: Lorenzo Brino, Giovanna Cassani, Alessandra De Cesaris, Pietro Jarre, Donato Ludovici, Vittorio Manassero, Matthias Neuenschwander, Moreno Pescara, Enrico Maria Pizzarotti, Tatiana Rotonda, Alessandra Sciotti and all the Scientific Committee members for their effort and valuable time.

SPONSORS

The WTC2019 Organizing Committee and the Editors wish to express their gratitude to the congress sponsors for their help and support.

Tunnels and Underground Cities: Engineering and Innovation meet Archaeology,
Architecture and Art, Volume 4: Ground improvement in
underground constructions – Peila, Viggiani & Celestino (Eds)
© 2020 Taylor & Francis Group, London, ISBN 978-0-367-46868-2

WTC 2019 Congress Organization

HONORARY ADVISORY PANEL

Pietro Lunardi, President WTC2001 Milan
Sebastiano Pelizza, ITA Past President 1996-1998
Bruno Pigorini, President WTC1986 Florence

INTERNATIONAL STEERING COMMITTEE

Giuseppe Lunardi, Italy (Coordinator)
Tarcisio Celestino, Brazil (ITA President)
Soren Eskesen, Denmark (ITA Past President)
Alexandre Gomes, Chile (ITA Vice President)
Ruth Haug, Norway (ITA Vice President)
Eric Leca, France (ITA Vice President)
Jenny Yan, China (ITA Vice President)
Felix Amberg, Switzerland
Lars Barbendererder, Germany
Arnold Dix, Australia
Randall Essex, USA
Pekka Nieminen, Finland
Dr Ooi Teik Aun, Malaysia
Chung-Sik Yoo, Korea
Davorin Kolic, Croatia
Olivier Vion, France
Miguel Fernandez-Bollo, Spain (AETOS)
Yann Leblais, France (AFTES)
Johan Mignon, Belgium (ABTUS)
Xavier Roulet, Switzerland (STS)
Joao Bilé Serra, Portugal (CPT)
Martin Bosshard, Switzerland
Luzi R. Gruber, Switzerland

EXECUTIVE COMMITTEE

Renato Casale (Organizing Committee President)
Andrea Pigorini, (SIG President)
Olivier Vion (ITA Executive Director)
Francesco Bellone
Anna Bortolussi
Massimiliano Bringiotti
Ignazio Carbone
Antonello De Risi
Anna Forciniti
Giuseppe M. Gaspari

Giuseppe Lunardi
Daniele Martinelli
Giuseppe Molisso
Daniele Peila
Enrico Maria Pizzarotti
Marco Ranieri

ORGANIZING COMMITTEE

Enrico Luigi Arini
Joseph Attias
Margherita Bellone
Claude Berenguier
Filippo Bonasso
Massimo Concilia
Matteo d'Aloja
Enrico Dal Negro
Gianluca Dati
Giovanni Giacomin
Aniello A. Giamundo
Mario Giovanni Lampiano
Pompeo Levanto
Mario Lodigiani
Maurizio Marchionni
Davide Mardegan
Paolo Mazzalai
Gian Luca Menchini
Alessandro Micheli
Cesare Salvadori
Stelvio Santarelli
Andrea Sciotti
Alberto Selleri
Patrizio Torta
Daniele Vanni

SCIENTIFIC COMMITTEE

Daniele Peila, Italy (Chair)
Giulia Viggiani, Italy (Chair)
Tarcisio Celestino, Brazil (Chair)
Lorenzo Brino, Italy
Giovanna Cassani, Italy
Alessandra De Cesaris, Italy
Pietro Jarre, Italy
Donato Ludovici, Italy
Vittorio Manassero, Italy
Matthias Neuenschwander, Switzerland
Moreno Pescara, Italy
Enrico Maria Pizzarotti, Italy
Tatiana Rotonda, Italy
Alessandra Sciotti, Italy
Han Admiraal, The Netherlands
Luisa Alfieri, Italy
Georgios Anagnostou, Switzerland

Andre Assis, Brazil
Stefano Aversa, Italy
Jonathan Baber, USA
Monica Barbero, Italy
Carlo Bardani, Italy
Mikhail Belenkiy, Russia
Paolo Berry, Italy
Adam Bezuijen, Belgium
Nhu Bilgin, Turkey
Emilio Bilotta, Italy
Nikolai Bobylev, United Kingdom
Romano Borchiellini, Italy
Martin Bosshard, Switzerland
Francesca Bozzano, Italy
Wout Broere, The Netherlands
Domenico Calcaterra, Italy
Carlo Callari, Italy

Luigi Callisto, Italy
Elena Chiriotti, France
Massimo Coli, Italy
Franco Cucchi, Italy
Paolo Cucino, Italy
Stefano De Caro, Italy
Bart De Pauw, Belgium
Michel Deffayet, France
Nicola Della Valle, Spain
Riccardo Dell'Osso, Italy
Claudio Di Prisco, Italy
Arnold Dix, Australia
Amanda Elioff, USA
Carolina Ercolani, Italy
Adriano Fava, Italy
Sebastiano Foti, Italy
Piergiuseppe Froldi, Italy
Brian Fulcher, USA
Stefano Fuoco, Italy
Robert Galler, Austria
Piergiorgio Grasso, Italy
Alessandro Graziani, Italy
Lamberto Griffini, Italy
Eivind Grov, Norway
Zhu Hehua, China
Georgios Kalamaras, Italy
Jurij Karlovsek, Australia
Donald Lamont, United Kingdom
Albino Lembo Fazio, Italy
Roland Leucker, Germany
Stefano Lo Russo, Italy
Sindre Log, USA
Robert Mair, United Kingdom
Alessandro Mandolini, Italy
Francesco Marchese, Italy
Paul Marinos, Greece
Daniele Martinelli, Italy
Antonello Martino, Italy
Alberto Meda, Italy

Davide Merlini, Switzerland
Alessandro Micheli, Italy
Salvatore Miliziano, Italy
Mike Mooney, USA
Alberto Morino, Italy
Martin Muncke, Austria
Nasri Munfah, USA
Bjørn Nilsen, Norway
Fabio Oliva, Italy
Anna Osello, Italy
Alessandro Pagliaroli, Italy
Mario Patrucco, Italy
Francesco Peduto, Italy
Giorgio Piaggio, Chile
Giovanni Plizzari, Italy
Sebastiano Rampello, Italy
Jan Rohed, Norway
Jamal Rostami, USA
Henry Russell, USA
Giampiero Russo, Italy
Gabriele Scarascia Mugnozza, Italy
Claudio Scavia, Italy
Ken Schotte, Belgium
Gerard Seingre, Switzerland
Alberto Selleri, Italy
Anna Siemińska Lewandowska, Poland
Achille Sorlini, Italy
Ray Sterling, USA
Markus Thewes, Germany
Jean-François Thimus, Belgium
Paolo Tommasi, Italy
Daniele Vanni, Italy
Francesco Venza, Italy
Luca Verrucci, Italy
Mario Virano, Italy
Harald Wagner, Thailand
Bai Yun, China
Jian Zhao, Australia
Raffaele Zurlo, Italy

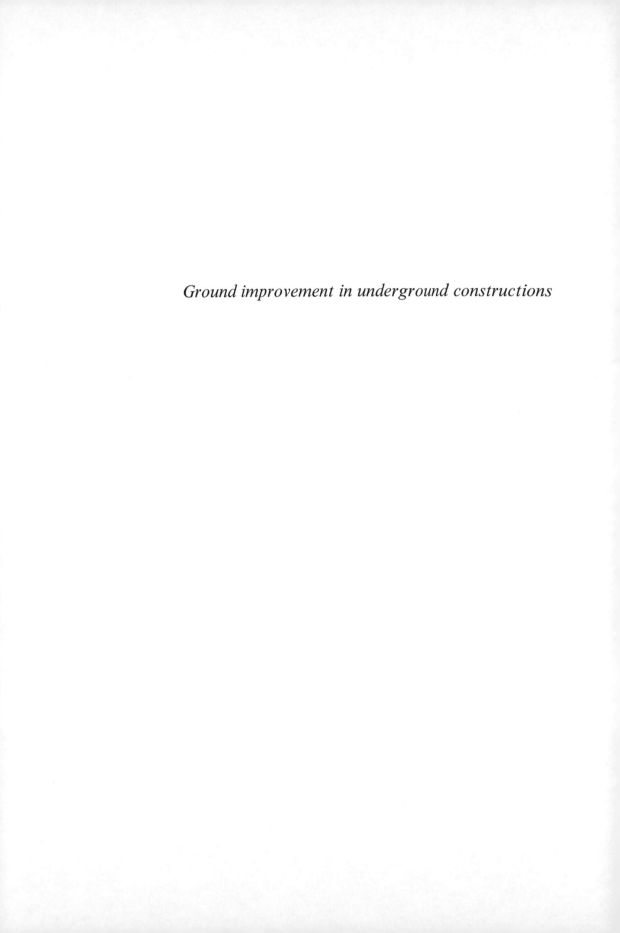

Ground improvement in underground constructions

Tunnels and Underground Cities: Engineering and Innovation meet Archaeology, Architecture and Art, Volume 4: Ground improvement in underground constructions – Peila, Viggiani & Celestino (Eds)
© 2020 Taylor & Francis Group, London, ISBN 978-0-367-46868-2

Permeation grouting and consolidation works to solve flooding problems in an old tunnel

A. Antonelli
MM S.p.A.

A. Balossi Restelli & E. Rovetto
Studio ingegneria Balossi Restelli e Associati

ABSTRACT: The raising water table in the Milan area affects some of the Metro lines constructed in the past without waterproofing. The problem should be faced by acting on the Metro lines in operation before water inflow could compromise the regular Metro service. The paper describes the two phases intervention studied for subway line M2 between the Piola and Lambrate stations. The first phase, was aimed at stopping water inflow and to repairing the invert strongly compromised. The second phase concerns water tightness and long-term stability of tunnel, through the creation of a shell of consolidated ground. Grouting operations will be executed both from inside the tunnel and from the surface, the last one requiring particular attention for the protection of the tunnel lining. Consequently a dedicated real-time monitoring system has been studied to avoid any damages to the lining in order to guarantee the metro service in safety.

1 THE AQUIFER AND ITS EVOLUTION IN MILAN AREA

1.1 *Hydrogeological characteristics*

The municipal territory of Milan occupies an area of sq. Km. 182, in central position with respect to the two important tributaries of the Po river: the "Ticino" at West and the "Adda" at East. It develops in an area with a very low slope of 2–3‰ with altitudes in m a.s.l. between 143–145 in the North and 102–103 in the South.

A thick alluvial and fluvio-glacial deposit characterizes the subsoil up to about 40 m from the ground level, with gravel and pebbles with particle size from medium to coarse sometimes intercalated with lenses of silts and clays, variable in size and depth. The Wurmian "gravelly-sandy" sediments outcrop on the surface. They are characterized by the proximity of the main water courses Olona, Seveso and Lambro whose contribution is represented by a continuous coarse sedimentation. At an average depth of 40 m there is instead a clayey level of some meters, which separates the most superficial coarse horizon, from the sandy one below and figures out diverse ways of water circulation of groundwater.

A dense network characterizes surface hydrography, only partly of natural origin (Lambro and Olona rivers, the Navigli and various irrigation ditches) which ensures drainage of the superficial waters and, at the same time, supplies the underground aquifer.

The superficial alluvial deposit have high permeability (about 5E-04 m/s), which decreases both with the depth and with the transition from North to South.

The first aquifer consisting of coarse deposits of the first 40 meters contains free groundwater; the name "second aquifer" identifies the sandy-gravelly succession with frequent clayey diaphragms between 40 and 110 m, almost always containing groundwater in pressure. Between the free groundwater circulation of the first aquifer and the water in pressure of the "second aquifer" there could be local intercommunications. The metropolitan lines of the city are in the range from the ground surface to about 35 m depth.

1.2 *Historical data and sources of water table levels*

The first data structured on the characteristics of the water table in Milan date back to 1889 with the construction of the first wells of the Milan aqueduct; to this day, the values of water depth in each of the 37 Drinking Water Plants are measured and recorded.

Since the 1950 two additional monitoring networks have been developed: - the first by MM, the Municipal institution dedicated since 1955 to metro lines implementation, along the directions of the metro lines that achieved n. 110 piezometers; - the second due to the sewerage Municipality of Milan that, starting from the 85 points of measurement in 1954, today has approximately 30 efficient piezometers.

Since 1998 the SIF (Aquifer Information System), that manages all the data related to the levels of the aquifer in the Milan area, has been set up at the Central Environment Management of the Province of Milan.

1.3 *Historical evolution of the aquifer in Milan territory*

Figure 1 summarizes the trend of the water depth in Milan from 1915 to today.

In the progress, the following phases could be found:

From 1915 to 1950: the groundwater level remained constant and very superficial for a long time even after the construction of the first Milan Aqueduct stations (the first pumping plant dates to 1889, in 1910 there were already 10 plants with 87 wells); the first signs of lowering date back to the years 1930–1940 and are linked to the intensification of industrial activity.

From 1950 to 1970 the massive industrial and urban infrastructures development and the increasing waters withdrawal accelerates the decreasing of the aquifer at a rate of one meter per year with tips of 2.5 m causing relevant subsidence phenomena, up to 22 cm in the Duomo area.

From 1975 to 1980 the trend abruptly inverts because many industries move to the North Milan area and many private wells close.

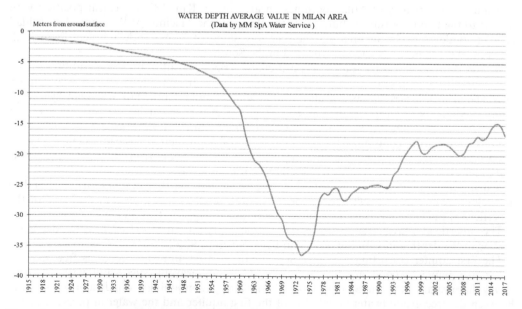

Figure 1.   Average value of water depth in Milan area from 1915 to 2017.

The following period 1980–1990 is characterized by the water table equilibrium until a new sudden increase, between 1990 and 1998, for the dismission of large plants (Pirelli, Breda, Falk, Montedison), responsible for the depression of the post-war years. Until 2007 the levels are still stable.

The levels return to increase from 2007 to 2015: a first and only partial interpretation may be due to the abandonment of several water extraction wells in the North of Milan due to the increasing concentration of contaminating agents.

2 UNDERGROUND LINES AND RELATIONSHIP WITH THE AQUIFER

2.1 Aquifer levels in the projects of underground lines

In the design practice since the beginning of its activities, MM has always referred to two conventional aquifer levels.

• The first level named "reference aquifer" that refers to the water table during the construction phase which affects the executive methodologies
• The second level named "project aquifer", the estimated "long term" level during the life cycle of the opera, that influences the sizing and the necessity of waterproofing

The choice of these levels, especially in the case of the project water table, is delicate since any long-term forecast is uncertain and full of consequences on the realization and maintenance costs of the opera.

During the design of the Line 1 developed in the 1956, they adopted the project aquifer level of the year 1958 exstimated on maximum levels registered in 1951 (year of flooding of Po valley) decreased by the average annual lowering from 1951 to 1958.

During the design of the Line 2 developed in the 1965, it was initially thought to adopt the same criterion, which was later changed based on a farsighted evaluation that was deduced from the MM archive documentation *". . ..The "aves" (as the aquifer was defined at that time) appears deeply modified with respect to the one of the past decades and is still in a rapid variation. The definition should therefore be formulated by examining all the facts that may occur in the years to come that could affect the aquifer evolution"*.

Thus, they considered the influences of many factors such as water supply, industrial water extraction, urbanization, pluviometry data, the forecast of new aqueduct plants as well as the announced decision to regulate and limit the concessions to private withdrawals. Therefore they assumed as project aquifer the maximum level recorded in 1964, increased by 3 meters as a margin of safety.

During the design of the Line 3, developed in 1979–1980, the evaluations were still more forward-looking by examining a broader hydrogeological context than the Milan area and thematic maps were elaborated that defined the groundwater flow lines and the possible aquifer evolution in the city area. The forecast of a recovery phase of the piezometric depression of the early 1970s is the reason why all the Line 3 is waterproofed.

2.2 Real aquifer evolution and effects on built lines

Compared to the hypothesis assumed in the design phases, the real evolution of the piezometric levels has produced over the years situations of potential crisis of underground lines where not adequately waterproofed. In 1997, following the first peak in the trend of ascent of the aquifer, the level of apprehension regarding the phenomenon was such as to lead to the establishment of a "Coordination Group for the water table raising in Milan Area" (participated by Region, Province, Municipality, MM, Villoresi Consortium, CAP, Authorities concerning the Po River). The purpose of the Coordination Group was to plan and implement interventions aimed at controlling the groundwater, including the feeding of surface water courses with first stratum wells, the tribute of new wells and the recovery of old ones.

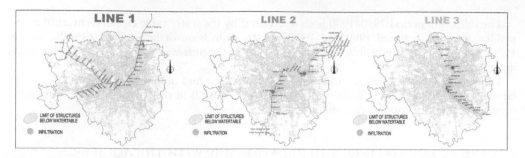

Figure 2.   Milan Metro lines: stretches below water table peak level (1999).

The plans of the metropolitan lines in Figure 2 highlight the sections below the phreatic level of those years (1999) and the areas affected by infiltration phenomena.

It is clear that the raising of the aquifer could represent a risk factor for the underground works only when it is accompanied by other items that contribute to the possibility of infiltrations such as the absence or insufficiency of the waterproofing, the deterioration of the structures and/or waterproofing membranes, the presence of constructions defects, points of weaknesses, the use of poor quality materials.

It should also be noted that in the case of metrolines, the risk of infiltration has a much greater impact than in other underground structures, due to the difficulty of maintenance with the line in operation and the excessive costs of accidental interruption of the metro service.

3   THE METRO LINE 2 CASE HISTORY

The Metro line 2 represents with its 40.4 km the most extensive metro line in the Milan area. Its realization, carried out in progressive steps, covers a wide time period, from 1969 to 2011 (see Figure 3), characterized by a very rapid evolution of the design tools, by a growing experience on the issues of aquifer fluctuation and in constructions under the water table, by significant qualitative evolution of building materials in general and of waterproofing technologies in particular.

Therefore, nowadays each part of the line in exercise shows different problems and maintenance requirements related both to the depth of the track and to the design and construction choices, peculiar of the realization period.

According to this it is not surprising that the most ancient part of the track is the one that has suffered mostly for the water table raising with infiltration problems more important where the trace gets deeper and the constructional defects or materials deterioration are more pronounced.

Focusing on the portion of the line 2 between Piola and Lambrate stations, infiltration problems related to the aquifer raising date to several years before the year 2000 and affected

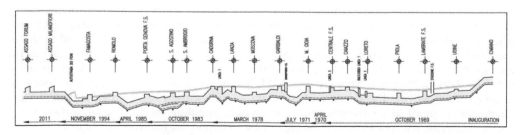

Figure 3.   Metro line 2: periods of execution.

particularly the uneven track tunnel. Already in 1999 the operator ATM adopted security measures prescribing the trains to slowdown in case of severe water inflow or service interruption in case of water level exceeding the track level in the tunnel. In 2011 ATM started a waterproofing intervention on the top surface of the inverted arch by using injections of acrylic bi-component resins. The intervention didn't get any appreciable result because of the flowing back of resins through the cracks between the inverted arch and the sides. Therefore starting from 2011, a series of investigations were carried out to check the conservation status of the lining in order to define the proper intervention.

3.1 Construction phases of tunnel's lining

The twin tunnels between Piola and Lambrate stations, extends for 670.2 m, have a distance from each other centre line of 17.60 m and a double slope of the rail profile with a minimum in the central zone at 104.607 m a.s.l.

The tunnels were not waterproofed because in the period of construction the water table was just below the excavation level.

The casting phases of the lining were the followings:
1. provisional lining consisting of steel ribs and shotcrete
2. casting of the lining of the crown arch and side walls
3. excavation and casting of the inverted arch.

The lining demonstrated to have a weak point in correspondence of the joint between the side walls and the inverted arch, where the water inflowed under pressure, when the water table raised.

3.2 Investigations

Starting from 2011, a series of investigations were carried out in the uneven track tunnel to characterize the lining, mainly the geometry and the construction materials, and to focus the weak zones of the lining that favoured the water inflow.

Archive drawings (see Figure 4) were found illustrating the geometry and materials used for the provisional and permanent lining.

Investigations were performed in 2011 by the Milan Polytechnic in order to check the thickness of the inverted arch and the strength and chemical characteristics of the concrete. Ten drillings and ten cores were performed to check the thickness of the inverted arch and 25 compressive strength tests were carried out, which provided an average resistance of 26.7 MPa. A georadar survey (GPR) along two alignments checked the thickness of the inverted arch on both the right and left side that resulted as follows: in 77% of the sections the thickness reduction was more than 25%, in 20% of the sections the thickness reduction was less than 25% and in 3% of the sections no reduction was observed. The costructions defects,

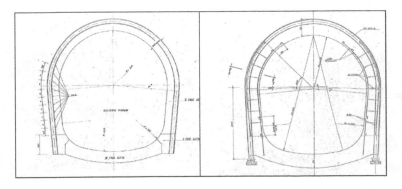

Figure 4.   Metro lines 2 tunnel lining sections: cast sequence and reinforcement bars.

detected in the invert concrete, could be related to water pumping during casting operation to lower the water table.

Moreover, in different areas, a detachment was observed between the filling concrete and the inverted arch through the water infiltrated.

3D Laser scanner and Thermographic survey conducted in 2016 by the company SINE-CO showed the deterioration of the lining (cracks, water inflow) affecting large portions of the tunnel.

4 FIRST PHASE: URGENT WORKS

Before carrying out the consolidation grouting it was necessary to execute urgent works for the safety of the tunnel and for the railway traffic security, aimed mainly at reducing the water inflow into the tunnel (see Figure 5).

In 2011, when the consolidation design was done, the flow was around 200 m^3/h and at the construction phase (2015) it had increased to 600 m^3/ h (the flow value refers to the entire length of the tunnel equal to 670,2 m).

The interventions carried out in this first phase were the followings:

- Execution of self drilling anchors bars on the sides and on the inverted arch. The anchors used were ARCO R 32x5,2 Termic CE tipe, F_{tk} 450 kN, F_{yk} 380 kN and working load 300 kN, 6 m long. The large water inflow during the drilling phase have initially made the cementation of the bars difficult, so that it was necessary to insert a conical rubber gasket at the top of the holes having the purpose of containing the water and cement mixture inflow. Once this technological measure was adopted, the cementation of the bars was more regular and the quantity of the cement mix injected into the self-drilling bars reduced from about 300 l/bar to 190 l/bar.
- Execution of cement grouting with a mix with cement/water ratio equal 2, through single-valve pipes. The project foresaw 3 linked pipes of different lengths, the first one to fill the obturator bag in order to prevent water inflow, the second to inject the cement mixture and the third for the possible injection of the silica mixture to reduce the setting time of the mixture (Joosten method). The project was however modified reducing the drilling diameter from 100 mm to 50 mm in order to reduce water inflow and using only one single-valve pipe for the injection of the cement mixture and subsequently for the injection of the silicate through the same pipe. The sealing of the hole was guaranteed by a rubber conical seal. This variant, used in order to realize smaller diameter drillings, nevertheless has entailed greater difficulties in controlling the setting time of the cement mixture which often flowed through the open cracks between the inverted arch and the sides. In the deepest zones of the tunnel it was consequently necessary to prepare metal formworks fixed to the tunnel lining, in correspondence of the joints between the inverted arch and sides, which allowed to slow down the flow velocity inside the tunnel thus favoring the setting of the cement.
- Realization of connection bars between the inverted arch and the filling by means of anchor bars R32x6 standard, about 1m length.

Figure 5.   Metro line 2 first phase works: before, during and after the intervention.

5 SECOND PHASE: WORKS FOR LONG TERM WATERPROOFING AND STABILITY

During the design stage of the two phases intervention, in 2011, was decided to ensure the long-term stability and waterproofing of the tunnel lining by creating a second inverted arch of consolidated ground just below the lining, realized by grouting injections executed from inside the tunnel with traffic interruption.

However during the first phase intervention execution, in 2015, the water table raised above the centre level of the lining (see Figure 6) so it was decided to extend the second phase consolidation all around the lining, to create a grouted continuous shell.

5.1 Numerical analysis

In order to evaluate the effectiveness of the intervention from both hydraulic and structural point of view, a finite difference analysis was performed with the numerical modeling software FLAC. The intervention on the existing structure was classified as a static improvement according to NTC 2008 so that the Knowledge Levels (KL) of the various items involved in the calculation model (geometry, construction details and materials) were identified, and the related Confidence Factors (CF) to be used in the analysis were defined. A limited Knowledge Level KL1 has been identified corresponding to a Confidence Factor equal to 1.35.

The materials used for the provisional and permanent lining and the respective resistance features are the followings:

- NP160 steel ribs 0,6 m spaced: Aq42, yield and tensile strength respectively $f_y = 230$ MPa $f_u = 420$ MPa;
- Ø30 mm 15 cm spaced smooth rebar: Aq50, yield and tensile strength respectively $f_y = 270$ MPa, $f_u = 500$ MPa;
- Concrete: $R_{ck} = 22.5$ MPa.

The static and filtration analysis were carried out considering 3 different boundary conditions, namely: 1) actual condition after the first phase consolidations works (self-drilling bars), and water table at 108 m a.s.l.; 2) long-term situation after first and second phase consolidations and water table at 113 m a.s.l.; 3) long-term situation in the hypothesis of executing only first-phase consolidations works and water table at 113 m a.s.l. Figure 7 shows the stress values in the self-drilling bars which, for the three cases described above, respectively correspond to 10 kN, 88 kN and 70 kN from which the maximum design force resulting equal to $P_d = 88 \cdot 1.3 = 114$ kN according to the design approach 2 - A1+M1+R3 of NTC 2008.

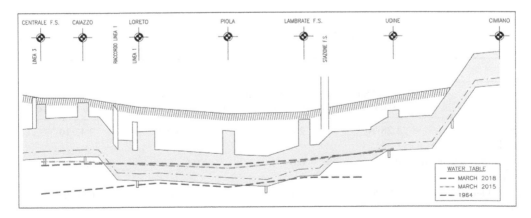

Figure 6. Metro line 2: water table levels at the construction stage, during the design of repairing intervention and today.

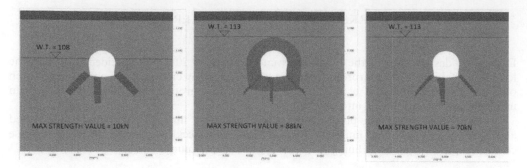

Figure 7. Strength in the bar anchors in the 3 different conditions considered in the numerical analysis performed.

The design force P_d was then compared with the anchors' total design tensile geotechnical resistance that resulted equal to 157 kN and with the structural resistance of the steel bar equal to 336 kN.

Flexural and shear resistance analysis were done in the long term with water table at 113 m a.s.l. in the hypothesis of consolidation of the ground and collaboration between inverted arch and filling.

The filtration analysis allowed to evaluate parametrically the inflow increase that could have occurred if no consolidation sealing grouting were carried out around the tunnel. In the numerical model it has been reasonably assumed that most of the water entered through the joint between inverted arch and sides.

In the analysis the following coefficient of permeability values were assumed: - natural soil mainly sandy silty gravel k=2E-05 m/sec; - consolidated ring, 2m thick, grouted with cement and chemical mix k=5E-07 m/sec; external consolidate ring, 1m thick, grouted with cement only k=5E-06 m/sec.

Table 1 shows the estimated flow in the different situations examined that are: 1) Actual water table level at 108 m a.s.l. before consolidation and waterproofing grouting; 2) Actual water table level at 108 m a.s.l. and consolidation grouting; 3) Long term water table level at 113 m a.s.l. and consolidation grouting; 4) Long term water table level at 113 m a.s.l. and no consolidation grouting.

The effectiveness of the grouted area around the existing tunnel is evident due to the reduction of soil permeability (k=5E-7 m/s). The inflow is reduced from 371 m³/h to 3 m³/h with the water table at 108 a.s.l. to then increase to 9 m³/h in the case of water table raising, up to 113 m a.s.l.

Table 1. Estimation of water inflow into the tunnel for different water table levels and ground consolidation conditions.

	Phase 1	Phase 2	Phase 3	Phase 4
	w.t. 108 m a.s.l. no cons. grouting	w.t. 108 m a.s.l. with cons. grouting	w.t. 113 m a.s.l. with cons. grouting	w.t. 113 m a.s.l. no cons. grouting
Flow rate [m/s]	$1.28 \cdot 10^{-4}$	$20 \cdot 10^{-6}$	$3.23 \cdot 10^{-6}$	$3.47 \cdot 10^{-4}$
Filtration section area [m²]	0.6	0.6	0.6	0.6
Tunnel length [m]	670.20	670.20	670.20	670.20
Total water inflow [m³/h]	371	3	9	1005

6 GROUND CONSOLIDATION AND WATERPROOFING

6.1 *Operating procedures for the shell execution and effects on line excercise*

Once defined the efficient shell section we had to evaluate how to execute the grouting injections.

The reinforcing and waterproofing intervention completely made from inside of the tunnel would require:

- in the hypothesis of only night-time processing, the duration of the works can be estimated as 4/5 years, intervening every night with two drills for about 4 hours useful: in this case indeed it's necessary to ensure the removal of all site interferences before the resumption of trains service with adequate safety margins. It is obvious that this solution is not workable due to the urgency of the intervention;
- in case of service interruption, a duration of about 4 months, counting 3 shifts per day, 7 days per week, with 4 advancement fronts. Such a prolonged closure of the service resulted from the estimates of the line operator that is incompatible with the needs of the users;

The traffic, in normal operating conditions on the route analyzed, is about 22,000 passengers/hour/direction with trains every 180 seconds. The organization of a substitute surface transport, considering an 18 minutes bus, would need 6–7 buses for each train.

Such a bus number, even if possible for the ATM operator, it would require operating on a sufficiently long route, assuming the closure of the section between the stations of "Centrale" and "Gobba" with discomfort for the users. Therefore, the interruption of the service can be foreseeable only in the period of August when the number of passengers drops to about 5500 passengers/hour/direction with trains every 360 seconds and the surface replacement service runs more easily on a reduced stretch. Figure 8 shows the routes of the substitute vehicles in the hypothesis of closing the line between the stations of "Udine" and "Loreto" in August: during this period the daily passengers of the route vary from a minimum of about 60,000 in the central week to a maximum of 110,000 in the first week.

Therefore, it was chosen a third mixed solution with execution of the injections part from the surface and part from inside the tunnel with service interruption limited to the period of August.

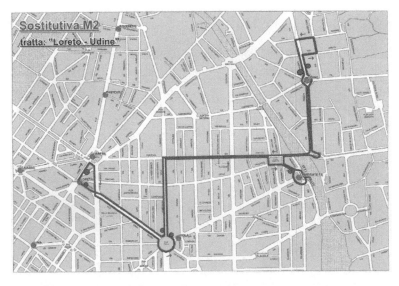

Figure 8.   Routes of substitute vehicles in the hypothesis of closing the line between "Udine" and "Loreto" stations.

6.2 Ground consolidation execution

Figure 9 shows the typical sections of consolidation grouting executed from the road and from inside the tunnel. Drilling sections are normally spaced 1.60 m from each other and for each section drilled from the road, 3 holes will be made in correspondence of the left side of the tunnel (tree side), 3 from above the crown arch and 3 in correspondence of the right side (houses side).

Before drilling the holes on the right side, due to the presence of underground pipelines close to the sidewalk, it will be necessary to perform a pre-excavation about 1.40 m deep, in order to locate the gas and the water pipelines.

Inside the excavation a PVC guide pipes ($\emptyset$ = 140 mm) will be placed, at a distance at least 25–30 cm from the pipelines. The guide pipes will be held in position fixing the top and bottom sides with a template, to ensure the correct inclination.

After filling the excavation and compacting the soil, drillings for the installation of valved pipes (TAM pipes) will be executed inside the guide pipes.

Drillings in proximity of the gas pipes will be preferably executed from mid-April to mid-September not to interfere with the period of domestic heating.

Drillings from inside the tunnel, for the consolidation of the inverted arch and in the stretch under the sewer main pipe at the crossing with Teodosio st. (see Figure 10), are planned in the month of August with total suspension of the service, and will be performed under the water table with an hydrostatic pressure of about 0.2–0.3 bar.

Preliminary to the grouting, the most of residual water inflow will be temporarily blocked by injection of high expansion polyurethane foam resin in order to limit any possible mix inflow into the tunnel that could occur during the injection of cement mixtures carried out from outside.

Drillings will therefore be carried out with the aid of the preventer device for the management of water inflow; once the drilling has been completed, the manchettes pipes will be inserted and the annular space between pipe and drilled hole filled with a plastic mix (fill grout).

The injections will be performed according to the "controlled volume" system, i.e. by injecting a preset amount of mixture into each valve based on the volume of soil to be treated in the surrounding area and the grain size of the soil.

Cement mixture stabilized with bentonite will be generally injected in 2 stages, while for the silicate mixture 1 stage only is foreseen.

The "preset allowable grout pressure" is preset at maximum value of 13÷15 bars in order to preserve both existing buildings close to the consolidating area and the tunnel lining. For the same reason the injection flow rate will also be limited to a maximum flow of 400 Lt/h.

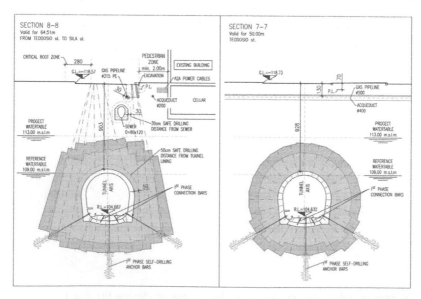

Figure 9.   Typical sections of consolidation grouting performed from the road and from inside the tunnel.

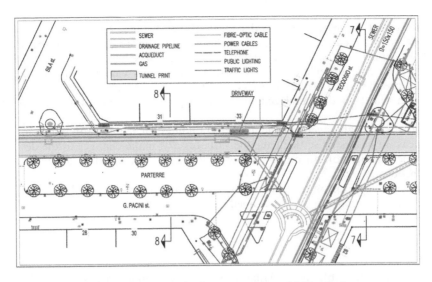

Figure 10.   Sewer main pipe, gas and water pipelines at crossing with Teodosio street.

7   ACTIVITIES PLANNING

The activities have been planned taking into account the following constraints: - the viability entering in Milan through Pacini street must always be guaranteed, providing the partial use of the road with a lane 3 m wide as minimum. The activities from the road will then be carried out in two separate phases, respectively with the closure of the left half of the road (trees side) for the consolidation of the left side and of the crown arch of the tunnel, and with the closure of the right half of the road (side houses) for the consolidation of the right side of tunnel; - works close to the gas pipelines, in particular as regards drilling, will be scheduled from mid-April to mid-September not to interfere with the period of buildings' heating. Injections can be scheduled outside this period also; - works from inside the tunnel will be carried out only in August months of two consecutive years with interruption of the trains service and use of substitutive surface vehicles. In these two months drillings and injections from the street will be suspended avoiding to create overlaps and interferences between surface's and tunnel's works; - the organization of the work was conceived to complete first the consolidation from above (crown arch and sides), at least in the area characterized by greater hydraulic head and water inflow, secondly the injections from inside the tunnel for the consolidation and waterproofing of the inverted arch.

This sequence has been identified with the purpose of exploiting the consolidation on the sides, already performed, as a lateral shielding to contain and limit any dispersion and wash-out of the mixture during the injections of the inverted arch, executed from inside the tunnel, and so improving the efficiency and performance of the consolidation itself.

8   MONITORING SYSTEM

During the works from the road with trains in service, the tunnel will be monitored by topographic convergence readings automatically performed by a total station and by means of the Railway Deformation System of Sisgeo (RDS) and topographic readings of the micro-prisms installed on RDS-T gauges. Night inspections will be organized with frequency to be defined according to the progress of the works to check the lining.

As the trains will be in service during works from the road, a control system has been designed to prevent any damage to the tunnel lining, with particular regard to possible piercing of the lining by the drilling rods. Therefore each drilling rig will be equipped with an alarm so that when reaching a preset depth, defined preliminarily for each hole, the drilling will be interrupted automatically in order to check the distance of the drilling rod from the tunnel's lining. In the case the distance from the lining is more than 30cm, the drilling may proceed, vice versa it will be interrupted and the hole re-drilled.

Moreover, a video surveillance system with fixed thermal cameras will be set up in the tunnel to ensure the safety of the train service. The images taken by the cameras of the monitoring system will be video analyzed in order to promptly report any event of breaking through of the lining.

Level staffs will be arranged on Pacini street buildings for manual topographic readings in order to measure heaves or settlements and, if necessary, make changes to the design specifications. Topographic monitoring of the sewerage is also envisaged to evaluate any heaves caused by grouting.

Environmental monitoring for noise measurement and for chemical release into the atmosphere and in the watertable is also envisaged.

9 CONCLUSIONS

The raise of the water table in the Milan area is creating many problems for the structures and infrastructures constructed between the 60s and 80s, when the water table reached the lowest historical levels and it was not expected to rise again in the long period. In particular, in M1–M2 and M3 subway lines there are some stretches that are currently below the water table that had not been waterproofed at the time of construction.

The left tunnel between Piola and Lambrate station is the one that suffered most of the problems both because of the high water level and because of criticalities detected in the structure of the inverted arch.

In addition to the technical problems related directly to the consolidation intervention, including the protection of the tunnel, buildings and underground pipelines, as well as of the railway service, the planning of the interventions took into account the effects that the works executed from inside the tunnel would have had on the interruption of the service and consequently on the road viability. The works have been planned in order to minimize the impact on the road system, by planning the interruptions of the train service only in the months of August of two consecutive years.

The numerical analysis allowed to estimate the reduction of the inflow due to the soil waterproofing and to evaluate the stresses in the long-term conditions, after the consolidation grouting around the tunnel, that were found to be compatible with the resources of the structure.

ACKNOWLEDGEMENTS

We wish to thank for their active collaboration Eng. L. Marraudino and Eng. A. Romano of ATM S.p.A. company who gave us documents regarding the I stage of works and metro line 2 traffic data.

We are also grateful to MM S,p.A. Water Service for the documentation regarding the water table.

We also would like to thank Mr. G. P. Mandruzzato of MM S.p.A. Engineering for the help given in the research of historical documents about aquifer of archive drawings regarding the construction phase.

REFERENCES

Balossi Restelli, A., Rovetto, E. & Sesini, A. 1999. L'imprevedibilità dell'innalzamento della falda nell'ambito della progettazione di una galleria metropolitana a Milano. *XX Convegno Nazionale di Geotecnica, Parma, 22–25 September 1999*: 317–325.

Conta, R & Mandruzzato, G.P. 2014. Falda acquifera e opera della metropolitana milanese. *Il Giornale dell'Ingegnere* (7).

*Tunnels and Underground Cities: Engineering and Innovation meet Archaeology,
Architecture and Art, Volume 4: Ground improvement in
underground constructions – Peila, Viggiani & Celestino (Eds)
© 2020 Taylor & Francis Group, London, ISBN 978-0-367-46868-2*

Dam rehabilitation for safety operation

O. Arghiroiu
Lecturer, University of Oradea, Oradea, Bihor County, Romania

S. Călinescu
Vice-President, Romanian Tunneling Association, Bucharest, Romania

ABSTRACT: Lesu dam is a rockfill dam, having a reinforced concrete mask in upstream, 61 m in height and it performs an accumulation of water for more than 30 million cubic meter. The water infiltration, during exploitation, reached today over than acceptable limits. So, it was decided, according to a Technical Expertise, to have refurbishment works of the dam, as follows: a) waterproofing of the upstream concrete mask; b) strengthening grouting performed in the underground gallery; c) strengthening grouting from the ground surface; c) dam foundation drainage system, after the underground grouting; d) underground grouting gallery rehabilitation; e) monitoring and warning alarm system rehabilitation; f) surface rehabilitation of the dam g) protection of riverbanks riverbed downstream of the dam; h) equipment rehabilitation.

1 INTRODUCTION

1.1 General information

Lesu Dam and storage reservoir is located in North – West region of Romania, Bihor County, in the Crisuri Hydrographical Basin River, placed on the Iadului Valley, a left hand side tributary to Crisul Repede River, about 70-km distance from Oradea City (Figure 1).

The dam is managed by "Apele Române" National Administration, specifically by Oradea Water Basin River of Crisuri Administration and is operated within Bihor Water Management System by the Crisul Repede Hydrographical System, Lesu operation and maintenance team.

It was built between 1969 and 1973. It is a rockfill dam, having a reinforced concrete mask in upstream, 61 m in height and it performs of water accumulation for more than 30 million cubic meter. It functions are: a) downstream water shortages for water supply of villages and Oradea City; b) production of electricity; c) flood control; d) recreational lake. According to Romanian norms, it has the "special importance class" and necessity for special tracking the behavior.

1.2 Problems occurred

After the first water filling of the storage reservoir in 1973, great seepages were recorded and this fact called for draining the storage reservoir for investigations and restoration works. The storage reservoir was drained repeatedly and the restoration works were carried out at the facing and at the connection between the cut-off and the bed rock.. Nevertheless, 1000 l/s of seepage through the dam were exceeded. After the last campaign of the restoration works (1992 ÷ 1994), the measured seepage discharges remained within the admissible limits. After 1999, the discharges started to increase again, being at present around the value of 300 l/s,

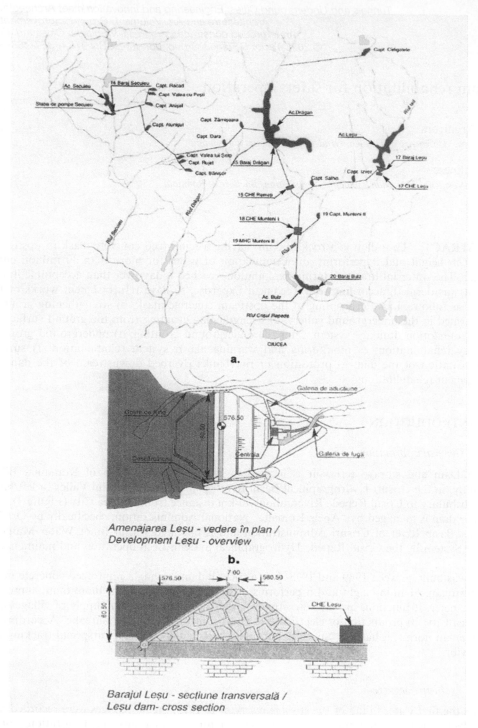

Figure 1.   Lesu Dam location.

under the condition of keeping the low water level in the storage reservoir around 555.00 mASL (Above Sea Level), in comparison with NRL = 574.50 mASL.

The role of Lesu Dam is a storage reservoir to supply with drinking and industrial water Oradea City and the downstream villages and towns, to generate electrical power through

Lesu Hydropower Plant (3.4 MW power at a discharge of 8 m³/s) and flood wave attenuation. The design discharge is 1% ($Q_{1\%}$ = 170 m³/s) annual exceeding probability and a check discharge value of 0.1% ($Q_{0.1\%}$ = 294 m³/s).

2 LOCATION BASIC DATA

2.1 *Climate, precipitation and wind regime*

Thermal condition of the moderate continental temperate type, characteristic to Bihor County, shows great diurnal, monthly, seasonal and annual fluctuation, they being determined by the circulation and frequency of air masses, by the presence of the main relief stages, orientation of the main mountain and hilly peaks, as well as by the orientation of the major valleys, abutment inclination and so on. The mean annual temperatures can vary between 2°C ÷ 7°C in the high mountain area, 7°C ÷ 8°C in the low massifs and 10°C ÷ 11°C in the hilly area.

The precipitation regime shows a quantitative increase with altitude, thus, in the plain area, the mean multiannual precipitation range between 500 ÷ 700 mm, in the hilly area they are between 700 ÷ 1000 mm, in the lower massif area they are between 1000 ÷ 2000 mm, while in the high zone these precipitation exceed 1400 mm.

Wind regime is determined by the nature, succession and frequency of the pressure system. At the level of the Bihor County, the highest frequency is recorded of the wind blowing from the Southern sector (about 28% in December and 17% in August)

2.2 *Geological and geotechnical data*

Both the dam and the storage reservoir are totally on an igneous massif, at the altitude of 800 m. The massif is characterized by heavy, massive forms and sedimentary lands, made up especially their lower part by limestone, it is marked by an exocarst relief and by structural relief which renders evident mostly the fault system in steps. The structure has a wide range of rocks, whose products were attributed of the alpine subsequent magmatism. The rock shows sometimes intense self-metamorphism, marked by albitization and plagioclase epidote and chlorotization of the formic elements. Rhyolites are the most frequent rock which occur in the location.

2.3 *Seismic conditions*

According to the Seismic Romanian Norms, Lesu Dam site lies in the micro-seismic zone I = 6 on the MSK scale. The seismic design zone is characterized by: the peak ground acceleration reaches to a_g = 0.8 m/s². For the main recurrence interval IMR = 100 years and the control period of response spectrum is T_c = 0.7 s.

2.4 *Dam description*

The flowing through and checking discharges at the high water outlet structure, the levels in the storage reservoir by taking into view a backup volume (at 2 m under the spillway crest) are:

- Design discharge: $Q_{1\%}$ = 170 m³/s,
- Attenuated design discharge: $Q^{attenuated}_{1\%}$ = 91 m³/s,
- Checking discharge: $Q_{0.1\%}$ = 294 m³/s,
- Attenuated checking discharge: $Q^{attenuated}_{0.1\%}$ = 216 m³/s.

Lesu Dam is rockfill dam, with portlandite semi-crystallized homogenous and compacted limestone. It has the following characteristics (Figure 2):

- Maximum dam height: 60.50 m,
- Crest elevation in axis: 580.50 mASL,
- Crest width: 7.00 m,
- Crest length: 181.00 m
- Spillway crest elevation: 576.50 mASL,
- Water intake invert elevation: 539.00 mASL,
- Bottom outlet invert elevation: 522.05 mASL,
- Foundation elevation: 520.00 mASL,
- Inclination of slopes:
  - Upstream: 1:1.3,
  - Downstream: 1:1.2; 1:1.3; 1:1.4 (two 2 m wide berms).
- Dam waterproofing: reinforced concrete face with variable thickness (0.3 m at crest to 0.8 m at the lower part), with M35 plastic strip and bituminous mastic between slabs, resting on a plain concrete cu-off along upstream dam foot, where the grouting and drainage are located. The gallery is located only on the right side of the dam, to allow re-grouting of the grout curtain.

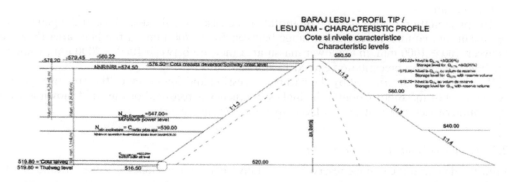

Figure 2.   Lesu dam cross-section.

For waterproofing the dam a grout curtain was achieved on a single row, in three stages, as follows:

- On the left bank, from a groin, at interaxis equidistance of 10 m/5m/2.5 m;
- On the right bank, grouting from the grouting gallery, performed in the concrete cut off, at interaxis equidistance of 5 m/2.5 m/1.25 m.

The grouting work was performed during the carrying out of the dam and immediately after the first filling up of the storage reservoir, on the areas with great seepages. Subsequently, re-grouting works carried out in two other campaigns, between 1978 ÷ 1979 and 1992 ÷ 1993. Due to the special geological structure of the right abutment it was decided that, even from the project phase, i.e. execution, during the storage reservoir operation, depending on the seepage evolution, to resume the grouting works.

On the left abutment and in the bed, the IIIrd stage (at 2.5 m) the grouting was generally rock – cut off connecting grouting. On the right abutment however, the resulting absorptions and high cement consumptions rendered evident open fissures in the rhyolites mass, in parallel to the valley that is why the grouting in the grout curtain was doubled with a row of 10 m long drillings in the rock at 2 m upstream of the main row.

The upstream cut off connects the face with the country rock. Fixing in the rock is variable (between 2 ÷ 7 m). The excavations expanded sideways since execution led to filling up the supplementary excavated volume with rockfill, injected with mortar in a repair campaign. The cut off is provided with an inside gallery which can be inspected for carrying out grouting works and for tracking the drainage system behind the grout curtain. The upstream reinforced concrete face has a variable thickness, between 30 and 80 cm and it is performed in slabs between which a waterproofing strip out of plastic material and bituminous mastic was provided for filling the developed joints. The Lesu concrete dam face, designed at the end of the 60's, accordance with the knowledge and possibilities of that time, underwent settlements and impairments of the joint waterproofing, which called for repairs for reducing the seepage discharges.

The two repairs concentrated on filling the hollows under the face and sand and cement mortar injection, on the one hand, and on the repair of the joint waterproofing, by filling with bituminous mastic and plugging (with wood) and covering them with rubber band stuck to the facing, on the other hand. The 1979 repair was concentrated on the left hand side of the facing (bottom of the valley – left abutment), while the 1992 one, on the right hand side of the facing. The repairs lasted a few years, the seepage discharge decreased immediately after repairs, each time, but than they returned to inadmissible high values.

2.5 *Works to be carried out as a preliminary step*

In order to provide work "in dry" conditions, during the period when the grouting works are carried out for consolidating/concreting the upstream floor, the water which remains stagnant between the cofferdam used at achieving the dam (between the bottom outlet intake) and dam will be discharged by pumping-out.

It is considered that for emptying this volume (estimated at $12,000 \text{ m}^3$) pumping equipment are required with a total discharged of about $200 \text{ l/s} = 720 \text{ m}^3/\text{h}$, which may draw out this volume in a period of about 17 h. The water which may be infiltrated through this cofferdam towards the downstream foot of the dam will be discharged by means of the same pumps throughout the empty storage reservoir period.

The volume of water coming from seepages through the cofferdam, estimated at about 10 % of the corresponding volume $Q_m = 2.2 \text{ m}^3/\text{s}$, will be discharged by pumping throughout the empty storage reservoir period. The volume of $Q_m = 2.2 \text{ m}^3/\text{s}$, will be discharged by pumping throughout the empty storage reservoir period. The volume of water drained in 365 days – 30 days (storage reservoir emptying) = 335 days will be of $6,367,680 \text{ m}^3$ and it could be taken out by means of a pump with $Q_{total} = 200 \text{ l/s} = 720 \text{ m}^3/\text{h}$ in 8,884 h. The total pump operating hours is of about 8,900 h, with a consumption E of about 356 MWh.

It will be taken into consideration also to install a pump during the construction period in the grouting gallery sump for discharging the water used at the drillings and grouting works in the gallery, as well as the seepage waters in the gallery, in the foundations, during the carrying out of the works of concreting the bridge and of possible expanding the present sump.

2.6 *Waterproofing upstream dam face*

Analysis of the solutions of the past, as well as of the seepage evolution, leads to the need of a radical intervention in this rehabilitation. The solution which is proposed in the project is the one of totally covering the reinforced concrete face with geomembrane well fixed on the marginal outline of the upstream dam foot (on the concrete cut-off on the right side of the dam and groin, on the left side) and on the upstream face meridians as far as level $Q_{1\%}$. For installing the geomembrane the place has to be totally cleared.

The waters remaining in the area between the bottom outlet invert elevation, the cofferdam in front of the bottom outlet and the dam will be discharged into the bottom outlet, by means of the dewatering throughout the works performed for laying the geomembrane and of the grouting works in the grout curtain (carried out from the gallery) and connection-consolidation grouting (performed from the surface, from a concrete board, upstream of the cut-off).

During the storage reservoir emptying and re-filling period, the discharge infiltrated in the dam and in the grouting gallery will be closely monitored to establish, when the geomembrane has been installed, its efficiency at the partial and final acceptance of the work.

During the emptying of the storage reservoir frequent readings will be made in the downstream measuring houses after laying the geomembrane and installing the flow meters in the grouting gallery, the continuously measured discharges will be pursued at the 4 flow meters installed in the gallery, throughout the whole storage reservoir re-filling period.

Preparatory works at the reinforced concrete face before laying the geomembrane will be:

- Cleaning the deposits from the facing by excavation, air blowing and washing with pressure water jet, the material gathered on the cut-off will be disposed of from the cut-off and groin area and it will be transported to an area upstream of the storage reservoir site, in the golf on the right bank;
- Carrying out Georadar test on the concrete surface, in order to detect the significant voids under the facing. Significant voids are considered the ones having the $\Delta Hg \geq 25$ cm depth and $S \geq 4$ m^2 site, namely $V_{voids} > 1 \div 2$ m^3.
- If the Georadar study, which will be carried out after cleaning the face and before laying the geomembrane, will render evident significant voids under the face, these voids will be filled with fluid concrete, with maximum 16mm grain and $12 \div 15$ cm settlement, similar to the pumped concrete, by drillings performed in the face. A number of 10 voids is estimated in the Technical Project to be filled, representing $V_{voids} = 20$ m^3.
- The geomembrane is laid on a firm and perfectly flat surface (provided by the existing reinforced concrete face, after its proper treatment, as it is shower below): without great humps and without protuberances; that is why, the humps between the boards ascertained after the cleaning of the face will be treated by: cutting the sharp edges, filling the space between the boards with concrete. The Project estimated L_{joints} impaired boards = 100 m.
- All the metal parts existing on the upstream face, including the level markers, will be cut and the surface will be smoothed. The displacement measurements on the upstream face will no longer be carried out because after 35 years (setting into operation of the dam = 1974) the dam has used up all its settlements.
- Treatment of the reinforced concrete face joints and significant cracks, visible on the areas established in the project, namely the joints on the face, on $d \sim 9 \div 10$ m as to the marginal outline of the face and on the significant void areas, detected by the Georadar, by means of:
  - ➤ Cleaning the joints of the existing material,
  - ➤ Achieving a 10 cm^3 groove, on the whole length of the joints to be treated;
  - ➤ Filling the joints and cementing them with cement mortar by adding Admix - C type material.

The cracks and open joints which created voids under the face, have to be closed by means of the above technology and adapted in situ. The geo composite (geomembrane) is a polyvinyl chloride material (PVC), laminated ("welded") together with a perforated, unwoven geotextile. The 2 m wide strips of material is packed by pre-welding in 6 m strips (minus the lapping of the strips in the welding area), so that the distance between two meridians axes is of 5.70 m.

The details of the top (horizontal), meridian (vertical- inclined), lower and lateral (marginal) fixing and the details of connecting to the side walls, as well as the details of draining/ collecting and discharging the water under the geomembrane are given in the project. The PVD structure of the geomembrane is based on additives and processing elements which increase its resistance to atmospheric agents (UV, ozone, heat, ice etc.), tensile strength and to give it excellent welding characteristics. The geomembrane is resistant to perforation and tear during its handling and during serviceable life.

Laying the geomembrane begins after emptying the storage reservoir, disposal of the deposits on the concrete face, the carrying out of the georadar study, possible filling with fluid concrete of the great voids and repairs of the face surface and cementing the joins. The geomembrane is laid from the crest, without demolishing the existing parapet, the geomembrane being mounted beginning from the elevation 578.00 m ASL.

The space below the geomembrane is drained, the infiltrated waters being led by the meridian drained towards a marginal drain which discharges the flows collected by two ϕ50 drains/ pipes in the grouting gallery, in its ending zone. At laying the geomembrane on the facing, if it turns out to be necessary, a third drain will be installed for collecting the waters below the geomembrane, in the face right abutment connection area (on the side inclined towards the marginal drain). The collected discharge which has thus entered the grouting gallery is measured by a flow meter after which it is discharged into the gallery sump. The sump collects also other discharges infiltrated in the gallery (through the gallery joints and walls or from drains – 14 drains inclined towards the downstream area, 8 short, vertical drains, 3 drains in the downstream gallery wall, in the dam body); these discharges are measured separately, so that the total flow discharged by pumps is the sum of all flows entering the gallery, knowing however the extent of each flow.

2.7 *Dam foundation treatment*

The Technical Project aims at waterproofing the dam foundation in the areas where great permeability were ascertained both at the initial study works and during the carrying out of the dam and at the subsequent intervention as well (concentrated in the right abutment), by providing the following works (Figure 3):

- **The works of supplementing the existing grout curtain by:**
  - ➤ Grout curtain thickening, performed in the grouting gallery and drainage in the cut-off, on 1 row, in 4 stages, with equidistance of 15 m/5.0 m/2.50 m/1.25 m, with lengths of 50 m/30 m/25 m/20 m, drillings and grouting arranged through the grouting gallery length. First the exploration drillings are carried out by recovering cores, at about 15 m distance one from the other. Then drillings are performed without core recovery. Stage I proper (at 5 m, $h_{drilling}$ = 30 m), stage II (at 2.5 m, $h_{drilling}$ = 25 m) and stage III (at 1.25 m, $h_{drilling}$ = 20 m). On the area towards the spillway, the drillings are carried out in fan mode, securing the closing of the grout curtain in the spillway. On the horizontal area of gallery, the drillings are vertical ones, while on the gallery inclined area, the drillings have an inclination of 15° as to the vertical ones, while on the gallery inclined area, to intercept transversally the cracks and faults, according to the conclusions/recommendations of the technical -geological survey. On a whole, 110 drillings are carried out which sum up 3826 m. Drillings and curtain grouting orientation and lengths was established so that the lower limit of the Stage III ($h_{drilling}$ = 20 m) drillings should be below the upper limit which supposed to be the rhyolite – clay schists contact.

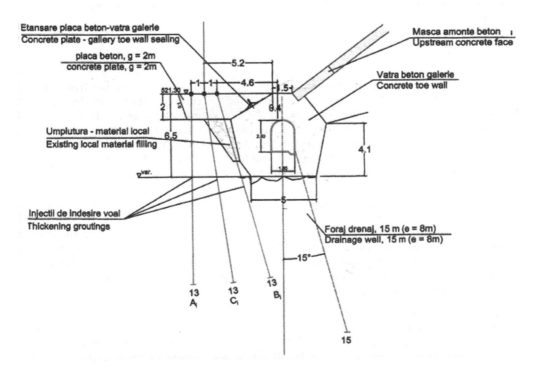

Figure 3.   Cross Section on Toe Wall and Grouting Gallery.

- ➢ Connection and consolidation grouting performed from the surface, from a concrete board adjacent to the cut-off, toward the upstream area, will be excavated, after which a 2 m and wide between 5 ÷ 7 m concrete board, 3 rows of drillings of consolidation and connection of the cut-off with dam foundation, $L_{drilling} = 15$ m (2 m in the concrete board and 13 m in the rock) will be carried out. Distance between the rows is of 1.00 m, equidistance between the drillings of one row is of 1.50 m the drillings on the three rows will be arranged in chess mode, so that the distance between two consecutive drillings, belonging to each to the other row should be of 0.75 m. The A upstream drilling row is vertical, B downstream row is inclined towards the downstream area at 8° as to the vertical line. These works can and shall be performed only with the storage reservoir empty. On a whole, 264 = 3 x 88 consolidation drillings are carried out, which sum up 3960 m.
- ➢ It is urgently mentioned that the connection and consolidation grouting was not provided in the initial project and the greatest cement consumptions were recorded in this area, in all the grouting works campaigns.

- **Tracking/monitoring works of the grout curtain efficiency:**
  - ➢ After grouting has finished in the gallery, 14 drainage drillings of 15 m length were carried out at about 8 m apart, towards downstream, inclined at 15°, along the grouting gallery length, $\phi = 76$ mm, fitted manometer and valves, cased with HDPE perforated pipe. At present there are 10 drillings, out of which 5 are operational and they are partially silted up. The 14 drains will have at the top HDPE pipes, fitted with valve, pressure manometer. The 10 drainage drillings are reinforced and are abandoned.
  - ➢ Such 8 short, vertical drains will be carried out of drilling = 6 m (2 m in concrete + 4 m in the rock), at the downstream limit, fitted as above;
  - ➢ 3 drains performed in the gallery, in the wall downstream of gallery, in the dam body, inclined at 30° as to the horizontal line, of 3 m length, up to the concrete face, to capture any possible seepage under the piezometer line, which remained stored in the space between the face and the cut-off level and the unexcavated promontory which advanced in the right abutment towards the valley.
  - ➢ Generally, drains are kept open, discharging freely the captured seepage into HDPE $D_{ext} = 110$ mm, $L = 116$ m, collecting pipe.

2.8 *Grouting works*

The grouting works are performed only on the right bank which far weaker from the geological point of view. From the geological survey of the right abutment (detailed survey of the face cut-off) it results a higher density of the breaking discontinuities, the faults having frequently openings between 30 ÷ 50 cm, being partially silted up with clayey material. The crack system has definite orientation and inclinations, reflecting the intense fragmentation of the rhyolite massive in the right abutment. In addition, the contact between sloughing shale, cretaceous intercalation and in rhyolite, are generally highly tectonic, inducing a great permeability of this area (3 ÷ 5 l/m/atm).

Generally, the major discontinuities have a great inclination nearly to the vertical line. It is recommended that their maximum interception should be made with drillings with inclinations of 15° towards the abutment. In the 1992 ÷ 1993 curtain grouting stage comparatively lower consumptions were obtained than the ones in the 1978 stage, the grouting works having been carried out only with cement suspension and they were on the contact sector.

For carrying out the connection and contact grouting, adjacent to the cut-off on its entire length the following preparatory works will be required.:

- Excavation of the filling material adjacent of the cut-off of the up stream rock and the elevation in the drawings
- Pouring of a 2 m thick and variable thickness between 5 ÷ 7 m concrete board.
- The join between the cut-off and the board will be waterproofed with sika strip.

– The connection and contact grouting will be carried out from the concrete board on 3 roads (2 m in concrete board and 13 m in the rock).

After the carrying out of the grouting works "post grouting" control drillings will be started up from the gallery.

2.9 *System of collecting and discharged waters infiltrated in the gallery*

At present, the drain assembly of the dam is made up of 10 drains cased with steel pipe of 50 mm in diameter. Only 5 are active, but partially silted up, the rest being inactive.

The water taken over by the active drains in discharged inside the gallery but from here on it reaches gravitationally in the pumping station sump. The drains proposed to be carried out have an important role of monitoring the efficiency in time of the grout curtain and measuring the discharges infiltrated in the gallery by distinctive seepage sources.

Due to the fact that Lesu dam is a rockfill dam, the drainage proper of the foundation to reduce sub-pressures is not justified, the drains which are provided aim only at pursuing the efficiency in time of the waterproofing works provided and deciding future minimum measures of intervention.

The drainage system is made out of:

– 14 drains inclined towards of 15 m length for collecting the discharged of infiltrated water to the deep grout Curtin, equipped with valves and manometer, in the area where grouting works are done.
– 8 short drains vertical ones for collecting the flows in the grout curtain contact area.
– 3 drains which will collect water from the geomembrane and discharge into a pipe and then in the sump.
– 3 drains performed from the gallery in the downstream gallery wall, in the dam body, inclined 30° as to the horizontal line of 3 m length up to the concrete face to capture any possible seepage under the piezometer line in the dam from the area in which the water remains stored upstream of the foundation promontory left unexcavated.
– Ditches on the both sides of the gallery which will gather the seepage through the gallery walls and joints, with discharges by means of pipes of the flows on the downstream ditched into the upstream ditch.

3 CONCLUSIONS

According to the existing Romanian Operating Regulation the normal operating state of the grouting gallery is "flooded gallery". The seepage is discharged through an overflow (metal pipe with discharges mainstream). In case of need, the discharge of the gallery by means of mobile pumps and keeping it to dry during access to its lower elevation.

The gallery rehabilitation works modifies this solution, by keeping the gallery permanently dry. So, it is possible to monitor separately all the discharges infiltrated in the gallery: under the geomembrane, from the drains and trough gallery joins and walls.

After measuring each of this discharges the total inflowing water is taken over by the pump systems installed in the sump and discharged through a pipe which goes up through the gallery as far as the crest elevation (in the chamber at the top of the gallery access tower) where after measuring the flow it discharges in the spillway.

REFERENCES

Aquaproiect SA Bucharest, 2003. Provisions of the Lesu Dam Feasibility Study.
Halcrow Group Ltd., 2009. Technical Report: Placing Lesu Dam under Safety Conditions.

Tunnels and Underground Cities: Engineering and Innovation meet Archaeology,
Architecture and Art, Volume 4: Ground improvement in
underground constructions – Peila, Viggiani & Celestino (Eds)
© 2020 Taylor & Francis Group, London, ISBN 978-0-367-46868-2

Horizontal directional drilling technique applications in ground improvement from low pressure injections to jet grouting

B.B. Bosco
Consultant., Pavia, Italy

S. Carraro
SOGEN srl, Padova, Italy

ABSTRACT: The paper describes some experiences about the use of HDD methodologies for consolidation and ground improvement in underground constructions. Traditional vertical or subvertical drilling sometimes become problematic for the presence of streets, sewers, underground utilities, existing buildings or for the particular shape of soil improvement. These obstacles constraints the executive choices. The possibility of using horizontal drillings, especially if directional, with the possibility to correct planimetric and altimetric model, makes free the imagination of the designer, who could experiments new applications. HDD technique makes possible to place T.A.M. ("Tube A Manchette"). The possibility to execute directional coring, with HDC (Horizontal Directional Coring), makes possible to carry out geotechnical (geognostic) surveys, also for tests in the hole: so we can have like a little pilot tunnel for geotechnical or in general drilling information. Finally the paper describes the field test, and the positive results of jet grouting consolidation performed by HDD technique.

1   INTRODUCTION

There is a great advantage in the use of sub-horizontal HDD drilling in waterproofing and/or soil consolidation work (Chirulli, 2011).

Through these controlled drillings it is possible to introduce TAM in the ground for low pressure injections or directly (soon it'll be possible) realize jet grouting.

2   LOW PRESSURE INJECTION TECHNIQUE AND JET GROUTING TECHNOLOGY

The technique of injections by low pressure permeation in loose soils, (preceded the jet grouting technique of a hundred years) allows TAM to introduce waterproofing or consolidating mixtures into the porosity of the soil.

In fact, with both cementitious and chemical mixtures it is a matter of reducing the number and the amount of the voids (even if in pitch) so as to modify the characteristics of compressive strength or permeability of loose soil (Bosco, 2013).

With low pressure injections we do not change the texture of the ground, but we fill in the voids and move the groundwater in order to transforming a loose soil into a cemented one.

Waterproofing is a function of the degree of voids filling. For this reason in loose soil, with the TAM, in a first phase the largest voids with the cement mixture are clogged, then gradually with cement but also with chemical mixtures and resins, which are the most penetrating mixtures.

By consolidation with low pressure injection, with the cement mixtures, a reduction in permeability is obtained, while the resistance achievable reaches values from 0.1 to 2.5 MPa.

The jet grouting technique (JG), born in Japan, by hydraulic/mechanic fracturing and mixing soil, introduces a cementitious and/or chemical mixture through pressures from 20 to 50 MPa, by special nozzles inserted in a monitor, placed geometrically upstream of the drilling tool, of a battery of rods (which pump in the joints the hydraulic seal) (Associazione Geotecnica Italiana, 2012).

The high pressure injection occurs with the rotation of the rods, and this pressure through the jet causes the breaking of soil and the mixing of the particles of which the soil is composed with the mixture.

The mixture consolidates and waterproofs a cylinder of soil, until it forms a column of variable diameter (depending on the applied technology, the depth, soil mechanical resistance, soil grain size).

At the same time soil resistance is a function both of the characteristics of the consolidating mixture and of the soil composition.

The resistance achievable with this technique reaches values from 2 to 25 MPa (the lower values are due to the presence of clay particles and silt or from organic materials).

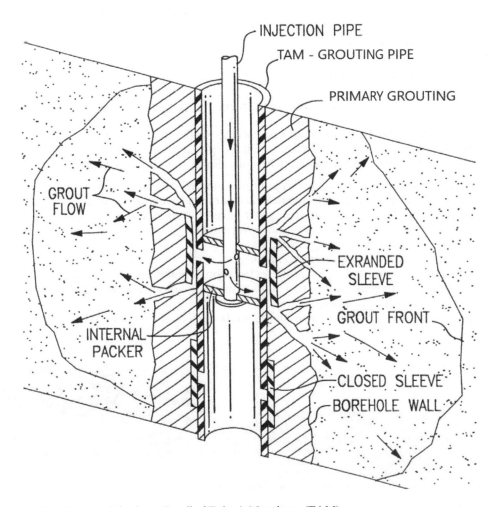

Figure 1.   Low Pressure injection – Detail of Tube A Manchette (TAM).

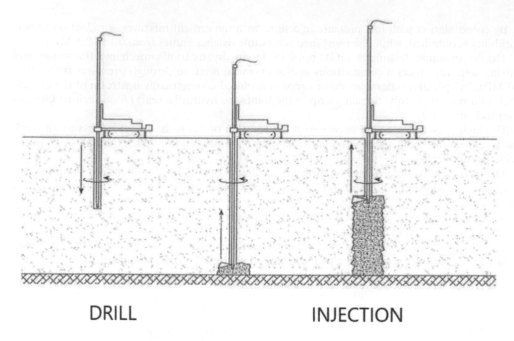

DRILL   INJECTION

Figure 2.   Jet Grouting Technology.

3   USE OF HDD PERFORATION

This technology derives directly from the drilling and consequent deviations of the oil well technique.

At the moment the HDD are used successfully, for the laying of telephone cables, electric cables, for piping of water, gas, district heating, oil pipelines, sewage systems, etc.

In recent years new technologies have been developed applied to HDD guided drilling:

- Micro-fissured pipes for drainage with the aim of stabilizing landslides;
- pipes connecting drainage wells to create deep draining networks;
- adherent pipes for the remediation of sewer pipes or other types;
- TAM, to provide low pressure permeation injections;
- TAM for compaction grouting technique;
- TAM for compensation grouting technique;
- TAM and micro-fissured pipes for the reclamation of polluted soils or landfills;
- steel tubes for freezing probes to realize temporary consolidations and waterproofing of loose ground in groundwater;
- steel pipes or structural elements positioned on the contour of the tunnel to be excavated safely.

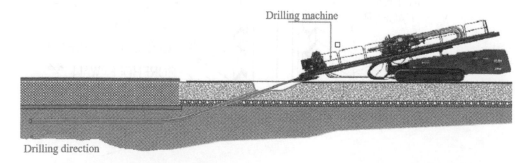

Figure 3.   HDD Technique for LP Injection - Drill.

Soon it'll be possible, according by HDD technique, associate the use of high pressure or jet grouting injections of the single or double-fluid type for the treatment of the contour of the shallow tunnel (canopy technique) to be excavated or for the excavation face treatment, directly from ground level.

In general use of HDD for jet grouting will permit a realization of bottom plugs for trench, immersed tunnels, landfills, etc.

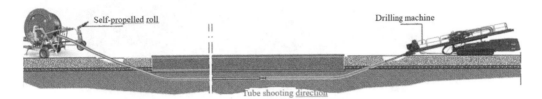

Figure 4.   HDD Technique for LP Injection – TAM placement.

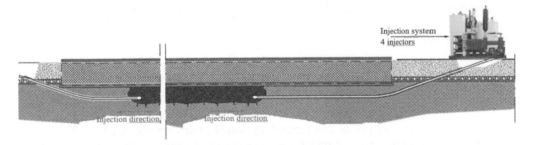

Figure 5.   HDD Technique for LP Injection – Injection.

Figure 6.   HDD Drilling machine on the entrance side.

Figure 7.   HDD on the exit side.

4   HDD AND LOW PRESSURE INJECTION BY PERMEATION. EXAMPLES

4.1   *Mestre – Metro Tunnel under railway station.*

One of the first projects involving this technology was the one commissioned by the company Ing. E. Mantovani for the crossing underground of the 12 platforms of the Mestre railway station (the double-track tunnel serves the trams that connect Mestre to Marghera, near Venice).

The project involved the waterproofing of a sufficient thickness of soil outside the line of tunnel: the r.c. precast tunnel, to be driven under railway lines, had the following dimensions: 120 meters long by 8 wide and 5 high.

After an accurate validation of the project, the difficulty of planning the jacking of over 5000 tons for a length of over 120 meters, they made us lean towards the construction of a traditional tunnel, with excavation in the middle of the tracks following the laying of sheet piles walls.

However, the preparation of this project gave us the opportunity to experiment with the technique and the materials to be used. A special test field was carried out at the workshop warehouse of the Anese company in Concordia Sagittaria (VE), to test methods and materials.

After performing a 150 meter HDD perforation with the Vermer D 220x300 equipment, a TAM was installed.

The drilling was performed in 2 hours and 30 minutes with a probe and two helpers, at the same time they were placed with the probe, TAM 26 bars 6 meters long, in the drilled hole of 160 mm diameter.

The hole was supported by bentonite mud, in fact in one hour the drilling and boring rods were extracted. The TAM consisted of a 75/50 mm PVC tube joined with metal sleeves, the same thickness as the manchettes. It was positioned in the TAM a 120 meter deep packer, to try to inject bentonite mud (this instead of using a cementitious or chemical mixture). We did not want to carry out the injection of manchettes with cementitious mixtures as the TAM extraction was planned.

The washing of the required TAM was also tested during the various injection phases to keep the injection tube accessible. After waiting for 3 hours, the TAM was easily extracted, proving that the hole protected by the bentonite sheath was not collapsed. In the real case, the bentonite primary grouting applied during drilling is replaced by injecting the cement mixture from a valve, that will surround the TAM.

Once the primary grouting has been taken, the manchettes tube was be made integral with the ground surrounding the hole so that it was possible to continue with the injection of the surrounding soil, valve by valve.

4.2 *Settlements control in Govone building (compensation grouting)*

To stabilize the subsidence under the foundations of a building in correspondence of Govone Barrier, Asti-Cuneo motorway, a preliminary consolidation treatment of the ground underlying the foundations was carried out. To minimize the extent of the interventions, it was assumed to perform a series of sub-horizontal drillings under the pillars, according to HDD technique. The drillings were equipped with valved tubes (TAM), to be able to improve soil mechanic characteristic and control settlements foundation. Through the combined use of an external guide system and directional hole bottom tools, holes were drilled according to pre-established paths with tolerances of the order of the decimeters. These drillings were performed along the whole length. Subsequently, the primary drilling battery was extracted and the injection tube was introduced; subsequently the battery of temporary casing was extracted and primary grouting immediately realized.

This is necessary to fill the annular space between the casing drilling and the injection tube and prevent displacement as a result of drillings.

This technique allowed the execution of a path with plano altimetric curves that can have radius of curvature of 100 meters: radius that is a function of the characteristics of the used equipment, of the ground in site and of the injection pipes to be installed.

The HDD boring involves the use of a special asymmetrical tip with a tilted surface provided with peaks and nozzles from which the drilling fluid, usually a bentonite mud, exits at programmed pressure. This tip, if rotated, allows the straight feed, while if pushed without rotation it exerts on the ground the pressure necessary to deviate according to the inclined surface, defined for the maneuver. The elevation drilling path along which the HDD drilling was carried out had some characteristic features consisting of an entrance ramp (lenght to descend), a curvilinear stretch with a radius of curvature 100 meters, below the foundations, with a useful lenght of injection about 20 meters.

The HDD boring were designed on 6 alignments, at depth between -0.50m and -8.50m from the ground level. The alignments that underpin the foundations at depth of 0.50m are made up of two parallel holes with an interaxis of 0.80 meters, injected at low pressure (max 1 bar) so as not to generate undesired lifting on the structures. The other five alignments were injected between 3 and 5 bars. Once the HDD boring was completed and the drilling head was extracted with the rods, the valve tube arranged for the injections (TAM) was introduced into the hole, while the hole was supported by the drilling mud. The drilling mud was then replaced with a special primary grouting, to allow the subsequent injection phases. Through these injection pipes, injections of cementitious and integrative mixtures were carried out. The combined effect of these two mixtures in the soil has led to a global improvement of the geotechnical characteristics of the soil.

The injection intervention was carried out in two successive phases:

- First phase: injection of low viscosity cementitious mixture and with improved penetrability, the variable injected volume of 5-8% of the Total Volume (injection in two successive shots).

- Second phase: injection of a silicate and inorganic reagent chemical mixture, 25-30% of the total volume (injection in two successive stages). The injections were performed with continuous control of the pressures, flow rates and injection volumes.

The injection volume for each pillar was about 80 cubic meters:

- 15.00 m^3 of cement mixture equal to 65 liters/valve for the deeper holes and 35 liters valve for the couple of holes under the foundations;

- 65m^3 of chemical mixture equal to 270 liters/valve for the deeper holes and 135 liters for the couple of holes under the foundations.

4.3 Field test of HDD+JG technique in Silvano d'Orba.

The possibility of ground improvement using the JG technique was considered using HDD drilling.

This technique could be used as a substitute for low pressure injections made from HDD technique, if higher mechanical parameters are required for the soil (compared with the results of LP injetion), as well as limited times execution.

This application could be used both for the bottom plug of immersed tunnels, and for the ground improvement during excavation for shallow tunnels, drilling directly from ground level (machine out of tunnel), avoiding works on the the face of tunnel (machine in tunnel).

The need to limit the use of TAM in a tunnel to be excavated gave us the opportunity to experiment in a special test field at the workshop warehouse of the Injectosond company in Silvano d'Orba the feasibility of creating jet grouting through the HDD driller.

With one special piece of equipment, one operator and three helpers, we drilled a 70-meter hole in one hour, after which with a special reamer, the hole has been widened in the presence of bentonite mud up to the diameter of 200 mm.

A third passage was then carried out with a 200 mm diameter spindle equipped with compressed air nozzles and a cement mixture; this time was powered by three high pressure flexible pipes.

The site was equipped with a 20,000 liter and 15 bar air compressor, a 7 liters per second jetting pump at 400/500 bar, a worm pump for the washing water at 25 bar.

The spindle equipped with nozzles was rotated like a monitor with the drilling rods, while a special tool was used to connect three flexible pipes (so that the pipes themselves could not rotate with the spindle. The experimentation allowed us to positively evaluate the results of system, including the size of the JG treatment tested with one series of cores after one months. The samples show a diameter of treatment from 1 to 1.20m.

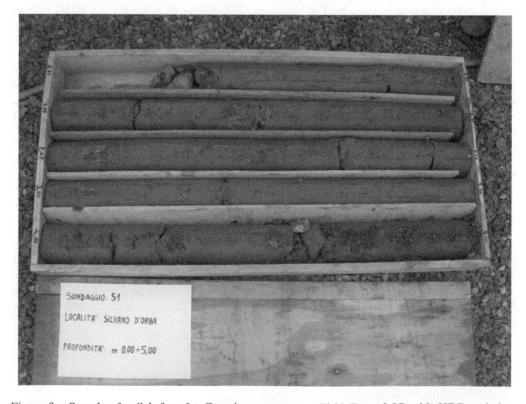

Figure 8. Sample of soil before Jet Grouting treatment – Field Test of JG with HDD technique application.

Figure 9. Sample of soil after Jet Grouting treatment – Field Test of JG with HDD technique application.

5 CONCLUSION

The development of HDD technology is in progress. For engineers it is a good time to continue experimenting with new applications to this technology.

As seen in the applications, with the HDD technique the low pressure injection can be carried out but it is possible to carry out the jet grouting that avoids the difficulty of injection of any single manchette, with the problem of executive times. Interesting then the execution of the coring in progress behind the face of tunnels, Horizontal Directional Coring (HDC), in particular on rock; it is therefore possible to know widely advance on the excavation the condition of the rock or the presence (and pressure) of water.

Loose soil can be sampled so that you can evaluate the injection treatments needed before the excavation reaches the survey progressives. In both these cases, in situ tests can be carried out that can guide the designer in the technical and economic choices.

REFERENCES

Associazione Geotecnica Italiana AGI 2012, *Jet Grouting Raccomandazioni*, Edizioni AGI, Roma – Italy
Bosco B.B. 2013, *Consolidamento e impermeabilizzazione dei terreni, la tecnica del grouting*, Maggioli Editore, Rimini – Italy
Chirulli R. 2011, *Manuale di tecnologie NO DIG*, Nodig.it, Milano - Italy

Tunnels and Underground Cities: Engineering and Innovation meet Archaeology,
Architecture and Art, Volume 4: Ground improvement in
underground constructions – Peila, Viggiani & Celestino (Eds)
© 2020 Taylor & Francis Group, London, ISBN 978-0-367-46868-2

Glassfiber reinforced polymer consolidation for enlargement of a railway underpass in Brandizzo, Italy

M. Bringiotti
MAPLAD S.r.l., Catania

S. Carraro & D. Stella
SOGEN S.r.l., Padova

E. Piovano
ITALFERR S.p.a., Torino

ABSTRACT: The paper describes the enlargement of an existing railway underpass in Brandizzo, Turin, Italy. Variable geometry of consolidation and reinforcement of Glassfiber Reinforced Polymer, with definition of the geotechnical parameters of the soil and the embankment, allowed to perform the complex excavation, with constant verification of design assumptions and operational control of the executive phases. Lateral to the existing underpass on both sides, on the body of the railway embankment, cemented forepoling were performed, defining as a discrete volume of soil improved, like an earth retaining wall. A series of subhorizontal nails made from inside the existing underpass, has represented a tie for the underpass. During jacking phase of new reinforced concrete element, bigger than the existing underpass, demolition has involved at the same time the old underpass, and the excavation of two lateral parts of embankment.

1 INTRODUCTION

The underpass, as part of the work of "Soppressione del passaggio a livello al km 22 + 871 e 23 + 114 of the Turin - Milan railway line", in the municipality of Brandizzo, Turin, Italy, was affected by a widening of the roadway. The existing underpass, with a net section of dimensions 4.00mx3.70m (Figure. 1) has been replaced by a new box-shaped underpass, formed outside the railway embankment, and subsequently driven by jacks under the railway line, with simultaneous demolition of the existing one.

The enlargement of the underpass required the construction of a new box in reinforced concrete (called "monolith") with net internal dimensions B x H = 8.5 x 4.9 m (Figure 1). The realization of the new structure was preliminarily accompanied by the execution of temporary support works for the excavation and the construction of the thrust area of the "monolith".

In order to avoid the interruption of the railway line, the new r.c. building was positioned under the tracks by means of a progressive thrust system, hydraulically applied with a battery of jacks.

The positioning of the "monolith" in its final order took place by means of excavation sequences from the inside of the work, demolition of the portions of the old sub-area, which are included in the new one, and its drive into the ground. At the end of every single advancement in the ground the system has stopped because of two reasons: the first one to reposition the cylinders of the jacks, and the second one to affix the extension elements, between the thrust structure and the jacks themselves, with the aim of creating the thickness sufficient to start a new step of advancement.

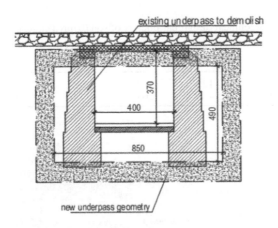

Figure 1. Geometry and dimensions of the existing underpass and of the new project monolith.

The original project based on the tender, contemplated the push of the project "monolith" and the complete demolition of the existing walls of underpass, during the pushing phase, after filling with concrete (with low percentage of cement) the volume of the vacuum constituting the existing underpass. The modification of construction design was necessary in order to adapt the lateral support works for the changes to the relative secondary municipal road network, but above all for the modification of the construction phases, in order to optimize and simplify the execution times of the thrust. The variant consisted in the consolidation of the railway embankment affected by the excavation, on the two sides of the new monolith shape, in addition to the use of horizontal anchors connecting the two portions of the existing underpass and the consolidated soil volumes behind the original walls of underpass.

The advancement of the new "monolith", for each single step of pushing, was preceded by the demolition of discrete portions of concrete, constituting the underpass to be demolished: portions of masonry/concrete and arched upper structure. Both the lateral consolidation at the embankment and the horizontal lateral anchors, perpendicular to the axis of the manufactured artefact, have been realized through a wide use of tubular elements in GRP, after drilling and casing, and have been cemented with a sequence of injections. Two subhorizontal coring were preliminarily executed to verify the geotechnical characteristics of soil and the railway embankment.

2 GEOLOGICAL, HYDROGEOLOGICAL AND GEOTECHNICAL CONTEXT

The geotechnical characterization of soil was referred to the original tender design documentation. This is stratigraphic and geotechnical information obtained from different series of geognostic surveys, carried out at other sub-areas within the same contract, and located at a certain distance. The underpass in question, except for a depth of about 130cm compared to the current shares, does not affect the ground level at great depth, according to a railway line above ground. This is the reason why, according to the detail design, a stratigraphy and a definition of geotechnical parameters were used, deriving from neighboring works within the same contract, during the "Variante Project", in order to characterize the natural soil geotechnic and the soil constituting the embankment, two horizontal coring were performed inside it. In general, the geology of the site consists of alluvial deposits characterized by the presence of gravelly-sandy materials, given the proximity to the torrent Malone. On the basis of the available data, the stratigraphic situation of the site has been defined as follows: from ground level and up to a maximum depth of about 12m there is a layer of gravels in a sandy matrix (formation A); from the base of this formation up to the maximum investigated depths there are well

Table 1. Geotechnical design parameters of soil (from Tender).

n.	Desc.	Depth (m)	γs (kN/m^3)	E (MPa)	c' (kPa)	Φ (°)
A	GW	0 - 12	19.0	50	0	34
B	GM	12 – 19.5	20.0	70	0	35

Figure 2. Sub-horizontal coring execution S1 on embankment.

Table 2. Geotechnical parameters of embankment.

n.	Desc.	H emb. (m)	γs (kN/m^3)	E (MPa)	c' (kPa)	Φ (°)
R	GM	4.0	20.0	30	0	34

thickened sand deposits (formation B). The design stratigraphy described below was then defined (Table 1), as derived from the original project based on the contract.

The water table is assumed at the altitude of + 180.50m, about 4 meters from the ground medium level, set at + 184.50. As regards the two sub-horizontal surveys carried out inside the railway embankment, S1 and S2, respectively on the West side shoulder and on the East side shoulder of the existing underpass (Figure 2), the geotechnical parameters used are reported (Table 2). The overall height of the embankment on ground level is about 4 m.

3 THE METHOD OF DRIVEN UNDERPASSES

The construction of railway underpasses represents something extremely complex due to the delicate nature of the intervention context. Over the years the need to realize underpass works at railway lines in operation has grown exponentially, so the need to reduce all types of inter-ference with railway traffic has progressively been highlighted and, at the same time, to reduce time of implementation of the interventions. Important is the effort of the proprietary com-panies of the railway lines (in this case "Rete Ferroviaria Italiana") in order to modernize the infrastructures, also eliminating level crossings, replacing them with these artefacts.

The works are carried out with modern techniques and are prevalently inserted in strongly urbanized areas, for which special measures are necessary. Since it is easy to understand how the lines to be upgraded are the busiest ones, it is evident that the search for less penalizing solutions for circulation is one of the first working hypotheses of any design. In addition to railway works, a considerable package of interventions to the urban mobility system that should allow the substantial improvement of urban traffic problems is added; there are also many road works in progress that intersect the railway lines, thus making necessary to solve the mutual interference.

In this context, the box-like sections underpass railway lines are inserted, whose main target is the elimination of level crossings and, in general, the improvement of urban traffic conditions. The technique widely used today for the construction and implementation of such structures is that of the "push box", which consists in the construction of an underpass by prefabrication (the entire r.c. underpass called "monolith"), in a special building site on the side of the embankment, and the subsequent placement of this, with hydraulic system, inside the embankment road or rail. During the execution of the work the track is stiffened by a set of beams parallel and connected to each other by crosspieces placed at a limited distance between centers for supporting the rail; this longitudinal stiffening structure rests transversely on steel beams having the function of supporting the whole by sliding on the extrados of the same "monolith" during the launching step. This technique has the following advantages:

- maintenance of the operation of the communication route affected by the underground crossing;
- significant reduction in the support structures of the communication line;
- operational speed;
- minimum site risks;
- installation of the tunnel simultaneously with the construction of the excavation;
- reduction of the environmental impact;
- lower costs than those that characterize traditional executive technology.

The need to guarantee rail traffic, to limit the failure of the infrastructure within very narrow limits, with particular regard to the skew of the rails, to support the side embankment under the tracks and sideways to the axis of the new building, to ensure the impermeability excavation and work during their development are very stringent requirements for designers and contractors.

One of the key elements to solve the problem of supporting the tracks is to adopt an adequate beam system. The Essen system, used here incorporates the concepts underlying the traditional method of rail bundles (stiffening beams in correspondence of the rails and maneuvering beams), and creates a system of constraints able to allow the transit of trains up to 80 km/h. A series of wooden piles, appropriately fixed in the body of the railway embankment, allows the support of the maneuvering beams placed orthogonally to the track. This system supports the Essen bridges, placed longitudinally to the rails throughout the area affected by the thrust works. The maneuvering beams are also constrained to particular beams called "bracing" which balance the frictions during the thrust phase.

The real problem to be solved in the design and execution of an underpass with pushed monoliths, however, is the support of the embankment which is lateral to the monolith itself. From a theoretical point of view, the infixation of the monolith is in close contact with the surrounding ground, with continuous thrust, digging from the front the ground penetrated and protected by the monolith itself, whose front structure is configured with a beak, with inclined spokes.

In reality, due to the sub-roofs made on plain ground, especially under the water table, the intimate connection between side walls and ground to penetrate is rather rare, and the excavation face, especially in the presence of scarcely cohesive soils, tends to collapse inside the monoliths thus involving the ground on the back of the rostrums. Consolidating the ground in which it is necessary to excavate, above all to the side of the rostra, under the rails and for the whole section to be excavated, from the stalls to the arrival of the monolith, is the most complicated challenge. This is because consolidations must be performed from outside the railway line, and often during night hours to eliminate interferences with rail traffic.

Moreover, in the presence of groundwater, these works must guarantee characteristics of absolute impermeability. In general, these are delicate processes, with limited productivity, which can induce lowerings (micropiles) or elevations of the railway tracks (jet-grouting, low pressure injections), which is why these are usually performed with constant topographic monitoring of the tracks.

4 PROJECT DESCRIPTION AND ON SITE ACTIVITIES

The solution proposed to Brandizzo, and then realized, foresees the preventive consolidation of the body of the embankment with threading (nailing) in GRP 60/10mm bars, of variable length from 12 to 17 meters and diameter of perforation equal to 180mm.

The nailing is made sideways to the axis of the future monolith on both sides, along two orthogonal directions, in order to create a grid of reinforcing elements that consolidate the loose material of the embankment.

The first direction is parallel to the axis of the railway line: the nails are made almost horizontal, with the drilling point placed inside the existing underpass, below the railway line; the nails are executed at different heights orders, in numbers of 5/6 nails for each order. They are obviously executed on both sides of the underpass.

The second direction is transversal to the axis of the line; the nails are made from points of perforation which are lateral to the embankment: the rivets are then guided by injection points, two on one side of the embankment and two on the other side, straddling the existing underpass. For each drilling point, 4 nail umbrellas with variable inclination are carried out inside the embankment under the railway line. In this way, a volume of surveyed ground is created, placed immediately behind the walls of the existing underpass. Therefore, there have been no instability phenomena towards the monolith front, nor disturbances to the sediment of the railway line. Some pictures of the intervention are shown in the previous pages. Specifically, Figure 4 shows the view of the heads of the bars in GRP at the exit of the embankment, for the direction of injections orthogonal to the railway axis; it is evident how the nailing point is external to the embankment, therefore without any disturbance for the railway line; the grooves are placed on the back of the wing walls of the existing subbase, thus consolidating the portion of soil which will be subjected to excavation when the monolith is pushed. Figure 5 shows a view from inside the monolith in the pushing phase, with simultaneous

Figure 3.   Details of GRP reinforcement during monolith driving.

Figure 4.   Demolition of existing underpass during the driving phase of the new one.

Figure 5.   The end of the driving phase of the new underpass, and the old one pulled out.

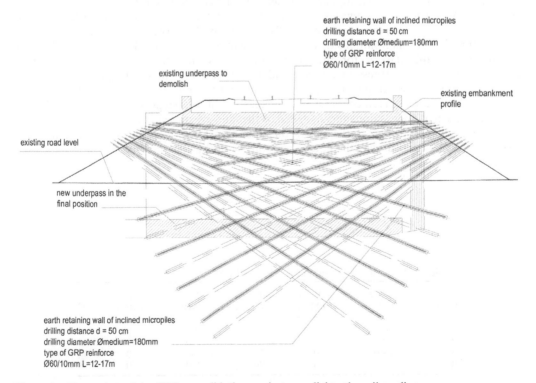

Figure 6.   Front view of the GRP consolidation project, parallel to the railway line.

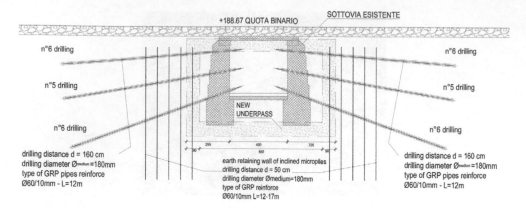

Figure 7. Front view of the GRP consolidation project, perpendicular to the railway line.

demolition of the existing underpass. The ground on the back of the old wall in r. c. is clearly visible, a terrain that presents an almost vertical excavation wall thus avoiding instability of the body of the embankment, which could have an impact on the operation of the above railway line.

5 NUMERICAL ANALYSIS

For the design of the steps work a calculation model was implemented under conditions of plane state, relative to the plane parallel to the railway axis of the embankment. The analyzes were conducted with the PLAXIS 2D calculation code (Brinkgreve and Broere, 2008), schematizing the soil with an elastoplastic model with a hyperbolic hardening (Hardening Soil Model, HSM). The various phases of construction of the building were modeled, starting from the existing initial state: positioning of the Essen bridge, execution of the nailings, progressive thrust of the monolith inside the embankment, removal of the Essen bridge. The excavation of the embankment for the positioning of the new monolith was simulated by decreasing the initial stiffness of the consolidated survey, with descending steps of 20% for each phase, and the simultaneous activation of the clusters relative to the transversal section of the monolith in reinforced concrete, with increasing stiffness steps of 20%, for each phase.

In Table 3 calculation parameters assigned to the land are presented.

For nailing perpendicular to the axis of the embankment, an increase in cohesion and stiffness has been associated with the equivalent soil. In detail, the value of cohesion increase is equal to

$$\Delta c = \frac{\Delta \sigma_3}{2} \cdot \sqrt{K_p} \tag{1}$$

Table 3. Geotechnical parameters used for FEM code PLAXIS.

Desc.	γs (kN/m^3)	E_{50ref} (MPa)	E_{50edef} (MPa)	E_{urref} (MPa)	c' (kPa)	Φ (°)
R	20.0	30	30	90	0	34
Rcons*	20.0	50	50	150	100	34
A	19.0	50	50	150	0	34
B	20.0	70	70	210	0	35

* Rcons=embankment reinforced with FRP

where $\Delta\sigma_3$ = confinement pressure acting on the excavation front and Kp = $\tan^2 (45 + \phi/2)$ = passive earth pressure coefficient. The value of $\Delta\sigma_3$ to be used is the lower value between the yield strength of GRP elements (pipes) and the resistance to lateral skin friction of the injected and cemented pipes.

$$\Delta\sigma_3 = \left(\Delta\sigma\frac{1}{3} = \frac{Avtr \cdot \sigma d}{A}; \Delta\sigma\frac{2}{3} = \frac{\pi \cdot D \cdot L \cdot \tau}{A}\right). \qquad (2)$$

where Avtr = fiberglass pipe area, σd = fiberglass pipe design strength (600MPa), A = area of competence of a single pipe, D = diameter of drilling, L = anchorage length, τ = tangential tension of adhesion mortar/soil (50kPa). In terms of safety, cohesion values have been used, equal to 100 kPa, while for the stiffness of the consolidated soil, whose increase is proportional to the area of the fiberglass elements, on the area of competence of each individual nail, a value of E50ref = 50 MPa was chosen. Fiberglass nails parallel to the railway embankment, in the FEM calculation, were considered as linear beam behavior with linear elastic behavior, with the following calculation stiffness values EA = 30 x 103kN/m and EJ = 10 kNm^2/m. A beam-type element was also considered for the Essen track-supporting bridge in the FEM calculation, with linear elastic behavior, with calculation stiffness values EA = 1 x 106kN/m and EJ = 3 x 105kNm^2/m. The following is a view of the geometrical calculation model at the intermediate excavation and final phase (Figures 8-9).

Some views of the results are shown in terms of displacements calculated in the model, in the two phases shown above. The results in terms of displacement were compatible with the work in progress: rail traffic weighed on supporting bridges of the Essen-type platforms, while in addition to being guaranteed the stability of the detections, the movements were limited and confirmed by the continuous topographic monitoring of the trains. These have undergone only and exclusively the movements induced by the passage of the train convoys, however, their load is not directly acting on the survey. Also the displacement value shown in Figure 11,

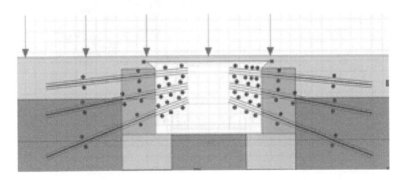

Figure 8.   FEM Geometry: excavation and demolition simulation reducing rigidity of excavated elements.

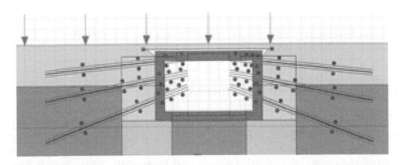

Figure 9.   Model FEM: excavation ultimated and underpass driven in a definitive position.

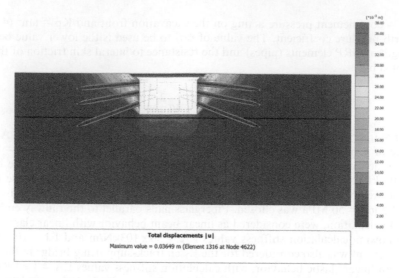

Figure 10. Displacements resulting from the FEM model: excavation and demolition simulation with reduction in rigidity of excavated elements: maximum total displacement 3.7cm.

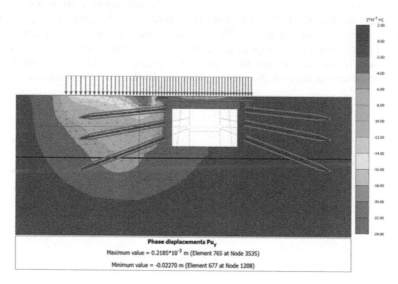

Figure 11. Displacements resulting from the FEM model: excavation occurred and monolith in final position, with railway overhead presence: increase in vertical displacement Puy = 2.3cm.

due to the railway load acting directly on when the work is completed, is compatible with the displacements of the ballast in operation.

6 NEW MATERIAL FOR GRP PIPE: BASALT FIBER

Basalt fiber is a material made from extremely fine fibers of basalt, which is composed of the minerals plagioclase, pyroxene and olivine. It is similar to carbon fiber and fiberglass, having better physic/mechanical properties than fiberglass (Table 4), but being significantly cheaper than carbon fiber (Manuele, Bringiotti, Laganà and Fumagalli, 2019). Basalt fibers are 100%

Table 4. Comparison between basalt rebar and fiber glass rebar.

Properties	Glass Rebar	Basalt Rebar	Unit
Elastic modulus	>30000	>50000	N/mm^2
Elongation at break	>2	>2.5	%
Fiber content	>60	>70	%
Shear strength	>16	>20	Ksi

natural and inert. Tested and proven to be non-carcinogenic and non-toxic and easy to handle. In contrary, fiberglass is made from a mixture of many materials, some of which are not environmentally friendly. Since basalt is the product of volcanic activity, the fiberization process is more environmentally safe than glass fiber process. Basalt continuous filament is a green product, abundant in nature so it can never deplete the supply of basalt rock. This kind of new product, under investigation and test by Maplad Srl, engineered also by Sogen Srl, could really be the material of the future in the field of the composite materials.

7 CONCLUSION

The present paper describes the expansion of an existing railway underpass in Brandizzo, Turin, Italy, on the Turin-Milan historic railway line. The push insertion of the new structure in r.c. ("Monolith"), the demolition of the existing structure, the need to combine the support of the tracks avoiding landslides and collapses for the railway embankment itself, in order not to jeopardize the railway operation, required a careful design of detail, with use of "intelligent" building technologies and sequences. In particular, variable geometries of consolidations and reinforcement in GRP, together with a correct and in-depth knowledge of the geotechnical parameters of the affected terrain, has allowed the delicate interventions to be carried out in safety, with constant verification of the project hypothesis and operational control of the executive phases. On both sides of the existing underpass, on the body of the railway embankment, a succession of VTR insertions were carried out, on perforations then sealed with a cementitious mixture, in form of sub-horizontal micropiles reinforced with VTR pipes. In the body of the embankment this geometry has defined a discrete volume of improved terrain, with supporting wall behavior. A series of nails perpendicular to it, subhorizontal and realized from within the existing sub-line, represented a constraint for them. During the push of the new monolith, having a shape with greater encumbrance compared to the existing underpass, there have been demolition phases of the old structure and demolition phases of part of the consolidated renovation, within which the monolith has found space. The close collaboration between the Works Management, with a daily presence on site, and the Company's Designers, in constant contact with each other, was fundamental for verifying the goodness of the project hypotheses and the on-site compliance of what was previously planned, as well as to solve every single unexpected or contingent problem.

REFERENCES

Brinkgreve R.B.J., Broere W. (eds) 2008. *Plaxis-2D User Manual.*
Manuele G., Bringiotti M., Laganà G., Fumagalli G. 2019. MPLD fiber glass and composite materials as structural reinforcement and systems; different applications and usages from Metro Milano up to long basis tunnels as Brenner and high speed train MI-Genoa, WTC2019, Napoli

Tunnels and Underground Cities: Engineering and Innovation meet Archaeology,
Architecture and Art, Volume 4: Ground improvement in
underground constructions – Peila, Viggiani & Celestino (Eds)
© 2020 Taylor & Francis Group, London, ISBN 978-0-367-46868-2

A novel low enthalpy geothermal energy system based on ground freezing probes

A. Carotenuto, N. Massarotti, A. Mauro & G. Normino
Università degli Studi di Napoli "Parthenope", Napoli, Italy

A. Di Luccio & G. Molisso
Ansaldo STS | A HITACHI GROUP COMPANY, Napoli (Na), Italy

F. Cavuoto
Ingegneria delle Strutture, Infrastrutture e Geotecnica, Napoli, Italy

ABSTRACT: Geothermal energy plants allow both heating and cooling of buildings by using the ground as a renewable energy source. The present work shows the numerical results obtained for the recovery of the freezing probes used for the Artificial Ground Freezing (AGF) technique of the two tunnels between Line 1 and Line 6 of the new Metro station in Piazza Municipio, Napoli. The AGF is a consolidation technique used in geotechnical engineering. The recovered probes are connected to a geothermal heat pump. The authors propose to develop an ad hoc innovative plant, in accordance with the principles of the Industry 4.0. In particular, the authors have developed a thermo-fluid dynamic numerical model, solved by means of finite elements, in order to design an innovative energy system, recovering the freezing probes installed for the excavation of the two tunnels.

1 INTRODUCTION

Human activities, especially in the last two centuries, have increased the greenhouse gases (GHG) emission level in the atmosphere, changing the natural balance of the environment and causing climate change. One of the consequences of these changes is the global warming of the planet. Temperatures in Europe have risen by about 1°C since 1850, the past decade has been the warmest ever recorded in Europe, Earth's surface temperature higher than 1.3°C above the average temperature in pre-industrial era. Extreme weather events (storms, floods, droughts and heat waves) have become more frequent and more intense. In order to tackle climate change, the European Union has set itself the objective of reducing, by 2050, GHG emissions, of 80%, compared to 1990 level. This objective must be achieved through a decarbonisation of the energy system and increasing energy efficiency.

A number of numerical and experimental works analysing the performance of energy galleries are available in literature. In particular, (Lee et al. 2012) propose the development of a textile heat exchanger on the ground, called "Energy Textile", positioned between a layer of shotcrete and geotextile draining. To evaluate the energy performance, they developed a prototype Energy Textile in a railway tunnel located in South Korea abandoned in Seocheon. In addition, they developed a finite volume 3-D analysis in FLUENT to simulate the operation of the heat exchanger enclosed in energy fabric, by varying the thermal conductivity of shotcrete, the velocity of circulation of fluid, the existence of groundwater and the geometrical configuration of the tubes.

(Lee et al. 2016) report a long period-monitoring of the system shown in the paper (Lee et al. 2012), in order to measure the thermal exchange capacity of each module by applying artificial textile energy heating and cooling loads on it. (Mimouni et al. 2014) propose to use anchors (anchors) as geothermal heat exchangers. The authors in this article illustrate the finite element thermo-hydraulic model using the finite element code Lagamine. Finally, the authors compare the heating capacity/storage extracted or placed into the ground using the anchors as heat exchangers, with powers extracted from energy piles made in Switzerland.

(Barla et al. 2016) investigate the possibility of thermal activation of a new section of a tunnel of line 1 of the Metro Torino (Italy) to heat and cool the adjacent buildings. The authors show the design and optimization of the geothermal plant, the quantification of usable heat and evaluation of any consequences on the surrounding land. The authors developed a 3D model to study the efficiency of the system, reproducing a 2D tunnel ring, while analysing a full-scale model of the groundwater in Turin to investigate the viability of technology in terms of effects on the surrounding environment. The results of these numerical models, developed through finite element analysis using the numerical code FEFLOW, foresee, that thanks to favorable conditions of ground water flow in Turin, the system would exchange 53 and 74 W per square metre of tunnel lining during winter and summer, respectively.

(Barla et al. 2018) in this study provide a guidance on computational methods for quantifying the cooling-power that one can extract from the ground and evaluate the consequences on the surrounding land, showing how the thermo-hydro and thermo-mechanical numerical analyses can be used to obtain a correct and efficient design of the tunnels. The authors develop a thermo-hydro model for using finite element software FEFLOW © (Diersch, 2009), while the Thermo-mechanical model is developed using the calculation code FLAC (Itasca, 2016). Also, show two examples of possible applications. The first concerns an application shallow and relates to the inclusion of GeoExchange probes in Turin Metro tunnel, particularly trafficking from Lingotto station to Piazza Bengasi. The tunnel under construction is located below the water table and groundwater flow direction is orthogonal to the axis of the tunnel. The authors plan to use the heat extracted from the soil to cool the offices in the region. The second case is related to the application of deep tunnels for cooling, it was decided to use an inexpensive alternative to the ventilation systems, as in the new base tunnel Lyon-Turin railway line. This case was developed before a preliminary analysis, using finite element modeling. In this application, the piping system is installed on the lower surface of the coating. To decrease the cost, the system is considered to be active in cycles. After the first activation of the cooling system is turned off and on at time intervals that keep the temperature between 25 and 30 °C. The authors affirm that as a preliminary assessment, compared to conventional ventilation, thermal activation of the lining of the tunnel would provide cost savings between €10.000 to €20.000/(km·year)$^{-1}$. In addition, the system would allow thermal exploitation and any users in the neighborhood of the portals could use the heat retrieved. The heat extracted for a kilometer of activation was calculated with reference to the balanced condition after the first 50 days of operation and proved to be about 6.500 (MW·h)/anno.

The authors of this article propose innovative low-enthalpy geothermal energy use for air conditioning of confined environments, recovering freezing probes used for Artificial Ground Freezing (AGF) to build two tunnels between the line 1 and Line 6 of the new Metro station in Piazza Municipio, Naples. Artificial freezing of the land is a technique that reduces the permeability of soils and consolidate them temporarily during the excavations for the construction of tunnels, shafts and tunnels to link. This technique is now used successfully for decades, particularly, since the '70s onwards, its use has spread particularly in the Nordic countries. Geothermal probes recovered from frostbite, were connected to a thermo/cooling plant called Energy Box, which contains a heat pump, two inertial storage tanks and circulation pumps. The proposed system is characterized by high energy performance and lower environmental impact, in accordance with the energy policies, which have the goal 27% of energy produced by renewable sources by 2030, moreover, it is in line with the objectives of the circular economy, as in common practice freezing probes are left to leak into the soil. In the following sections, the authors describe the characteristics of the energy system in the tunnel, section three shows the numerical model developed. In section four describes the numerical and experimental results obtained, while the conclusions can be found in the last section.

2 DESCRIPTION OF THE ENERGY TUNNEL SYSTEM

The energy geo-structures, as seen in the previous section, can be coupled in traditional foundation works or underground tunnels, by integrating internal heat exchangers for air conditioning and the harnessing of low temperature geothermal energy infrastructure. This innovative technology comes in response to the needs to achieve a costs and space saving installation. The thermoactive geo-structure are systems that match a structural role with the ground heat exchanger and therefore the operating depth depends on the location and the size of the structure needed.

This technology requires only a mass temperature generally constant throughout the year, in fact, already about ten metres, seasonal fluctuations can be neglected, and the temperature settles around the mean superficial temperature value. The soil temperature at depths greater than 5–8 m typically varies between 8 °C and 18 °C, but remains constant throughout the year, up to about 50 m. Under such conditions, soil works as heat source for the buildings during the winter, and as a heat sink in summer when cooling is required.

In the case of tunnels, the heat traded at tunnel level can be transferred to the surface by inserting tubes into the shell of the gallery itself or through portals. In the galleries, the air temperature is relatively low (around 15 °C) throughout the year and the frequency of trains that cross is moderate, so it will not significantly increase the temperature in the tunnel.

Hot galleries, on the other hand, will usually show high internal temperatures. Urban tunnels (underground railways) with an inner diameter of about 7 m can have typical summer temperatures around 30 °C. Many stations and a rapid cycle frequency of trains lead to an additional heat input due to braking and commissioning. This increases the air temperature in the tunnel, which heats the ground.

2.1 Case study

The galleries treated in this article fall into the category of local galleries and, more specifically, the focus is on two connecting tunnels built between the town hall station, located near the moat of the Angevin, and 6 line TBM extraction pit. The two galleries with polycentric shape, have a length of 40 m following to a path slightly divergent and cross partly tufa formation (at the invert and piers downs) and partly the loose above. The two connecting tunnels have a section 6 m wide and high digging about 7 m.

For the construction of the two tunnels, it has been used the Artificial Soil freezing (AGF) technique for the top of the piers and the upper cover, in order to cope with the sash of water (about 16–17 m), while for the piers and the bottom, physical and chemical injections used MPSP TECHNIQUE (Multiple Packer Sleeved Pipe). The technique of freezing of the soil is characterized by the need for drilling into the ground to accommodate the freezing probes (a tube steel probe holder with diam. 114 mm and 10 mm thick).

Before inserting the freezing probes inside the holder, the exact location was found using the topographic map; for holes on which it was not possible to measure with the topographic method, to the end, it was used as an equipment developed for sub-horizontal hole deviation measurements, the Maxibor. All probes have been installed inside the pipe freezer steel holder with a diam. 114 mm and 10 mm thick, allied himself to the ground using cement grout. Freezer probes used for AGF technique are characterized by an outer pipe diameter. 76 mm steel and copper inner tube with a diam. 28 mm.

Once installed, the probes and the holders were linked by filling with cement grout the space between the tubes. Finally, within these probes liquid nitrogen or brine was circulated in order to form the frozen wall to perform the digging of the tunnels. After working for the construction of the tunnels, the authors have replaced the copper pipe with diam. 28 mm with a high-density polyethylene pipe of diam. 32 mm, by means of appropriate flanges, geothermal probes were connected to collectors and finally coupled to the heat pump installed in the Energy Box. Activities related to such intervention can be considered highly innovative, because they allow both to reuse and leverage equipment (freezer probes) that otherwise would be left in the ground, either to withdraw renewable thermal energy from the soil using geothermal heat exchangers and systems to match with freezer as heat pumps, with considerable energy savings and reduced environmental impact for the conditioning of the rooms. In Figure 1 shows the location of the shipyard located in the historical centre of the city of Naples, and the sight of the galleries and extensions to the collectors of GeoExchange probes.

Near the yard, there is not a confined room to be air conditioned, therefore one tunnel has been used to serve as pickup shaft of heat and simulates the winter condition, while the second simulates the summer condition.

The setup is shown in Figure 2 made at the construction site of the Naples metro, from this section it can seen the connection between the borehole and the Energy Box, the latter contains a heat pump, two inertial storage tanks and pumps circulation. Yet it is possible to observe the ten probes used as heat exchangers.

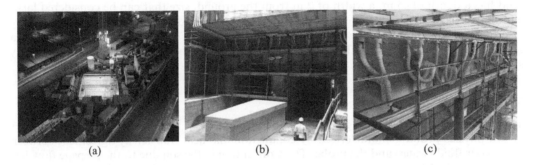

Figure 1.   (a) Construction yard; (b) Front view of the two tunnels; (c) Details of the geothermal probes.

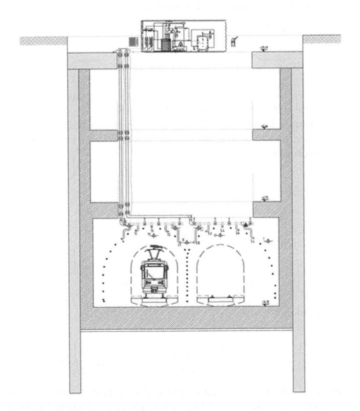

Figure 2.   Cross section of the experimental setup.

The links between the borehole and the Energy Box take place by means of geothermal collectors made of steel, while the pipes to connect the boiler to the collectors are made of high density polyethylene with a diameter from 50 mm.

3 NUMERICAL MODEL

The authors developed a transitional numerical model based on a bi-dimensional model considering heat and mass transfer in the pipe and in the annulus of the probe, and only heat between the probe and the surrounding terrain. The two-dimensional model is used to evaluate the influence of the concrete domain and its boundary conditions. The mathematical model employed has been developed in (Carotenuto et al. 2017). The numerical model was implemented within the commercial software Comsol Multiphysics. The domain of land has a depth of 45 m, a width of 2 m from the ground, size that can be considered large enough to neglect thermal noise, plus a thickness of 1 m to the outer shell of the gallery in concrete armed, and a length of 5 m and. Moreover, (Carotenuto et al. 2016) show a review on models for thermo-fluid dynamic phenomena in low enthalpy geothermal energy systems.

3.1 *Governing equation*

The thermal problem of freezing probes can be reproduced by the solution of the heat transfer in soil physics problem characterized by conduction and heat transfer by agreement between the carrier fluid (water) and the probe. The Convention in the soil due to the seepage flow has been overlooked because the barriers of jet-grouting, make it negligible. The authors developed a two-dimensional model to evaluate the thermal Exchange. Heat transfer in the domain of land was addressed through the following equation:

Transient conduction heat transfer

$$\rho_s c_s \frac{\partial T}{\partial t} - \nabla(k_s \nabla T) = 0 \tag{1}$$

where ρ_s = concrete or soil density (kg/m^3), c_s = specific heat capacity (J/kg·K), k_s = thermal conductivity (W/m·K) and T = temperature distribution (K).

A model of heat and mass transfer model to simulate in the behavior of the working fluid flowing in the probe. As a consequence, all the input and output variables correspond to the average value along the cross section:

Continuity equation

$$\frac{\partial(A\rho)}{\partial t} + \frac{\partial(A\rho u)}{\partial z} = 0 \tag{2}$$

Momentum conservation equation

$$\frac{\partial(\rho u)}{\partial t} = -\frac{\partial p}{\partial z} - f_d \frac{\rho}{2D_h} u|u| \tag{3}$$

Energy conservation equation

$$\frac{\partial\left(\rho_f A c_f T\right)}{\partial t} + \frac{\partial\left(\rho_f A c_f u T\right)}{\partial z} = \frac{\partial}{\partial z}\left(A k_f \frac{\partial T}{\partial z}\right) + f_d \frac{\rho}{2D_h}|u|^3 + Q_{wall} \tag{4}$$

where ρ_f = fluid density (kg/m^3), c_f = specific heat capacity at constant pressure (J/kg·K), k_f = fluid thermal conductivity (W/m·K), T = temperature (K), f_d = friction factor that depends on both Reynolds number, Re, and ratio between thickness and hydraulic diameter of the probe, ε/D_h, u = velocity (m/s), A = area of the cross section of the probe (m^2).

The last term on the right hand side of equation (4), $\dot{Q}_{wall}$ (W/m), represents the thermal power per length unit exchanged through the walls of the probe, and has been used to thermally connect the one-dimensional heat transfer model of the probe to the three-dimensional heat transfer model of the external probe and surrounding ground, as it has been calculated as follows:

$$\dot{Q}_{wall} = K_T(T_e - T) \tag{5}$$

$$K_T = \frac{2\pi}{\frac{1}{r_{int}h_{int}} + \frac{\ln\left(\frac{r_{ext}}{r_{int}}\right)}{k}} \tag{6}$$

where T_e = external temperature (K) (i.e. the external probe temperature), K_T = thermal transmittance of the probe (W/K), h_{int} = convection heat transfer coefficient (W/m^2·K), respectively, r_{ext} and r_{itn} = are the internal and external radii of the probe (m), L = probe length (m) and k = thermal conductivity of the probe (W/m·K). The second term on the right hand side of equation (4) represents the distributed pressure drops due to friction with the inner walls. However, the pressure losses in the fluid due to curve of the probe have been valued using the following equation:

$$\Delta_p = \frac{1}{2}k_{Df}\rho u^2 \tag{7}$$

where k_{Df} = shape factor related to the curvature of the probe.

Table 1 show the thermos-physics characteristics of the materials used for numerical modelling.

The thermophysical properties for concrete, HDPE and AISI4340 are defined by the manufacturers.

3.2 Boundary and initial conditions

The boundary conditions used in this model are reported to the domains of land, lens of the gallery and probe. The upper soil surface was set equal to the soil temperature, while the temperature of the soil was assumed as constant parsing, equal to the average annual temperature of the site and from measurements taken on site at the depth to which are installed geothermal probes (18° C). Where T_∞ is equal to 20° C and h_∞ is equal to 15 W/m^{2}K. Regarding the probe, the values of speed and temperature at inlet section were imposed, whereas no change of temperature is set to the output section. Figure 3 shows a schematic illustration of the domains that are in the modeling and the physical elements, while the boundary conditions described above are shown in Figure 4.

3.3 Mesh sensitivity analysis

The authors performed a sensitivity analysis of the mesh, to obtain numerical results independent from the grid. All computational grids are made of triangular elements and have been refined near the probe. Figure 5 shows the distribution of the mesh in the domains and condensing near

Table 1. Physical properties of the materials.

Materials	Thermal conductivity W/m·K	Density kg/m^3	Specific heat capacity J/kg·K
AISI4340	44.5	7850	475
Reinforced concrete	1.63	2500	837
HDPE	0.42	1100	1465
Tuff*	1.48	2713	1800

*(Cavuoto et al. 2017)

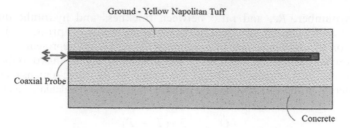

Figure 3.   Computational domain.

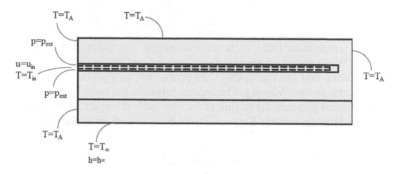

Figure 4.   Boundary conditions employed.

Figure 5.   Mesh along domain.

GeoExchange probe. Figure 6 shows a detail of the top of the geothermal probe for four cases of links considered. The volumetric flow rate considered in this case is equal to 800 l/h.

Figure 7 show the temperature diffusion in the domains for the different mesh considered. Table 2 shows the details of the four grid computing, along with a summary of the main numerical results. In particular, considering an inlet temperature of 35 °C, the temperature at the exit of the water was calculated and presented in Table 2, together with the computational time required to reach convergence. Based on this sensitivity analysis, the grid to use for calculations is the third with 130401 elements, because the percentage difference between the results obtained using this grid and those obtained using the densest grid is less than 1%.

Figure 7 shows that the efficiency of the probe is not influenced by the concrete domain and its boundary conditions. Therefore the authors can consider the axisymmetric model

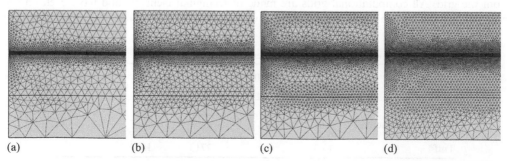

(a)   (b)   (c)   (d)

Figure 6.   Details of the meshes used for the grid independency study.

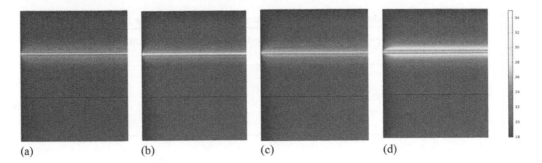

Figure 7. Temperature diffusion.

Table 2. Mesh details and summery of output data for the grid sensitivity analysis.

Mesh code	Number of elements	Computational time min	Outlet temperature °C
1	58139	23	28.67
2	68842	35	28.94
3	130401	36	30.65
4	174176	42	30.97

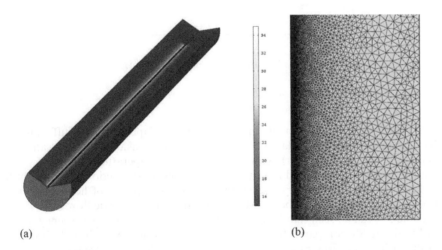

(a)   (b)

Figure 8. (a) Axisymetric model; (b) Mesh.

developed in (Cavuoto et al. 2018), neglecting the concrete domain. The Figure 8a and 8b show the axisymetric model and mesh considered.

4 NUMERCAL RESULTS

Given the importance of the validation procedure for numerical codes (Arpino et al. 2016, Massarotti et al. 2016), the mathematical model was validated in (Carotenuto et al. 2017), and here used to perform a parametric analysis in order to assess the volumetric flow rate that guarantees the best energy performance of the probe. The volumetric flow rate is the only parameter on which it is possible to intervene, because the size of the probes, lengths and diameters are defined in the technique of artificial ground freezing. The simulations were

Table 3. Numerical results for different values of the volumetric flow rate.

Volumetric flow rate m³/h	T_{out} °C	ΔT °C	Heat transfer W	Heat transfer per unit length probe W
0.2	28.13	6.87	1597	39.92
0.4	30.13	4.87	2273	56.82
0.6	31.17	3.83	2671	66.77
0.8	31.79	3.21	2982	74.55
1.0	32.25	2.75	3203	80.07

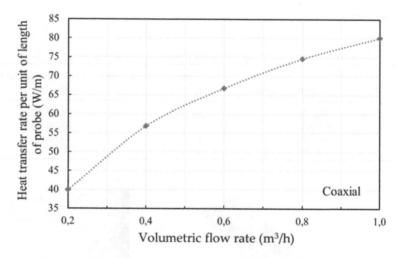

Figure 9. Heat transfer rate per unit of length of probe.

performed assuming a single layer of soil that is yellow Napolitano for tuff, this was determined by geological surveys available for authors. It was considered an operating period of 10 hours in cooling mode (summer), with incoming water temperature of 35 °C.

The main findings analyzed and reported in the following subsections are: output temperature and heat transfer along the probe. Table 3 shows the values of heat transfer and heat transfer per unit length of the probe measured for different volumetric flow rate.

From Table 3 it can be seen that by using a flow rate of 1.0 m³/h the efficiency of the system increases by about 6% compared to the volumetric flow rate 0.8 m³/h and about 50% compared to the volumetric flow rate 0.2 m³/h. The Table 3 shows the trends of the outlet temperature from the probe at different flow rates. The results obtained by varying a parameter, that is, the volumetric flow rate of the fluid flowing in the probe, were studied. The results in terms of heat transfer rate distribution at the exit of the probe, shown in Figure 9, clearly show that the increasing volume flow corresponds to a smaller temperature difference between inlet and outlet, due to increased output temperature, this leads to an increase in the thermal power extracted from the soil.

5 CONCLUSIONS

This work presents an analysis of heat transfer performance for retrieving freezing probes, firstly it was used an efficient calculation through a numerical model developed by the authors, verified with literature. The model, based on a 2D approach for heat and fluid flow in the probe and to conduct the heat into the ground, was used for parametric analysis to evaluate the best energy performance through the variation of fluids flow rate, as it is the only parameter on which the authors could intervene, because the parameters of shape and choice of the materials used were conditioned by the artificial ground freezing technique.

The main results of this paper can be summarized as:

- It is possible to retrieve material like freezing probes;
- Recovering the probes let the cost of drilling and installation of geothermal probes to be covered, allowing to overcome one of the problems of the development of low enthalpy geothermal systems which is the high drilling and construction costs related to the probes;

Retrieving freezing probes represent a precious opportunity to pair the heat extracted from the soil with heat pumps, with consequent benefits in terms of improved efficiency and use of renewable energy sources. However, numerical studies based on efficient and effective models are required in order to understand clearly the actual behaviour of these systems, based on the entire design and operation parameters involved.

ACKNOWLEDGMENTS

The academic authors gratefully acknowledge the financial support under the POR Campania FESR 2014/2020 O.S. 1.2 Az. 1.2.2 "Distretti ad alta tecnologia, aggregazioni e laboratori pubblico privati" program for the Project GEOGRID, CUP B43D18000230007.

REFERENCES

Lee, C. Park, S. Won, J. Jeoung, J. Sohn, B. & Choi, H. 2012. Evaluation of thermal performance of energy textile installed in Tunnel. *Renewable Energy* (42): 11–22.

Lee, C. Park, S. Choi, H.J. Lee, I.M. & Choi, H. 2016. Development of energy textile to use geothermal energy in tunnels. *Tunnelling and Underground Space Technology* (59): 105–113.

Mimouni, T. Dupray, F. & Laloui, L. 2014. Estimating the geothermal potential of heat-exchanger anchors on a cut-and-cover tunnel. *Geothermics* (51): 380–387.

Barla, M. Di Donna, A. & Perino, A. 2016. Application of energy tunnels to an urban environment. *Geothermics* (61): 104–113.

Barla, M. & Di Donna, A. 2018. Energy tunnels: concept and design aspects. *Underground Space.*

Cavuoto, F. Marotta, P. Massarotti, N. Mauro, A. & Normino, G. Artificial Ground Freezing: Heat and Mass Transfer Phenomena. ICHMT International Symposium on Advances in Computational Heat Transfer, May 28 - June 1, 2017, Napoli, Italy.

Cavuoto, F. Massarotti, N. Mauro, A. Normino, G. 2018. Recovery of freezing probes for the exploitation of geothermal energy in urban environment: a numerical analysis, Fifth International Conference on Computational Methods for Thermal Problems ThermaComop2018, July 9–11, 2018, Indian Institute of Science, Bangalore, India.

Carotenuto, A. Marotta, P. Massarotti, N. Mauro, A. & Normino, G. 2017. Energy piles for ground source heat pump applications: Comparison of heat transfer performance for different design and operating parameters. *Applied Thermal Engineering* (124): 1492–1504.

Carotenuto, A. Ciccolella, M. Massarotti, N. Mauro, A. 2016. Models for thermo-fluid dynamic phenomena in low enthalpy geothermal energy systems: A review, *Renewable and Sustainable Energy Reviews*, (60): 330–355.

Arpino, F., Carotenuto, A., Ciccolella, M., Cortellessa, G., Massarotti, N., Mauro, A. 2016. Transient Natural Convection in Partially Porous Vertical Annuli. *International Journal of Heat and Technology*, (34), S512–S518.

Massarotti, N., Ciccolella, M., Cortellessa, G., Mauro, A. 2016. New benchmark solutions for transient natural convection in partially porous annuli. *International Journal of Numerical Methods for Heat & Fluid Flow*, (26): 1187–1225.

Tunnels and Underground Cities: Engineering and Innovation meet Archaeology, Architecture and Art, Volume 4: Ground improvement in underground constructions – Peila, Viggiani & Celestino (Eds)
© 2020 Taylor & Francis Group, London, ISBN 978-0-367-46868-2

Field experiment on freezing rate of marine clay by artificial ground freezing method

H-J. Choi, D. Lee, H. Lee & H. Choi
Korea University, Seoul, Republic of Korea

Y.-J. Son
SK Engineering & Construction, Seoul, Republic of Korea

ABSTRACT: The Artificial ground freezing (AGF) method has been used in many geotechnical engineering applications such as temporary excavation support and underpinning, and groundwater cutoff, especially for constructing cross passages of a subsea tunnel through highly-fractured fault zones. The freezing rate is one the most important factors governing the applicability of the AGF method. This paper performed a series of field experiments to evaluate the freezing rate of marine clay in application of the AGF method. The field experiments consisted of the single freezing-pipe test and the frozen-wall formation test by circulating liquid nitrogen, which is a cryogenic refrigerant, into freezing pipes. The temperature of discharged liquid nitrogen was maintained through the automatic valve, and the temperature change induced by the AGF method was measured at the freezing pipes and in the ground with time. According to the experimental results, the single freezing-pipe test consumed about 11.9 tons of liquid nitrogen for 3.5 days to form a cylindrical frozen body with the volume of about 2.12 m^3. The frozen-wall formation test consumed about 18 tons of liquid nitrogen for 4.1 days to form a frozen wall with the volume of about 7.04 m^3. The radial freezing rate decreased with increasing the radius of frozen body because the frozen area at a given depth is proportional to the square of radius.

1 INTRODUCTION

Nowadays, the artificial ground freezing (AGF) method has been used in many geotechnical engineering applications such as temporary excavation support, underpinning, and groundwater cutoff (Crippa & Manassero, 2006; Papakonstantinou et al. 2013; Russo et al. 2015). Notably, the AGF method has recently been applied to tunnel construction in soft ground (Qin et al. 2010; Sun & Qiu, 2012; Song et al. 2016).

The AGF method conducts the freezing process by employing a refrigerant circulating through a set of embedded freezing pipes to form frozen walls serving as excavation support and cutoff wall. Freezing is a reversible process to improve the hydro-mechanical properties (strength, stiffness, and permeability) of the soil and to provide a closed arch, the frozen wall, around the excavated area. In other words, freezing fuses the soil particles together, significantly increasing soil strength, and making it impervious (Pimentel et al. 2012).

Currently, two refrigerants, i.e. brine and liquid nitrogen, are being used in geotechnical applications (Andersland & Ladanyi, 2004). Brine freezing is a closed system circulating the chilled brine at -20 to -40 °C through freezing pipes installed in the ground using refrigeration plants. On the other hand, liquid nitrogen freezing is an open system. The liquid nitrogen, which is an extremely low temperature of -196 °C, rapidly freezes the ground. Since the liquid nitrogen increases in volume by 800 times as it vaporizes, the phase-changed nitrogen gas is discharged into the atmosphere.

This study performed a series of field experiments to evaluate the freezing rate of marine silty clay in the application of the AGF method. The field experiments consisted of the single freezing-pipe test and the frozen-wall formation test by circulating liquid nitrogen, which is a cryogenic refrigerant, into freezing pipes. The temperature of discharged liquid nitrogen was maintained through the automatic valve, and the temperature change induced by the AGF method was measured at the freezing pipes and in the ground with time.

2 FIELD EXPERIMENT

2.1 *Test bed description and soil property*

The test bed for conducting the field experiment of AGF method is in Sinan-gun, Jeollanam-do, South Korea. A boring investigation and the standard penetration test were conducted to identify accurate geological features around the test bed. The boring investigation was performed three times with an interval of 3 m. As the results of boring investigations, the ground of the test bed consisted of a fill layer composed from the ground surface to the 1 m depth, a deposit layer composed of silty clay from a depth of 1 to 19.5 m, and the weathered rock appears to be deeper than 19.5 m (Figure 1).

The groundwater appeared at 1.0 m below the ground surface. In this field experiment, 1 m thick fill layer above the groundwater level was removed to simulate the AGF method for the saturated marine clay. As the results of the standard penetration test, the section at depth 1.0~4.2 m below the ground surface where freezing pipes were installed was evaluated as very soft ground with N values of 2~3.

The soil specimens were sampled with the in-situ ground condition from at a depth of 2 m and 3 m and the geotechnical properties of silty clay were evaluated by laboratory tests (Table 1).

Since geographical characteristics of the test bed adjacent to the sea, the salinity concentration of the pore water was measured to be about 17.7‰, which is about 50% of the average salinity of 35‰ in seawater.

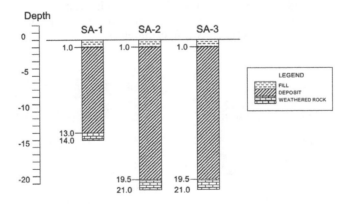

Figure 1. Geologic columnar section of test bed.

Table 1. Results of laboratory test for silty clay.

Depth (m)	Dry unit weight (g/cm³)	Water content (%)	Grain size analysis %					Atterberg limit		USCS
			#4	#10	#40	#200	2μ	LL(%)	PI(%)	
2.0~2.8	1.263	42.5	100	100	100	99.8	24.0	51.4	26.4	CH
3.0~3.8	1.212	46.0	100	100	100	99.8	26.5	52.8	27.5	CH

Since the thermal conductivity of the ground has a direct effect on the freezing efficiency of the AGF method, the evaluation of thermal conductivity in the unfrozen state and the frozen state should be preceded. In this field experiment, the thermal conductivity of the ground was evaluated using QTM-500. The QTM-500 uses the transient hot-wire method with the flat probe (PD-13 model) of 95 mm × 40 mm size. Measuring range of thermal conductivity is 0.023~12 W/m·K, and accuracy is ±5%. In the unfrozen state, the thermal conductivity was measured at room temperature of about 15 °C. In the frozen state, the thermal conductivity was measured in a freezer after a frozen specimen prepared in a freezer at -10 °C for 24 hours. Laboratory tests were shown in Figure 2, and the results of laboratory tests were represented in Table 2.

2.2 *Overview of field experiment*

The freezing pipes made of stainless steel were applied to the field experiment. The dimensions of the outer freezing pipe are 3.8 m in length and the 89.1 mm in diameter, and the diameter of the inner freezing pipe is 21.7 mm, as shown in Figure 3a. The freezing pipe was vertically constructed to reach the depth of 3.2 m in the ground. The upper part of the freezing pipe that was exposed to the ambient was insulated to maximize the freezing efficiency.

In this field experiment, the temperature of the liquid nitrogen that was applied as the cryogenic refrigerant is at -196 °C. As shown in Figure 4a, first, the liquid nitrogen stored in the tank was injected into the freezing system. Second, the liquid nitrogen that has passed through the freezing system was injected into the end of the freezing pipe through the inner pipe and then heat exchanges with the ground while moving to the upper part of the freezing pipe (Figure 4b). Nitrogen, which was phase-changed into the gaseous state through heat exchange with the ground, was discharged to the freezing system through the outlet located in the upper part of the freezing pipe. At this time, the injection of liquid nitrogen was managed by maintaining the discharged-nitrogen temperature constant through the automatic valve installed in the freezing system. Finally, nitrogen was discharged into the atmosphere.

In this paper, two cases of field experiments were performed. First, the freezing rate of marine clay was evaluated by performing a single freezing-pipe test through one freezing pipe. Second, to simulate the AGF method applied to marine clay, a frozen-wall formation test was conducted to form a frozen wall between two freezing pipes.

(a) unfrozen (15 °C)   (b) frozen (-10 °C)

Figure 2.   Measurement of thermal conductivity through QTM-500.

Table 2.   Results of laboratory test for thermal conductivity.

		2.0~2.8m	3.0~3.8m
Thermal conductivity	unfrozen	1.592	1.486
(W/m·K)	frozen	2.669	2.827

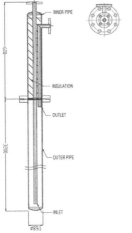

(a) specific configuration of freezing pipe

(b) construction of freezing pipe

Figure 3.   Profile and construction of freezing pipe.

(a) process of liquid nitrogen flow

(b) liquid nitrogen flow in freezing pipe

Figure 4.   Flow of liquid nitrogen.

2.2.1 Single freezing-pipe test

To measuring the temperature changes of the ground during the freezing process, three temperature-measuring holes were drilled in the ground at 0.25 m intervals from the freezing pipe (Figure 5). The thermocouples were installed at intervals of 0.5 m along the temperature-measuring hole from the ground surface to the depth of 4.2 m. Moreover, the thermocouples were also installed at intervals of 0.5 m along the freezing pipe from the ground surface to the depth of 3.2m. The ground surface was insulated to minimize the effect of the ambient air temperature (Figure 5c).

2.2.2 Frozen-wall formation test

In the frozen-wall formation test, the AGF method was simulated by forming the frozen wall between two freezing pipes. The spacing of the freezing pipes was designed to be 1 m, which was applied to the AGF method (Konrad, 2002; Crippa & Manassero, 2006; Li & Wang, 2010). To measuring the temperature change of the ground during the freezing process, seven temperature-measuring holes were drilled between two freezing pipes. 0.25 m and 0.5 m were

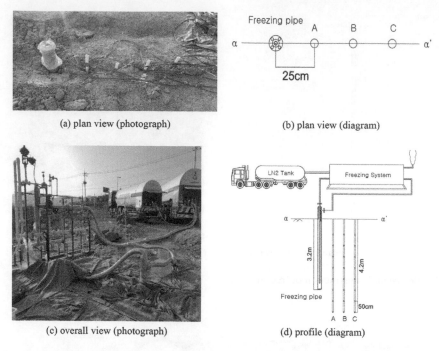

(a) plan view (photograph)

(b) plan view (diagram)

(c) overall view (photograph)

(d) profile (diagram)

Figure 5.　Arrangement of freezing pipe and thermocouples (single freezing-pipe test).

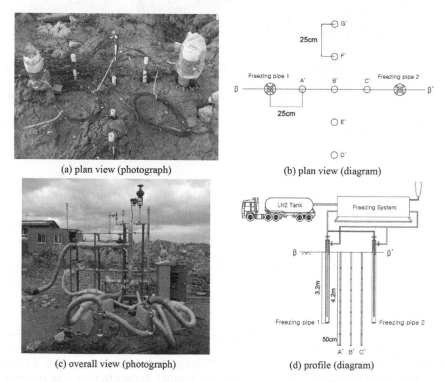

(a) plan view (photograph)

(b) plan view (diagram)

(c) overall view (photograph)

(d) profile (diagram)

Figure 6.　Arrangement of freezing pipe and thermocouples (frozen-wall formation test).

applied at intervals of the temperature-measuring hole and thermocouple, respectively, which were the same as the single freezing-pipe test.

3   RESULTS OF FILED EXPERIMENT

The salinity of pore water decreases the freezing point of pore water, which affects the freezing rate and target freezing temperature of the ground. In this test bed, the salinity concentration of the pore water was measured at 17.7‰, and the freezing point of the pore water was estimated to be -0.96 °C (Arenson & Sego, 2006). Based on the freezing point depression, the freezing time required was estimated.

3.1   Single freezing-pipe test

In the single freezing-pipe test, the temperature of discharged liquid nitrogen was maintained at about -170 °C through the automatic valve installed in the freezing system (Figure 7a). Figure 7b shows the temperature changes of the outer wall of the freezing pipe according to the time measured through the thermocouple.

The size of the frozen soil formed by the single freezing pipe was increased radially from the center of the freezing pipe over time (Figure 8). The field test was performed until the radius of the frozen ground was reached 0.5 m. The freezing time required at each step were summarized in Table 3.

As the result of the single freezing-pipe test, it took 3.5 days (84 hours) to form a cylindrical frozen soil having a volume of about 2.12 m^3 (radius of about 0.5 m, the height of about 2.7 m) around the freezing pipe. At this time, about 11.9 tons of liquid nitrogen was consumed.

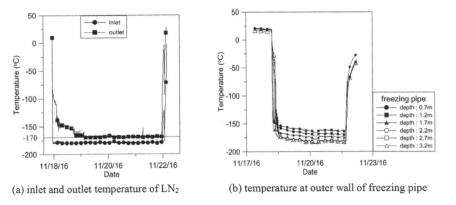

(a) inlet and outlet temperature of LN$_2$ (b) temperature at outer wall of freezing pipe

Figure 7.   Temperature change of liquid nitrogen and freezing pipe with time (single-freezing pipe test).

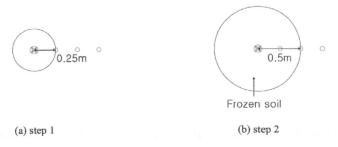

(a) step 1   (b) step 2

Figure 8.   Step of freezing process (single freezing-pipe test).

Table 3. Summary of freezing time required at each step (single freezing-pipe test).

Depth	Step 1	Step 2
0.7 m	20.1 hours	84.0 hours
1.2 m	19.9 hours	78.8 hours
1.7 m	20.6 hours	76.3 hours
2.2 m	18.7 hours	68.8 hours
2.7 m	17.2 hours	58.7 hours
3.2 m	33.0 hours	none

3.2 Frozen-wall formation test

In the frozen-wall formation test, the temperature of discharged liquid nitrogen was maintained at about -180 °C (Figure 9a). Figure 9b and 9c show the temperature changes of the outer wall of the freezing pipe.

The process of forming the frozen wall between two freezing pipes can be classified into four steps as shown in Figure 10. First, a frozen soil having a radius of 0.25 m was formed from each freezing pipe. Next, the two frozen bodies formed from each freezing pipe are attached. That is, the frozen wall starts to be formed. Next, the thickness of the frozen wall increased to 0.5 m. Finally, the thickness of the frozen wall was increased to the target thickness of 1.0 m, at which point the field experiment was terminated. The target thickness applied to this field experiment was applied to the AGF method (Konrad, 2002; Crippa & Manassero, 2006; Li & Wang, 2010).

The freezing time required at each step were summarized in Table 4.

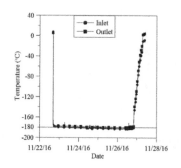

(a) inlet and outlet temperature of LN₂

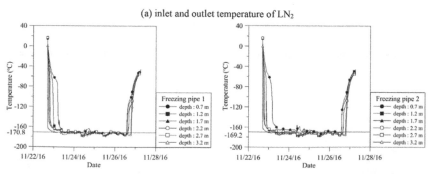

(b) temperature at outer wall of freezing pipe 1   (c) temperature at outer wall of freezing pipe 2

Figure 9. Temperature change of liquid nitrogen and freezing pipe with time (frozen-wall formation test).

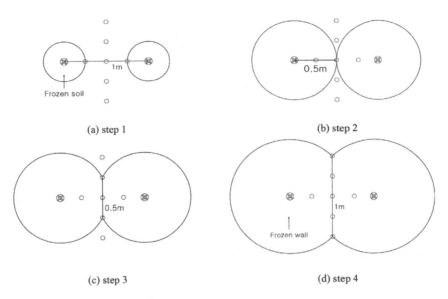

(a) step 1   (b) step 2

(c) step 3   (d) step 4

Figure 10. Step of freezing process (frozen-wall formation test).

Table 4. Summary of freezing time required at each step (frozen-wall formation test).

Depth	Step 1	Step 2	Step 3	Step 4
0.7 m	16.5 hours	39.1 hours	53.5 hours	79.1 hours
1.2 m	11.8 hours	37.9 hours	52.1 hours	80.0 hours
1.7 m	10.1 hours	36.0 hours	56.6 hours	81.8 hours
2.2 m	9.5 hours	31.2 hours	54.9 hours	75.6 hours
2.7 m	9.2 hours	29.4 hours	57.5 hours	80.8 hours
3.2 m	10.7 hours	40.1 hours	71.2 hours	98.2 hours

As the result of the frozen-wall formation test, the frozen wall with a volume of 7.04 m^3, about 2.2 m in width, 1.0 m in length and 3.2 m in height, was formed between the two freezing pipes. Formation of the frozen wall took 4.1 days (98 hours), and about 18.0 tons of liquid nitrogen was consumed.

4 RESULTS ANALYSIS

In the frozen-wall formation test, AGF method was performed until step 4 by setting the target freezing thickness of the frozen wall to 1 m. As shown in Table 4, it took 98.2 hours to form the frozen wall with a thickness of 1 m between two freezing pipes, and this is the freezing time required at the end of the freezing pipe, i.e. the thermocouple installed at a depth of 3.2 m. However, the freezing time required at the upper points of the freezing pipe were estimated to from 75.6 hours to 81.8 hours, which is less than that at the end of the freezing pipe. Similar results were obtained in the single freezing-pipe test, as shown in Table 3. At the end of the freezing pipe, the ground temperature below the freezing pipe acted as a boundary condition, so the freezing time required was estimated to be bigger than that at the upper part of the freezing pipe. The AGF method is conducted until the temperature falls below the target freezing temperature at the entire depth of the target frozen-wall thickness. Therefore, it is essential to monitor the end of the freezing pipe, which is the critical point of the AGF method.

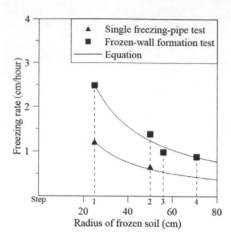

Figure 11.   One-dimensional freezing rate evaluated at 1.7 m depth.

The one-dimensional freezing rate can be estimated by the radius of the frozen soil and freezing time required (Equation 1).

$$v_{f_{1D}} = \frac{r_f}{t_f}$$

(1)

where $v_{f_{1D}}$ = one-dimensional freezing rate; r_f = radius of the frozen soil; and t_f = freezing time required.

Figure 11 shows the one-dimensional freezing rate in the radial direction at a depth of 1.7 m, which is the midpoint of 3.2 m with the freezing pipe embedded.

As shown in Figure 11, the one-dimensional freezing rate decreased as the radius of frozen soil increases in both experiments. The reason for the decrease in the freezing rate as the radius of frozen soil increases is that the area of the frozen soil is not proportional to the radius of the frozen soil but is proportional to the square of the radius of the frozen soil. In other words, when the radius of the frozen soil increases by n times, the area of the frozen soil increases by the square of n times. If the thermal energy injected through the freezing pipe is constant and frozen soil does not absorb or release the thermal energy after freezing, the one-dimensional freezing rate decreases by n times (Equation 2, Choi et al. 2018).

$$v_{f_{1D_n}} = \frac{1}{n} v_{f_{1D_0}}$$

(2)

5   CONCLUSION

In this paper, the freezing rate of marine clay was evaluated by field experiment of AGF method. The field experiments were conducted by circulating liquid nitrogen, which is a cryogenic refrigerant, into freezing pipes embedded at a depth of 3.2 m below the ground surface. The temperature of discharged liquid nitrogen was maintained through the automatic valve, and the temperature change induced by the AGF method was measured at the freezing pipes and in the ground with time.

The field experiments consisted of the single freezing-pipe test and the frozen-wall formation through two freezing pipes. According to the experimental results, the single freezing-pipe test consumed about 11.9 tons of liquid nitrogen for 3.5 days to form a cylindrical frozen body with a volume of about 2.12 m^3. The frozen-wall formation test used about 18 tons of liquid nitrogen for 4.1 days to form a frozen wall with a volume of about 7.04 m^3.

The freezing time required at the end of the freezing pipe was estimated to be less than that at the upper part of the freezing pipe. This is because the ground temperature below the freezing pipe acted as a high-temperature boundary condition at the end of the freezing pipe. Therefore, when the AGF method applied to temporary excavation support or groundwater cutoff, monitoring of the end of the freezing pipe is essential.

The one-dimensional freezing rate can be estimated by the radius of the frozen soil and freezing time required. Since the area of the frozen soil, which increase concentrically around the freezing pipe, is proportional to the square of the radius of frozen soil, not the radius of the frozen soil, the one-dimensional freezing rate decreases as the radius of the frozen soil increases.

ACKNOWLEDGMENTS

This research was supported by a grant (Project number: 18SCIP-B066321-06 (Development of Key Subsea Tunnelling Technology)) from Infrastructure and Transportation technology Promotion Research program funded by Ministry of Land, Infrastructure and Transport of Korean government.

REFERENCES

Arenson, L.U. & Sego, D.C. 2006. The effect of salinity on the freezing of coarse-grained sands. *Canadian Geotechnical Journal* 43(3): 325–337.

Anderson, O.B. & Ladanyi, B. 2004. *Frozen ground engineering (Second edition)*: 363. New York: John & Wiley Sons.

Choi, H-J., Lee, D., Lee, H., and Choi, H. 2018. Evaluation of freezing rate of marine clay by artificial ground freezing method with liquid nitrogen. *Journal of the Korean Society of Civil Engineers* 38(4): 555–565.

Crippa, C. & Manassero, V. 2006. Artificial ground freezing at sophiaspoortunnel (The Netherlands) - Freezing parameters: Data acquisition and processing. *Geo Congress 2006, Atlanta, 26 Feb- 1Mar 2006.*

Konrad, J.-M. 2002. Prediction of freezing-induced movements for an underground construction project in Japan. *Canadian Geotechnical Journal* 39(6): 1231–1242.

Li, D. & Wang, H. 2010. Investigation into artificial ground freezing technique for a cross passage in metro. *GeoShanghai International Conference 2010, Shanghai, 3–5 June.*

Papakonstantinou, S., Anagnostou, G., and Pimentel, E. 2013. Evaluation of ground freezing data from the Naples subway. *Geotechnical Engineering* 166(3): 280–298.

Pimentel, E., Sres, A., and Anagnostou, G. 2012. Large-scale laboratory tests on artificial ground freezing under seepage-flow conditions. *Geotechnique* 62(3): 227–241.

Qin, W., Yang, P., Jin, M., Zhang, T., and Wang, H. 2010. Application and survey analysis of freezing method applied to ultra-long cross-passage in metro tunnel. *Chinese Journal of Underground Space and Engineering* 6(5): 1065–1071.

Russo, G., Corbo, A., Cavuoto, F., and Autuori, S. 2015. Artificial ground freezing to excavate a tunnel in sandy soil: Measurements and back analysis. *Tunnelling and Underground Space Technology* 50: 226–238.

Song, H., Cai, H., Yao, Z., Rong, C., and Wang, X. 2016. Finite element analysis on 3D freezing temperature field in metro cross passage construction. *Procedia Engineering* 165: 528–539.

Sun, C.-W. & Qiu, P.-Y. 2012. Research on the freezing method applied to tunnel cross passage of the Guangzhou metro. *Modern Tunnelling Technology* 49(3): 161–165.

Tunnels and Underground Cities: Engineering and Innovation meet Archaeology,
Architecture and Art, Volume 4: Ground improvement in
underground constructions – Peila, Viggiani & Celestino (Eds)
© 2020 Taylor & Francis Group, London, ISBN 978-0-367-46868-2

Ground freezing and excavation of the Museum Island metro station under a river in central Berlin – challenges and experiences

J. Classen
Implenia Construction GmbH, Munich, Germany

P. Hoppe
Implenia Construction GmbH, Berlin, Germany

J. Seegers
Projekt-Realisierungsgesellschaft U5, Berlin

ABSTRACT: The extension of the U5 metro line in central Berlin closes the gap between the existing metro line in East Berlin and the so called Chancellor Line U55 running between the Berlin main station and the Brandenburg gate. The last part of the project is the construction of the 110m long Museum Island station near the Berlin cathedral and the newly constructed Berlin castle. The station lies beneath the Spree canal in fully saturated sands with underlying marl.

Freezing is used to stabilize the ground for the subsequent sequential excavation of the platform tunnels. A new method for horizontal directional drilling has been used to install the freeze pipes from shafts on either side of the future station. Extensive monitoring of ancient and new buildings and bridges was carried out during the freeze period and excavation. Sequential excavation and lining was used to limit movements of both the tunnel and the surface. The paper deals with the challenges and experiences during freezing, excavation and concrete works.

1 INTRODUCTION

Construction of the Berlin subway system has always been challenging due to the high water table and the many rivers and canals in the city center. Since the re-unification in 1990 a number of old lines and stations has been revived, and various new lines were established, one of which was the so-called Chancellor Line U55, running from the newly constructed Berlin main station, former Lehrter station to the Brandenburg gate, with only one intermediate stop at the German parliament, the so-called Reichstag. In former East Berlin the U5 today ends at the Ale-xanderplatz station. The new U5 line closes the approx. 2,2km gap running parallel under the street Unter den Linden for most of its alignment, but also passing under the new Berlin city castle and two shipping waterways. Three new stations, Berliner Rathaus (Berlin Town Hall), Museum Island and Unter den Linden, where a change-over to the existing line U6 is possible, will be built along the line.

2 THE PROJECT

Lot 1 comprises the train change-over facility near the town hall, the station Museum Island, the cross-over station Unter den Linden, and the connection to the existing Brandenburg Gate station. The parallel running tunnels were excavated by a slurry TBM of 6,7m diameter,

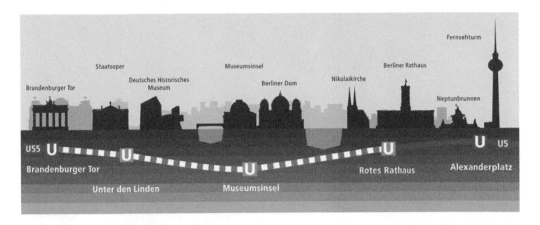

Figure 1. Alignment of the U5.

starting from the start shaft in the change-over facility. They first pass under the new Berlin city castle before running under the Spree river and the Spree canal, an artificial shipping waterway. After that they follow the street Unter den Linden up to the existing Brandenburg Gate station. After finishing the first drive, the TBM was dismantled underground, leaving the shield skin and cutterhead in the ground. It was transported back to the start shaft and finished the second tunnel drive.

The change-over facility and the station Unter den Linden were constructed using the open cut method, while the station Museum Island will be excavated from two shafts at the east and west end. The station is situated under the Spree canal, therefore open cut is not applicable.

In 2012 the works were assigned to Bilfinger Construction, which in late 2014 was acquired by Implenia Construction of Switzerland.

Works started in April 2012 and are planned to finish late 2019.

3 THE MUSEUM ISLAND STATION

The Museum Island station is the most demanding part of the works, since it is situated directly under the Spree canal, and therefore not accessible from the surface. It is also situated partly beneath the old castle bridge which is founded on wood piles, and the now called Bertelsmann building. Two shafts at the east and west end allow access for the necessary freezing operation. The ground consists of water saturated sands with underlying marl. The total length of the station will be approx. 180m, of which 110m will be excavated conventionally in the ice body.

3.1 *Excavation of the running tunnels*

Both TBM drives passed the later station placing precast concrete segments during excavation. These segmental tunnels will be demolished during excavation of the station, but needed to be supported during the process of ground freezing and excavation of the central station tunnel. This was achieved by placing precast concrete invert segments as foundation for triangular steel structures at distances of 1m. they carry circular steel rings which are connected to the tunnel segments by concrete filled bullflex hoses.

These support structures as well as the precast concrete segments will be dismantled during the excavation of the platform tunnels.

The tunnels also house the brine transfer lines and the measuring installation to control the forming and maintenance of the freeze body.

1315

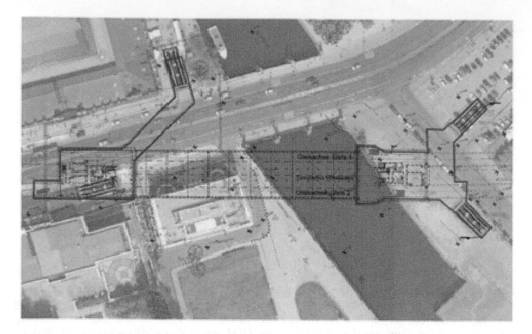

Figure 2.   Plan view of station and shaft.

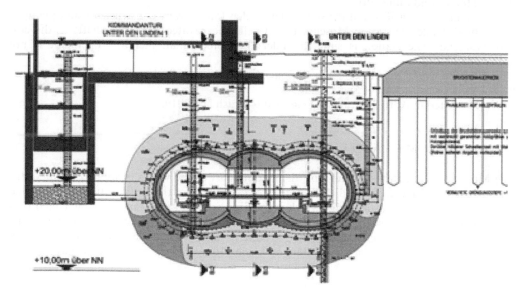

Figure 3.   Neighboring structures.

3.2 *Excavation of the entrance shafts*

Two access shafts were excavated on the east and west end of the future station. During construction they serve as access for the installation of the freeze pipes, and as starting and end point of the excavation of the platform tunnels.

During operation they will house the staircases, elevators and lifts to access and exit the station.

They were constructed using diaphragm walls and an injected invert. Subsequent excavation was performed in various levels.

Figure 4.   Segmental lining support and freeze pipe layout.

Figure 5.   Support structure in running tunnel.

4   GROUND FREEZING

4.1   *Geology and hydro-geology*

The geology in the area of the freeze body is dominated by saturated sands, with minor inter-calations of gravel, peat and big blocks. The underlying strata is marl, which was encountered by the freeze pipe borings at the western end of the station.

Ground water level is more or less at the surface, maximum water pressure therefore around 1,8bar.

Minimum cover of the station crown below the Spree canal water body is approx. 4m, the average flow speed of the ground water is around 1,5m/hour.

4.2 *Freeze pipe installation*

A total of 95 freeze pipe and temperature measurement borings with a length of 105m had to be performed from the eastern shaft. Short counter borings of 5 to 10m lengths were bored from the western shaft to ensure a secure connection of the freeze body to the diaphragm wall. They were situated circular around the two already existing tunnel tubes with a height of approx. 10m and a width of approx. 24m.

The borings started in the upper part of the headwall, which was followed by the excavation of the lower shaft section. After that the invert borings were carried out to allow faster installation of the freezing pipes. Lastly the borings in the idle section were carried out.

In addition measuring sections were also installed in the existing tubes to control the development and the maintenance of the freeze body.

Test borings were carried out in the area of the change-over facility in order to check the reliability and accuracy of the chosen boring equipment.

4.3 *Boring technique*

Due to the long length which required high pressures of up to 1200kN, and the heterogeneous ground conditions, standard boring techniques were not applicable. The contractor chose the press boring method, which was further developed for the removal of soil samples by flushing.

So the positive characteristics of the flushing method were combined with the higher advance power of the pressing method.

A specially developed drill bit allowed directional drilling by turning the inner drill rod in the wanted direction. It also allowed to drill past existing boulders, and to enter safely from the sand section into the much harder marl section.

Due to the high water pressure it was also necessary to use preventers at each bore hole head. It allowed to extract the bored ground safely while maintaining the counter pressure.

The boring pipes were also a new development since they had to follow tight curves when boring around obstacles.

Tests carried out before the actual installation confirmed tightness and stability of the tubing for radii less than 50m.

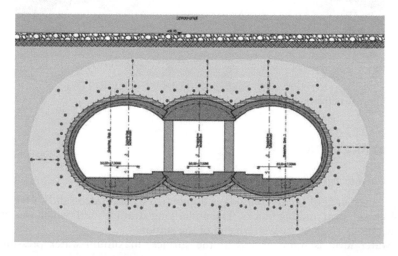

Figure 6.   Planned freeze body and borings.

Figure 7. Bending test on freeze pipe.

The pipes were installed by a boring rig installed on a working platform, which was verti-cally adjustable to allow multiple borings from a single point of installation. The thrust forces were distributed via horizontal steel beams to specially for this purpose installed diaphragm wall elements inside the shafts, serving as an abutment.

To increase further the safety against uncontrollable loss of brine, a welded pipe with spe-cial cold toughness was inserted in the outer pipe. The created void between the pipes was filled with a binder. This multi-layer construction formed a high security level against brine emissions.

4.4   *Execution of the borings*

The execution of a single boring, without encountering obstacles, took around 2 working days. Work schedule was 24/7, performed by 3 shifts. Results and further proceedings were coordinated in a weekly 'boring conference' which involved experts from the client, the designer and the contractor. This proved to be very efficient and showed positive results.

Figure 8. Drill rig on working platform.

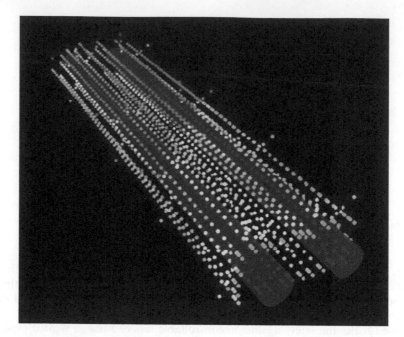

Figure 9. Visualization of the 3D model.

All 3 boring campaigns were performed in time. The deviation of single borings was maximum 10cm and always stayed within contractual limits. Only 3 borings did not reach the designated length, but were well within the area of the overlapping boring from the opposite shaft.

4.5 *Temperature control system*

A total of approx. 2000 temperature sensors were installed to monitor the development and state of the ice body. All results are fed into a specially designed 3D-model which allows analysis of single points or series of points at any given time and location. Results are web-based and can be followed by mobile phones or computers. Various warning levels are defined with relevanactions to be taken by the responsible persons.

The development of the ice body also defined the starting point for the excavation of the central tunnel.

5 EXCAVATION AND INNER LINING

The large finished section of the station requires sequential excavation, which starts with the central tunnel. This is again sequenced, starting with the crown area, followed by the invert. Excavation is performed by an excavator equipped either with a shovel or a rotating head.

Rock support consists of shotcrete, wire mesh and ring beams. In addition prefabricated rebar cases are installed for the later connection to the platform side tunnels.

For design reasons the inner lining of the central tunnel need to be installed prior to the excavation of the platform tunnels. Since there is not continuous wall between central and side tunnels, temporary steel columns are installed to carry the crown section of the inner lining. After having excavated the platform tunnels, the inner lining of all tunnel sections is connected, and the final concrete columns installed.

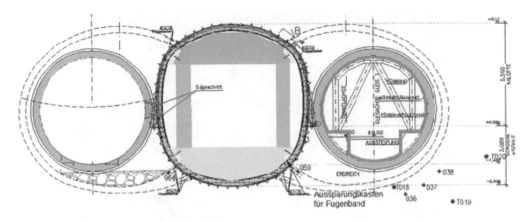

Figure 10. Excavation sequence of the central and platform tunnels.

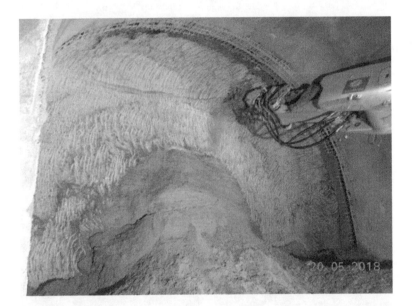

Figure 11. Excavation in the ice body.

6 CONCLUSIONS

At the time of preparation of this paper the central gallery is fully excavated and lined. The state of the ice body is as designed, and excavation proved to be as expected. At the end of November 2018, the excavation of the side platform tunnels started. The pre-fabricated boxes of the steel rebar cages to connect the shotcrete shell of the central gallery to the side platform shotcrete reinforcement proved to be a useful improvement, saving costs and time. All works are performed according to program, and so far without any accidents.

The excavation of the side platform tunnels will continue until end of February 2019, followed by the inner lining works until July 2019. Handover of the civil works for the Museums Island Station is scheduled for December 2019.

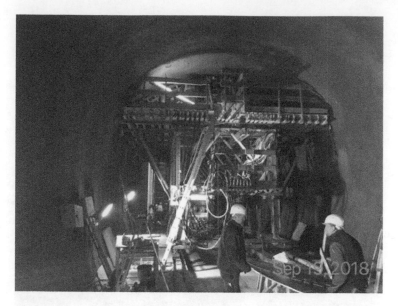

Figure 12.   First crown segment ready for concrete.

Figure 13.   Finished inner lining of the central gallery.

Tunnels and Underground Cities: Engineering and Innovation meet Archaeology,
Architecture and Art, Volume 4: Ground improvement in
underground constructions – Peila, Viggiani & Celestino (Eds)
© 2020 Taylor & Francis Group, London, ISBN 978-0-367-46868-2

Influence of the fibreglass reinforcement stiffness on the mechanical response of deep tunnel fronts in cohesive soils under undrained conditions

C. di Prisco & L. Flessati
Politecnico di Milano, Milano, Italy

G. Cassani
Rocksoil S.p.A., Milano, Italy

R. Perlo
Officine Maccaferri S.p.A., Zola Predosa, Italy

ABSTRACT: The fronts of tunnels excavated under particularly difficult ground conditions are commonly reinforced by inserting fibreglass pipes. This technique is particularly popular since it is very simple to adapt the reinforcement number/length according to the nature of soils encountered. In this paper, the authors illustrate the results of a 3D FEM numerical campaign aimed at analysing the influence of the reinforcements on the system response, conveniently summarized by employing a suitably normalized front characteristic curve. The numerical analyses were performed by assuming the material to be isotropic, homogeneous and characterized by an elastic-perfectly plastic constitutive relationship. Only undrained conditions are taken into account: the failure locus is defined according to the Tresca criterion and the flow rule is assumed to be associated. The numerical results show that the effectiveness of the inclusions is not related to the absolute value of the inclusion stiffness but to the value of a suitably defined non-dimensional variable. By employing this non-dimensional variable the authors show that it is possible to tailor the reinforcement stiffness according to the nature of the soil encountered.

1 INTRODUCTION

When tunnels are excavated under particularly difficult ground conditions, the front is commonly supported by inserting fibreglass reinforcements in the advance core. This technique had progressively gained popularity in tunnel engineering, since it is very simple to adapt the spatial reinforcements distribution (number, length and pattern), according to the nature of soil encountered.

In the past, numerous authors analysed the mechanical response of reinforced fronts, by employing experimental, numerical and theoretical approaches.

The mechanical response of reinforced tunnel fronts was experimentally investigated by means of small-scale models, by both performing centrifuge (Calvello and Talor (1999), Kamata and Mashimo (2003), Juneja et al. (2010)) and 1g model tests (Yoo and Shin (2003), Shin et al. (2008) and di Prisco et al. (2018a)).

From a numerical point of view, this problem was analysed by performing both FEM (Peila (1994), Yoo & Shin (2003), di Prisco et al. (2018c)) and DEM (Chen et al. (2013)) analyses.

An interesting theoretical approach was proposed in Wong et al. (2000 and 2004), where the front response is assimilated to the one of a spherical cavity excavated in an infinite soil domain subject to a uniform and isotropic state of stress. The presence of the reinforcements is took into account by defining a suitable homogenized material. Even if this approach allows the estimation of the front displacements, it is not very popular in the tunnel design.

In contrast, in practice, it is common to estimate an "equivalent cohesion" (Grasso et al. (1989)) and to employ the theoretical solutions obtained for unreinforced fronts by employing either the limit equilibrium method (e.g. Horn (1961)) or the limit analysis theory (e.g. Davis et al. (1980), Mühlhaus (1985), Leca & Dormieaux (1990), Augarde et al. (2003)).

An alternative theoretical approach based on the limit equilibrium method is proposed, for granular materials, in Anagnostou & Perazzelli (2015) and, for cohesive materials, in Perazzelli & Anagnostou (2017).

All the theoretical approaches based either on the limit equilibrium method or on the limit analysis theory, are however not suitable for reproducing the mechanical response of deep tunnel fronts in cohesive soils (di Prisco et al. (2018b)). Moreover, these approaches cannot take into consideration the experimentally observed (di Prisco et al. (2018a)) influence of the reinforcement stiffness on the mechanical response of the front.

In this paper, the authors intend to numerically investigate the influence of the reinforcement stiffness on the front response. To this aim, the authors considered a deep tunnel excavated in a homogeneous cohesive soil stratum behaving under undrained conditions and they performed a series of elastic-plastic numerical analyses. The failure condition is defined according to the Tresca criterion and, to avoid volumetric deformations, an associated flow rule is employed. The numerical results are discussed by employing the non-dimensional front characteristic curve, relating the mean value of stress applied on the front and the mean value of the front displacements, proposed in di Prisco et al. (2018b).

The paper is structured as it follows: in §2 the numerical model is described, whereas in §3 and in §4 the numerical analyses results are presented. Finally, in §5 a practical discussion of the influence of the reinforcement stiffness is reported.

2 NUMERICAL MODEL

To study the mechanical response of reinforced tunnel fronts, the authors performed a series of 3D FEM numerical analyses by employing the commercial code Midas GTS NX (http://en.midasuser.com/). The geometry of the numerical model is represented in Figure 1. The tunnel cross section is circular and its diameter is hereafter named D. For the sake of simplicity, the tunnel is assumed to be excavated in a homogeneous cohesive soil stratum, characterized by a constant saturated unit weight (γ_{sat}). The excavation operations are modelled as a reduction in the horizontal geostatic pressure initially applied on the front. For the sake of brevity, the results of only one reinforcement distribution (reported in Figure 1b) and one reinforcement number ($n = 100$) are discussed.

The numerical results are obtained by imposing the following simplifying hypotheses:

(i) the soil mechanical behaviour is assumed to be isotropic elastic-perfectly plastic. Undrained conditions are taken into account: the yield locus is defined according to the Tresca criterion (the undrained strength is hereafter named S_u). To prevent any volumetric deformation, the Poisson's ratio ν_u is imposed equal to 0.5 and the flow rule is assumed to be associated (the dilatancy angle is imposed equal to zero). Both the elastic undrained Young modulus (E_u) and S_u are assumed to be constant along depth;

(ii) only deep tunnels are taken into consideration: the cover/diameter ratio (H/D ratio) is sufficiently large to prevent the plastic domain to get the ground surface;

(iii) the reinforcements are assumed to be characterized by an equivalent circular cross section: their diameter (d) is the diameter of the boreholes drilled to install the reinforcements. The reinforcements are assumed to be constituted by an elastic equivalent material composed by both the reinforcement itself and the concrete grout. The stiffness of this equivalent material is hereafter named E_r. The reinforcements are connected to the soil by means of an elastic-perfectly plastic interface. The failure of the reinforcements is disregarded;

(iv) all the reinforcements are characterized by the same length. Since the actual excavation phases are not modelled, the reinforcement length L represents the minimum length of the reinforcements (before the installation of a new set of reinforcements);

(v) all the reinforcements are assumed to be parallel to the tunnel axis.

The geometrical dimensions of the numerical model (Figure 1) were varied by the authors to assess the reliability of the numerical results. The corresponding numerical results are here omitted for the sake of brevity.

The domain is subdivided in more than 144000 elements: mainly constituted of 8 node hexahedrons and, when necessary for geometrical reasons, of 6 and 5 node pentahedrons (Figure 1). The meshes employed are unstructured and the size of the elements is not uniform within the spatial domain: the mesh size is finer in the part of domain close to the tunnel lining and within the advance core (for a length equal to 2D). To assess the reliability of the numerical results, the authors took into consideration the influence of the spatial discretization on the results. The corresponding numerical results are here omitted for the sake of brevity.

The reinforcements are modelled by means of elastic embedded beam elements (Sadek & Shahrour, 2004). As was previously mentioned, to take into account the soil-reinforcement interface failure, the nodes of the beam elements are connected to the nodes of the soil elements by means of interface elements characterized by an elastic-perfectly plastic behaviour. To reproduce a "quasi rigid"-perfectly plastic interface. The elastic stiffness of the interface elements is significantly larger with respect to the soil stiffness. The maximum bonding stress along the tangential direction between the soil and the beam is assumed to be equal to S_u.

Nil are imposed vertical displacements on the lower boundary, as well as both normal and shear stresses on the upper boundary and stresses on the interior face of the tunnel lining. On the lateral boundaries, horizontal displacements are not allowed, whereas vertical displacements are permitted. The tunnel lining is schematized by means of a rigid plate perfectly constrained to the nodes of the soil domain.

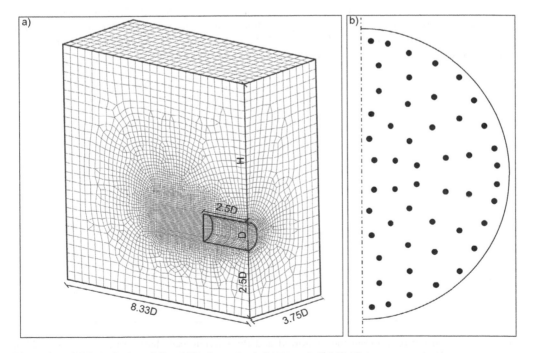

Figure 1.   a) Numerical model and b) adopted reinforcement distribution.

The numerical analyses were performed according to the three following phases:

(i) the initial state of stress is obtained by numerically simulating the stratum formation by keeping constant the geometry and by progressively increasing the gravity. The displacements of the tunnel lining owing to the initial stress conditions are not nil but they are nullified at the beginning of the numerical tests. During this phase, the elements modelling the reinforcements are not introduced in the numerical model;

(ii) the elements corresponding to the reinforcements and to the soil-reinforcement interfaces are activated;

(iii) the initial geostatic linear horizontal stress distribution is progressively nullified.

3 NUMERICAL RESULTS

The numerical results are discussed in the non-dimensional Q_f-q_f plane, introduced in di Prisco et al. (2018b). The non-dimensional stress Q_f and the non-dimensional strain q_f are respectively defined as:

$$Q_f = \left(1 - \frac{\sigma_f}{\sigma_{f0}}\right)\frac{\sigma_{f0}}{S_u}$$ (1)

$$q_f = \frac{u_f}{u_{fr,el}}\frac{\sigma_{f0}}{S_u},$$ (2)

where σ_f is the mean value of the stress applied on the front, σ_{f0} is the initial (geostatic) value of σ_f, u_f is the average front displacement and $u_{fr,el}$ is the elastic displacements for $\sigma_f = 0$ in the unreinforced case. According to di Prisco et al. (2018b) the employment of these non-dimensional variables is particularly convenient since, in the Q_f-q_f plane the front characteristic curve does not depend on the geometry, on the soil mechanical properties and on the initial state of stress.

The authors performed numerous numerical analyses by considering different reinforcements stiffness values, but, for the sake of clarity, in this section only one reference case ($D = 12$m, $H = 69$m, $\gamma_{sat} = 20$kN/m^3, $S_u = 120$kPa, $E_u = 50$MPa, $n = 100$, $E_r/E_u = 600000$, $d = 0.15$, $L = 18$m) is initially discussed. The results of the parametric study are reported in §4.

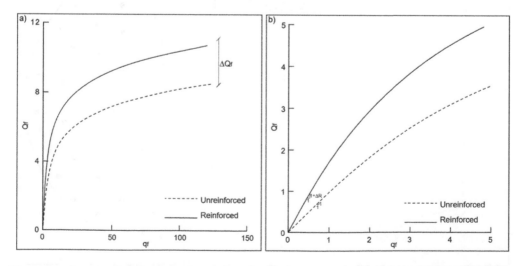

Figure 2. a) comparison between the reinforced and the unreinforced characteristic curve ($D = 12$m, $H = 69$m, $\gamma_{sat} = 20$kN/m^3, $S_u = 120$kPa, $E_u = 30$MPa, $n = 100$, $E_r/E_u = 600000$, $d = 0.15$, $L = 18$m) and b) detail of Figure 2a for $Q_f < 5$ and $q_f < 5$.

In Figure 2a the front characteristic curve obtained for the reference case (solid line) is compared to the corresponding one obtained without reinforcement (dashed line). To better appreciate the initial response, a magnification of Figure 2a is reported in Figure 2b.

The numerical results clearly highlight that the reinforcements severely influence the front characteristic curve. In particular:

- the initial slope of the reinforced curve is significantly larger than the unreinforced one. The initial slope of the unreinforced characteristic curve is 1 (Equations 1 and 2), whereas the initial slope of the reinforced characteristic curve is $1+\Delta R$, where ΔR is the increment in the initial slope with respect to the unreinforced case;
- the amplitude of the initial linear branch of the reinforced curve is larger;
- the final branch of the reinforced curve is almost parallel to the final branch of the unreinforced curve. The vertical distance between the two curves is hereafter named ΔQ_f.

4 PARAMETRIC STUDY

In this section, the influence of the reinforcements stiffness on the front response is numerically investigated. A complete discussion on the influence of the reinforcements (length, number and diameter) on the front mechanical response is reported in di Prisco et al. (2018c). For the sake of generality, a non-dimensional reinforcement relative stiffness is introduced:

$$\bar{E} = \frac{E_r d^2}{E_u L^2}.$$ (3)

This non-dimensional relative stiffness is chosen by assimilating the reinforcements mechanical response to the one of axially loaded piles embedded in an infinite elastic soil domain (Fleming et al. 1985). The characteristic curve in the Q_f-q_f plane is unique for any value of $\bar{E}$. For this reason in Figure 3 each characteristic curve corresponds to a different $\bar{E}$ value. The front characteristic curves obtained for different $\bar{E}$ values are reported in Figure 3a. To better appreciate the initial part of the characteristic curves, a magnification of Figure 3a is reported in Figure 3b.

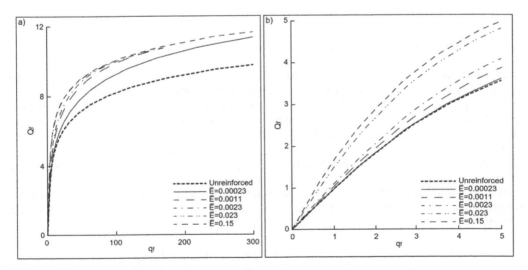

Figure 3. a) influence of the reinforcement stiffness on the front characteristic curve ($D = 12$m, $H = 69$m, $\gamma_{sat} = 20$kN/m^3, $S_u = 120$kPa, $E_u = 30$MPa, $n = 100$, $d = 0.15$, $L = 18$m) and b) detail of Figure 3a for $Q_f < 5$ and $q_f < 5$.

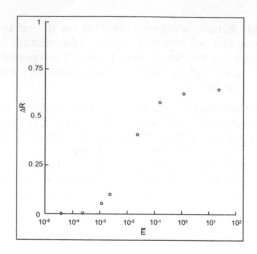

Figure 4. Variation of the initial stiffness of the front characteristic curve with the non-dimensional reinforcement relative stiffness.

The numerical results of Figure 3a and 3b highlight that the reinforcement stiffness significantly influences the initial slope of the characteristic curve: by increasing $\bar{E}$, ΔR increases. On the contrary, the final branch of the curves is not influenced by $\bar{E}$ and, therefore, for the sake of simplicity the influence of $\bar{E}$ on ΔQ_f is disregarded.

The numerical results of Figure 3 (along with some others omitted in Figure 3 for the sake of clarity), can be summarized by plotting the variation of ΔR versus $\bar{E}$ (Figure 4).

As is evident from Figure 4, the results are aligned on a monotonically increasing curve characterized by two horizontal branches.

The horizontal branch for $\bar{E} \to +\infty$ associated with "rigid" reinforcements testifies that for large $\bar{E}$ values increments in the reinforcement stiffness do not provide significant improvements in the global system response.

The horizontal branch for $\bar{E} \to 0$ is associated with reinforcements characterized by a stiffness comparable to the soil one and, therefore, the inclusions do not significantly improve the system response.

5 PRACTICAL IMPLICATIONS

In this section the numerical results reported in §4 are employed to discuss, with reference to E_r values typical for practical cases, the influence of the reinforcement stiffness on the front response and in particular on ΔR. As it was previously mentioned, ΔR is influenced by $\bar{E}$ and, therefore, not only by the reinforcement stiffness E_r but also by d, L and E_u. However, for the sake of brevity, d, L and E_u are hereafter kept constant ($d = 0.15$m, $L = 18$m and $E_u = 30$MPa) and only E_r is parametrically varied. The E_u value is chosen to be representative for the case of a tunnel characterized by $D = 12$m and $H = 69$m excavated in a normally-consolidated clay.

The reinforcements commonly adopted in practice are fiberglass pipes grouted in boreholes (Figure 5) and their equivalent stiffness E_r can be calculated by employing the following expression:

$$E_r = \frac{E_g(d^2 - d_{FG,e}^2) + E_{FG}(d_{FG,e}^2 - d_{FG,i}^2) + E_g d_{FG,i}^2}{d^2} \qquad (4)$$

where E_g and E_{FG} are the stiffness of the material constituting the grouting and the fibreglass pipes, respectively, whereas $d_{FG,i}$ and $d_{FG,e}$ are the internal and the external fibreglass

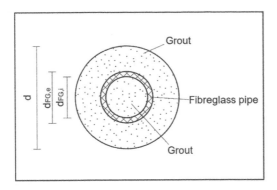

Figure 5.   Fibreglass pipe grouted in a borehole.

Table 1.   Values for the parametric study.

$d_{FG,i}$ [mm]	$d_{FG,e}$ [mm]	E_{FG} [GPa]	E_g [GPa]
$0 \div 40$	$60 \div 150$	$40 \div 200$	$25 \div 35$

pipe diameter (Figure 5). To study the influence of E_r on the system response, the authors considered different $d_{FG,i}$, $d_{FG,e}$, E_{FG} and E_g values (Table 1).

The influence of $d_{FG,i}$, $d_{FG,e}$, E_{FG} and E_g is discussed in the $\Delta R\text{-}\bar{E}$ plane in Figures 6a, 6b, 6c and 6d, respectively.

As is evident from Figure 6, in all the cases considered, the $\bar{E}$ values do not belong to the horizontal branches of the numerical curve. This implies that, an increment in the $\bar{E}$ can improve the front response. However, the increase in $\bar{E}$ associated with the variations in $d_{FG,i}$ are practically negligible, whereas by increasing $d_{FG,e}$, E_{FG} and E_g, $\bar{E}$ increases abruptly. Nevertheless, the variation in ΔR, by considering the mechanical and geometrical constraint imposed, is not very marked. This statement is strictly related to stiffness of the soil taken into consideration. Obviously, in case of quite large values of soil/rock stiffness, the variation in ΔR could be quite larger.

Therefore, the numerical results of Figure 6 can also be employed, in a performance based design perspective, to tailor the reinforcements according to the nature of soil encountered. For the sake of simplicity, it can be assumed that the system performance can be summarized with a given ΔR value, which, in its turn, is directly related to one single $\bar{E}$ value. For a given E_u value, E_r can be "designed" (by considering a proper combination of $d_{FG,i}$, $d_{FG,e}$, E_{FG} and E_g) to obtain the target $\bar{E}$ and the corresponding ΔR value.

6   CONCLUSIONS

In this paper, by performing a series of 3D FEM numerical analyses, the authors analysed the influence of the reinforcement stiffness on the mechanical response of deep tunnel fronts in cohesive soil strata under undrained conditions. For the sake of simplicity, the excavation process is modelled as a reduction in the pressure applied on the front. The soil behaviour is modelled by employing an elastic-perfectly plastic constitutive relationship.

The numerical results are analysed by employing a convenient non-dimensional front characteristic curve, relating macroscopic stress and strain variables.

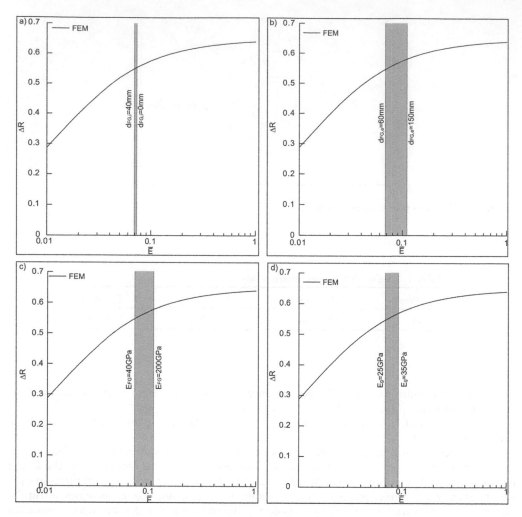

Figure 6. a) influence of $d_{FG,i}$ ($d_{FG,e}$ = 60mm, E_{FG} = 40GPa and E_g = 25GPa), b) influence of $d_{FG,e}$ ($d_{FG,i}$ = 40mm, E_{FG} = 40GPa and E_g = 25GPa), c) influence of E_{FG} ($d_{FG,i}$ = 40mm, $d_{FG,e}$ = 60mm and E_g = 25GPa) and d) influence of E_g ($d_{FG,i}$ = 40mm, $d_{FG,e}$ = 60mm and E_{FG} = 40GPa) for L = 18m, d = 0.15m and E_u = 30MPa.

The numerical results put in evidence that the reinforcement stiffness may significantly influence the front response and, in particular, the initial slope of the front characteristic curves.

Finally, the author show that, in a performance based design perspective, the reinforcements can be tailored on the soil properties and tunnel geometry.

REFERENCES

Anagnostou, G., & Perazzelli, P. (2015). Analysis method and design charts for bolt reinforcement of the tunnel face in cohesive-frictional soils. *Tunnelling and Underground Space Technology* 47:162–181.

Augarde, C. E., Lyamin, A. V., & Sloan, S. W. 2003. Stability of an undrained plane strain heading revisited. *Computers and Geotechnics* 30(5): 419–430.

Calvello, M., & Taylor, R. N. 1999. Centrifuge modelling of a pile-reinforced tunnel heading. *Proc. of Geotechnical Aspect of Underground Construction in Soft Ground*. Rotterdam: Balkema

Chen, R. P., Li, J., Kong, L. G., & Tang, L. J. 2013. Experimental study on face instability of shield tunnel in sand. *Tunnelling and Underground Space Technology*. 33: 12–21.

Davis, E. H., Gunn, M. J., Mair, R. J., & Seneviratine, H. N. 1980. The stability of shallow tunnels and underground openings in cohesive material. *Géotechnique* 30(4): 397–416.

di Prisco, C., Flessati, L. Frigerio, G., Castellanza, R., Caruso, M., Galli, A., Lunardi, P. 2018a Experimental investigation of the time-dependent response of unreinforced and reinforced tunnel faces in cohesive soils. *Acta Geotechnica* 13(3): 651–670.

di Prisco C, Flessati L, Frigerio G, Lunardi P. 2018b A numerical exercise for the definition under undrained conditions of the deep tunnel front characteristic curve. *Acta Geotechnica*, 13 (3): 635–649.

di Prisco, C., Flessati, L. & Porta, D. 2018c Deep tunnel fronts in cohesive soils under undrained conditions: a displacement-based approach for the design of fibreglass reinforcements. *Acta geotechnica*, under review.

Fleming, W. G. K., Weltman, A.J., Randolph, M.F., Elson, W.K. 1985. *Piling engineering.* Surrey University Press, Glasgow and John Wiley, New York.

Grasso, P., Mahtab, A., & Pelizza, S. (1989). Reinforcing a rock zone for stabilizing a tunnel in complex formations. *Proc. Int. congr. Progress innovation in tunnelling*, Toronto, 2, 671–678.

Horn, N. 1961. Horizontaler erddruck auf senkrechte abschlussflächen von tunnelröhren. *Landeskonferenz der ungarischen tiefbauindustrie*, 7–16.

Juneja, A., Hegde, A., Lee, F. H., & Yeo, C. H. 2010. Centrifuge modelling of tunnel face reinforcement using forepoling. *Tunnelling and Underground Space Technology*, 25(4),377–381.

Kamata, H., & Mashimo, H. 2003. Centrifuge model test of tunnel face reinforcement by bolting. *Tunnelling and Underground Space Technology* 18(2–3): 205–212.

Leca, E., & Dormieux, L. 1990. Upper and lower bound solutions for the face stability of shallow circular tunnels in frictional material. *Géotechnique* 40(4): 581–606.

Mühlhaus, H. B. (1985). Lower bound solutions for circular tunnels in two and three dimensions. *Rock Mechanics and Rock Engineering* 18(1): 37–52.

Peila, D. 1994. A theoretical study of reinforcement influence on the stability of a tunnel face. *Geotechnical & Geological Engineering* 12(3): 145–168.

Perazzelli, P., & Anagnostou, G. (2017). Analysis Method and Design Charts for Bolt Reinforcement of the Tunnel Face in Purely Cohesive Soils. *Journal of Geotechnical and Geoenvironmental Engineering* 143(9): 04017046.

Shin, J. H., Choi, Y. K., Kwon, O. Y., & Lee, S. D. (2008). Model testing for pipe-reinforced tunnel heading in a granular soil. *Tunnelling and Underground Space Technology*, 23(3): 241–250.

Wong, H., Subrin, D., & Dias, D. (2000). Extrusion movements of a tunnel head reinforced by finite length bolts—a closed-form solution using homogenization approach. *International journal for numerical and analytical methods in geomechanics*, 24(6),533–565.

Wong, H., Trompille, V., & Dias, D. (2004). Extrusion analysis of a bolt-reinforced tunnel face with finite ground-bolt bond strength. Canadian geotechnical journal, 41(2): 326–341.

Yoo, C., & Shin, H. K. (2003). Deformation behaviour of tunnel face reinforced with longitudinal pipes—laboratory and numerical investigation. *Tunnelling and Underground Space Technology*, 18(4): 303–319.

Tunnels and Underground Cities: Engineering and Innovation meet Archaeology, Architecture and Art, Volume 4: Ground improvement in underground constructions – Peila, Viggiani & Celestino (Eds)
© 2020 Taylor & Francis Group, London, ISBN 978-0-367-46868-2

Analysis of jet-grouting-consolidation in wide and deep shafts in alluvial soil

M. Ferrero, S. Torresani & R. Zurlo
BBT SE, Bolzano, Italy

ABSTRACT: In order to be able to execute ground freezing works, the construction of 4 shafts was required. Due to the geometrical set up of tunnels and geological conditions, the construction of the shafts is particularly challenging because they are wide (800-1600 m^2, ellipse with main axis 35-55 m), deep (up to 28 m) and subject to high water pressure (up to 18 m of water table above bottom). Each shaft was built by means of ground consolidation with 700-1000 jet grouting columns (d = 2000 mm) and underpinning excavation with final r.c. lining, in order to obtain a reasonably waterproof structure able to adequately counter the huge forces acting especially on temporary structures during excavation phases. The following paper will present an analysis of the experience acquired during the excavation of the first shaft, with particular reference to comparisons between design predictions, actual findings and critical issues encountered.

1 INTRODUCTION

The works described in this paper are located in the northern part of Italy, 30 km from the Austrian border, and are related to the southern part of the Brenner Base Tunnel, one of the most challenging works within the Scandinavian-Mediterranean corridor; once completed, the BBT high-speed railway tunnel will be the longest underground infrastructure, 64 km in length.

The area in which the works are located is marked by a strong variability in geo-logical/geo-technical conditions, through which the tunnels are to be built: the area is a fairly narrow valley and the Isarco river (known as the Eisack in German) runs through it; the Brenner Base Tunnel will cross the valley. As a result, in a relatively short stretch of approximately 2 km, the tunnels will have to be excavated through hard rock (Brixner Granite) in the more southern part, through debris/alluvial soil with the underpass of the Isarco river and again through hard Brixner Granite rock (with a fault under the Rio Bianco, a tributary of the Isarco). All the tunnels will be excavated with the full-face method using different techniques: drill & blast in hard rock, ground consolidation with jet grouting/injections in debris/alluvial soil and ground freezing for the underpass of the Isarco river.

The length of the stretch to be excavated with the aid of ground freezing is approximately 70 m. In order for the works to be successful, the drilling needed to install the freezing probes will be carried out from both sides of the Isarco river so to have a maximum horizontal drilling length of 35-38 m and limited drilling deviations. This paper will focus on the preliminary works required in order to be able to properly carry out the ground freezing for the underpass of the Isarco river, with particular reference to the construction of one of the four shafts.

The main characteristics of the shaft under discussion are its dimensions and the fact that it is located in alluvial soil with high water pressure: as a comparison, we can consider that the excavation for the foundations of a bridge pier (piles foundation, d = 1500 mm) in a river area is approximately 200 m^2 and located at a depth of 6-10 m, whereas the shaft under discussion is approximately 1,200 m^2 and 27 m deep with 18 m of hydrohead above the bottom of the excavation.

2 SHAFT DESCRIPTION AND GEOLOGICAL/GEOTECHNICAL CONDITIONS

As mentioned above, 4 shafts, 2 north and 2 south of the Isarco river, are needed. Four tunnels will run under the Isarco: 2 for the Brenner Base Tunnel (for the high speed railway) and 2 for the interconnection between the high speed railway and the local "normal" railway lines. In Figure 1 a general setup of the shafts is shown: on the left we have a schematic aerial view with a clear picture of the position of each shaft in the area of the Isarco river. We can see from this how the drilling for the freezing probes will be carried out from both sides of the river; on the right, a hypothetical setup of the shaft during the works in which the right-hand tunnel is ready to be excavated (freezing probes and ground consolidation are in place and operational) whereas the left-hand tunnel is already completely excavated and lined.

At the time of this writing (August 2018) the POBPN shaft was nearly completed, POBDN shaft was at 50% of excavation and construction of the final lining and the shafts in the southern part (POBPS and POBDS) were at 15% of excavation and construction of the final lining.

The shafts are under construction in the Isarco valley, which is characterized by the presence of fairly heterogeneous alluvial soils (spanning from fine sands, gravels, cobbles up to blocks and boulders; fine soils are substantially absent, while blocks of metrical dimensions are present, especially from ground level to 20 m depth) extending to a depth of no less than 70 m; the water table in the area is 6-8 m below ground level and its flow is quite fast, up to 15 m/day in the shaft area. As the bottom of the shaft is located at 27 m from ground level, the hydrohead will be 18 m, which means that a bottom plug will also be required in order to deal with the hydraulic pressure on the bottom of the excavation.

The shafts have been built by means of ground consolidation with jet grouting columns and underpinning excavation with a final reinforced concrete lining; the construction of each shaft has been executed according to the following sequence:

- preliminary excavation of approximately 2 m below the original ground level;
- execution of jet grouting columns, both for wall support and for the bottom plug; jet grouting columns are not reinforced with steel elements. During the execution of jet grouting drilling and injection, the relevant parameters are continuously monitored and recorded; due to the frequent presence of blocks and boulders, holes for jet grouting are usually predrilled with a separate equipment;
- execution of quality control tests in order to check the adequacy of the columns;
- excavation of a layer 2.5 m in depth;
- construction of final r.c. lining to bear all long-term stresses;
- the latter two steps will be repeated until the bottom of the excavation is reached (with underpinning excavation, the stretch of consolidated ground without the final r.c. lining will never exceed a depth of approximately 3 m: this provides stability to the ground which has been consolidated with jet grouting, but without steel elements);
- construction of the bottom reinforced-concrete slab.

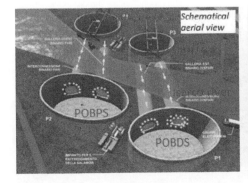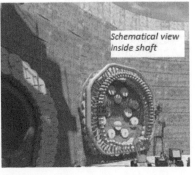

Figure 1.   General setup of shafts across Isarco river.

Of course, at the design stage, proper geological/geotechnical investigation campaigns and an *ad hoc* trial field to determine adequate injection parameters for jet grouting and columns meshes were performed; the description of these activities is not included in this paper.

The main roles of the jet grouting columns within the construction of shafts are as follows:

- to consolidate the ground in order to allow underpinning excavation;
- to reduce permeability of the side shell and the bottom plug in order to limit water infiltration during excavation;
- to consolidate the ground at the bottom plug level in order to counteract bottom upswell during excavation (up to completion of the bottom r.c. slab).

3 DESCRIPTION OF JET GROUTING INTERVENTION

3.1 *Introduction*

At the design stage and during an ad hoc trial field the injection parameters and other design parameters were defined, as follows:

- massive consolidation with double fluid jet grouting columns;
- nominal column diameter: 2000 mm;
- jet grouting columns' specific energy: 77 MJ/m;
- unconfined compressive strength of consolidated soil: 5 MPa;
- residual permeability of consolidated soil: $1 \cdot 10^{-7}$ m/s;
- residual flow in consolidate soil: 5 l/s/1000 m^2;
- column mesh: 1.65m × 1.43m;
- column length for walls: 33 m (2-3 rows of columns for walls);
- column length for bottom: 8 m (with 25 m drilled without injection).

Figure 2 shows the general setup of one shaft: the shaft is shown at its most critical construction phase, that is when the excavation reaches the bottom before completion of the lateral r.c. final lining and final r.c. slab; in this situation, the jet grouting must counter the highest stresses both on walls and on the bottom plug (with particular reference to the upswell of the base plug due to water pressure).

The shaft known as POBPN, which is the first shaft completed, will now be discussed; as follows a brief summary of relevant information related to the POBPN shaft:

- total number of jet grouting columns executed: 913 (319 for the lateral walls, 594 for the base plug);
- execution of jet grouting columns: from February 2017 to July 2017;
- execution of quality control tests: from September 2017 to November 2017;
- excavation and lining of the shaft: from December 2017 to August 2018;
- excavated area: 1203 m^2;
- water level above bottom excavation: 18 m;
- maximum excavation depth: 27 m.

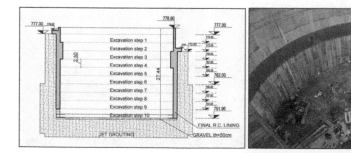

Figure 2.   General setup of one shaft (at the most critical construction phase).

3.2 Main challenges during construction of the jet grouting columns

Consolidation with jet grouting columns, nowadays, is part of common practice in geotechnical works but each project has its peculiarities: a project can be successful only if project peculiarities have been carefully analyzed starting with the design stage. In the area of the Isarco river, the major difficulties were due to:

- the natural soil characteristics;
- the deviation of drilling due to the length of the columns;
- the wear and tear of the drilling material.

The first two items will be discussed below, whereas the third is mainly an equipment issue and will therefore not be presented in this paper. All these aspects had been very well known since the preliminary study phase and were dealt with at the design stage.

3.2.1 Natural soil characteristics

As mentioned, the natural soil here is a mix of alluvial material, from the Isarco river, and debris from the surrounding mountains. The dominant lithology in the area is Brixner Granite, a very hard and abrasive rock: it is easy to understand how this lithology can affect the wear and tear of drilling material. Moreover blocks and boulders are fairly frequently found in natural soil: during geological studies a first layer was identified, from the surface to a depth of approximately 20-22 m, in which the presence of blocks and boulders was potentially more frequent, whereas at greater depths the soil was mainly composed of coarse and fine sand with a limited amount of blocks and boulders.

As mentioned, all parameters were monitored and recorded during jet grouting, including the presence of blocks and boulders: as a result it was possible to have a clear picture of the presence of blocks and boulders in the areas in which jet grouting columns were executed.

In Figure 3 the distribution of blocks and boulders encountered during the execution of jet grouting columns for the lateral walls is reported in detail:

- the blue dots represent the position in which blocks and boulders were recorded;
- the brown line represents the bottom of the excavation;

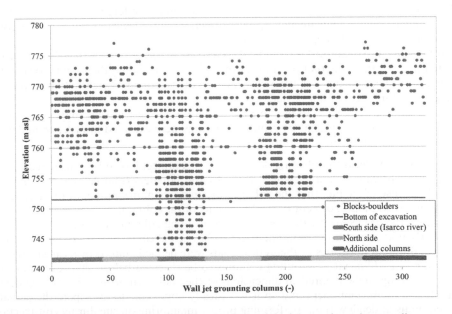

Figure 3. Blocks and boulders recorded during execution of the jet grouting column walls.

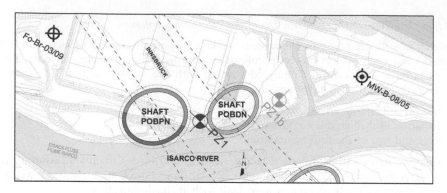

Figure 4.   Shaft POBPN – In green the north side of the shaft, in red the south side of the shaft. Piezo-meters are also shown (in black the historical piezometers, in colors recent piezometers), as described later.

- as the execution of jet grouting columns have been performed to different ground levels (due to the evolution of construction site) the depths recorded have been reported as eleva-tion: the actual ground level during execution of columns varied from 775 m asl to 779 m asl; the ground water level varied from 770 m asl to 771 m asl;
- in the lower part of Figure 3 the lines in different colors represent the plan position of the columns: in red the columns on the south side of shaft (that is the side close to the Isarco, see also Figure 4), in green the columns on the north side of the shaft and in violet the add-itional columns (most of them realized on the northern side of the shaft).

Analyzing Figure 3, it is interesting to note how proper geological/geotechnical investigations can provide accurate information that, such as in this case, was confirmed during execution (i.e. a higher concentration of blocks and boulders in the upper layer). Data related to the inner area of the POBPN shaft are consistent with data related to the lateral wall described above.

3.2.2 Drilling deviation
Data collected during the execution of jet grouting columns have been analyzed in terms of drilling deviation, in particular:

- drilling deviation can affect the final result with particular reference to water-tightness; if two adjacent columns deviate in opposite directions, the design overlap of the columns cannot be achieved with increasing depth;
- as drilling deviation cannot be completely avoided, the jet grouting shell has been recon-structed in 3D with the actual position of the columns (including their deviation) and their nominal diameter in order to have a clear picture of the potential areas in which the proper overlap of columns could not be achieved;
- after the analyses reported above, in areas where a correct overlap may potentially not have been achieved, additional jet grouting columns were executed in order to make up for the potential defects of the jet grouting shell.

It should be underlined that drilling deviation in this project has to be considered a challen-ging issue both because of the presence of blocks and boulders (that can cause drilling devi-ation) and because of the fact that the columns are quite long, meaning that drilling deviation can cause significant misplacement of the lower part of the column.

In Figure 5 the 3D reconstruction of actual jet grouting columns as executed is reported: it can be clearly understood how deviation affected the theoretical layout. As reported on the left of Figure 5, different categories of theoretical voids were defined, and recovery measures (basic-ally consisting in additional jet grouting columns to be executed before shaft excavation) were designed according to the theoretical dimensions of the voids. Unfortunately, drilling deviation has a highly random component and so it is basically impossible to anticipate the actual pos-ition of columns at depth without performing proper monitoring on site during construction.

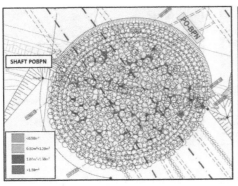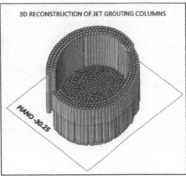

Figure 5.   Shaft POBPN – 3D reconstruction of actual jet grouting columns as executed: left, section at 30.25 m depth; right, global axonometric view.

Figure 6 analyzed drilling deviation, the values span from 0.12% to 2.94% with an average of 1.34%; the drilling specifications prescribed deviation of no more than 2%. In particular:

– Figure 6 left: the columns from 1 to 267 are related to the lateral walls of the shaft, from 268 to 853 to the bottom plug and from 854 to 913 to the additional columns. It is possible to see some differences in drilling deviation for the lateral wall (average 1.03%) and the bottom plug (average 1.48%), the additional columns have an average drilling deviation of 1.39%. The difference in drilling deviation, at first glance, does not seem to have any justification; on the contrary, the columns related to the lateral walls were the first to be executed and so, theoretically, should have been the "worst" ones because the entire system was still in a learning period. With an in-depth analysis of the jet grouting execution, it should be noted that during the execution of the lateral-wall jet grouting columns, the personnel was able to work in a wider and cleaner area (and at the very beginning with only one piece of equipment operating) while the execution of the columns related to the bottom plug was done in a confined area with 2-3 equipments working at the same time meaning that, most probably, the operations were negatively affected due to the limited space (see also Figure 7);
– the total number of jet grouting columns with drilling deviation above the prescribed specification is 63 (out of a total of 913 columns, that means 6.9% of the total): 5 columns from the lateral walls (out of 267, that means 1.9% of the lateral wall columns), 55 columns from the bottom plug (out of 586, that means 9.4% of the bottom plug columns) and 3 from the additional columns (out of 60, that means 5.0% of the additional columns). In this case again, it seems that the limited working space negatively affected drilling deviation;
– Figure 6 right: drilling deviation was plotted vs. length of jet grouting columns, and it can be noted that most of the columns with drilling deviation above specs are in fact the longest columns, as expected.

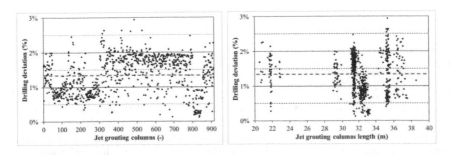

Figure 6.   Shaft POBPN – Left: drilling deviation of columns; right: drilling deviation vs column length; red dotted line: average drilling deviation 1.34%.

Figure 7. Shaft POBPN, execution of jet grouting – Left: lateral walls; right: base plug.

4 ANALYSIS OF SHAFT PERFORMANCES: JET GROUTING INTERVENTION

4.1 *Introduction*

As mentioned above, the main role of jet grouting is to ensure structural resistance and water-tightness in the short term, basically until the final lining in r.c. has been cast in situ. Although the time lapse until this operation occurs is relatively short, this role is essential to allow the physical construction of the shaft. During the excavation of the shaft, a monitoring system was installed, made up of the following equipment:

- 6 inclinometers, 30 m in length, to monitor deformations along the depth of the shaft;
- 12 topographical targets located on the top of the shaft to monitor displacement;
- 2 piezometers located outside the shaft to monitor the water level outside the shaft during excavation (and verify if the external ground water level is affected);
- 1 water flow meter system, to monitor the amount of water pumped out of the shaft;
- strain gauges on the steel reinforcement of r.c. lining at different depths to monitor the stresses inside the final lining.

Besides these instruments, the construction site has an extensive legacy piezometric monitoring system (approx. 30 piezometers, most of them installed during the early stages of the preliminary design and in part no longer available due to interferences with on-site works) to monitor water level in areas affected by the tunnel construction; this can be considered a truly valuable information asset as some piezometers date back to 2005, meaning 13 years of reliable information related to groundwater levels and water quality.

As follows several observations regarding the two main goals according to which jet grouting intervention was designed:

- structural resistance during shaft excavation;
- water-tightness during shaft excavation.

4.2 *Structural performances of the shaft during construction*

Thanks to the underpinning excavation, although the jet grouting columns are not steel-reinforced, no structural problems were encountered during the excavation of shaft POBPN. The concept behind the successful methodology adopted is related to the fact that any excavation step exposes 2.5 m of consolidated soil. Since the lateral wall of the shaft is built by means of ground consolidation with 2-3 rows of 2000 mm diameter jet grouting columns, these are able to counter the stresses induced during excavation; immediately after any excavation step, the r.c. lining is cast in situ, and from that moment on the resistant structure is a r.c. lining ring able to resist stress thanks to the combination of r.c. mechanical resistance and the ring shape itself.

Several observations appear useful, with regards to the shaft bottom plug: this structure should be considered particularly sensitive and proper care should be taken both during the design and the construction of this element, for the following reasons:

- the plug, if not properly designed and constructed, can, potentially, suffer catastrophic instant-aneous breakage due to the upswelling groundwater pressure: unfortunately there is usually no significant early warning, both in terms of time and magnitude, of this kind of breakage;
- monitoring of the bottom plug during excavation is practically impossible, mainly because of the lack of indicators to be monitored that might provide reliable early warning signals and also because of the limited space usually available on site;
- recovery measures to be performed during excavation in order to reinforce the bottom plug are practically impossible both in terms of accessibility of the area with proper equipment and because of the fact that during excavation, a groundwater flow is established causing an increase in water velocity that increases the difficulty of any kind of intervention (any material injected runs a high risk of being washed out);
- in case of critical situations the only reliable safety option available is the backfilling of the shaft: it is easy to understand the impact on the project.

4.3 Water-tightness performance of the shaft during construction

The watertightness of the shaft structure plays a relevant role for the following reasons:

- significant water infiltration during excavation will negatively affect works on site both in terms of safety and efficiency of the works even if, up to certain limits, the presence of water on site is unavoidable and can be easily managed;
- water infiltration can cause a lowering of the water table outside the shaft structure: for POBPN this is not a significant issue, as it is located in an area without sensitive structures nearby but it could be a relevant issue in urban areas. The shaft POBPS, where excavation has just started, is located close to the Verona-Munich railway line, so in this case any lowering of the water table could be an issue (see also Figure 8);
- significant water infiltration can activate transport of solids within the flow, causing local erosion with the potential of damage to the jet grouting structures;
- the infiltrated water must be properly treated before its discharge into a final receptor.

Figure 8.   Shafts close to the Isarco river (aerial view, July 2018).

In Figure 9 relevant data related to groundwater vs. excavation steps are presented on two ordinate axes: all data refers to the left axis (elevation) with the sole exception of water flow (on the right axis). In detail, Figure 9 shows the following information:

- brown line: excavation steps over time. Each excavation step has a depth of 2.5 m, Figure 9 shows the lowest level of excavation for each step;
- empty symbols refer to two legacy piezometers located in the northern shaft area (see Figuer 4), piezometers have been used to monitor the groundwater level of the Isarco valley since 2005 (MW-B-08/05) and 2009 (Fo-Br-03/09); as shown in Figure 9, seasonal groundwater level fluctuation has a magnitude of 2-3 m;
- full symbols refer to piezometers located close to the shafts (see Figure 4), the piezometers were installed (PZ1 in December 2017, PZ1b in April 2018) in order to verify if water pumped out of the shafts was affecting the water level outside the shafts;
- green line: in order to excavate the shaft, the groundwater inside the shaft itself has to be pumped out and the green line represents the average daily water flow pumped out from shaft POBPN. This data has been monitored since March 2018, when excavation reached a depth significantly below groundwater level (4-5 m); monitoring was suspended for some weeks (excavation step 6 was also slowed) for a significant improvement of the pumping system which was necessary to safely execute the following excavation steps.

Analyzing Figure 9 we can make the following observations on jet grouting consolidation:

- the first six excavation steps (step 1 is not reported as it was completely above groundwater level) did not affect groundwater levels and the amount of water pumped out was within 25-30 l/s, according to design predictions. A reduction in water level for piezometer PZ1 (located close to shaft POBPN) was only recorded during steps 2 and 3: most probably this was due to an anomalous (higher) level in PZ1 before excavation. PZ1 was installed after completion of the jet grouting columns and before excavation in the POBPN shaft, and in that area (PZ1) a significant number of jet grouting columns had been executed, as it is also close to shaft POBDN and to the tunnels to be excavated northwards (these tunnels were also consolidated with jet grouting columns executed from ground level before the excavation of POBPN). At the excavation stage, the reduction in groundwater level in PZ1 caused some concern, as at a first glance the impression was that the ground water level was decreasing at the same rate as the excavation; with more in-depth analysis and as

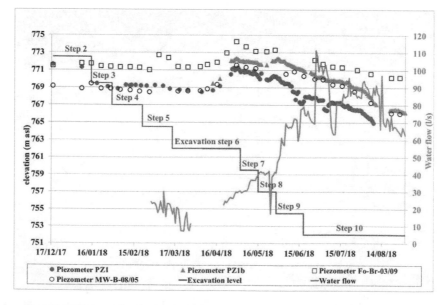

Figure 9. Shaft POBPN, monitoring of data related to groundwater during excavation.

excavation and monitoring proceeded, it became clear that the natural seasonal groundwater level in the area was similar to the one recorded in piezometer MW-B-08/05 (empty circle in Figure 9, elevation approx. 769.0 m asl) and not to the level in piezometer Fo-Br-03/09 (empty square in Figure 9, elevation approx. 771.5 m asl). This was also confirmed analyzing another legacy piezometer (no longer available) that was located in the same area as PZ1. Further confirmation came from piezometrical readings up to step 8, in late May 2018, when piezometer PZ1 showed basically the same readings as the legacy piezometer MW-B-08/05; moreover its qualitative fluctuations are absolutely consistent with groundwater level fluctuations in the Isarco valley (see also the fluctuations in piezometer Fo-Br-03/09); the same observations also apply to PZ1b;

- a small increase in water flow was recorded during excavation steps 7 and 8 but this was not significantly reflected in groundwater levels. Based on the observations above, it is possible to conclude that up to excavation step 8, that is an excavation of 20 m with 13 m of hydrohead, the jet grouting consolidation with underpinning excavation performed very well: no structural problems were reported and the water infiltrations recorded were reasonable;
- excavation steps 9 and 10 had several problems due to water infiltration. In Figure 9 we can see that with a relevant increase in water flow, the groundwater level in PZ1 no longer followed the general groundwater fluctuations for the Isarco valley but suffered a significant reduction (up to 2-3 m compared to MW-B-08/05). Step 9 experienced significant water flow and some additional injections were performed; during the excavation of step 10, a heavy increase in water flow, including also the transport of fine material and sinking levels in PZ1, was registered at the end of June 2018. During that event (which lasted approx. 5 hours), it was considered necessary, as a safety measure, to backfill part of the shaft which had already been excavated and, in the following weeks, to execute additional injections in order to limit groundwater flow and allow completion of the excavation works, the wall lining and the casting of the bottom r.c. slab. Additional injections were not easily performed because of the increased velocity of the groundwater due to the significant hydrohead, but several techniques and materials were applied in order to successfully manage the issue.

The increase in water flow described above can be explained with an increase in the permeability of the jet grouting shell at that depth (steps 9 and 10); this is also the depth at which a change in stratigraphy had been noted during investigations (see 3.2.1). Drilling deviations, as discussed in 3.2.2, and actual diameter (lower than nominal) of the columns, combined with specific stratigraphy, might provide an adequate explanation for the increase of permeability.

5 CONCLUSION AND LESSONS LEARNED

This paper has presented the analysis of jet grouting as used to consolidate wide and deep shafts in alluvial soil. The shaft in question, known as POBPN, has been built for the underpass of the Isarco river, by means of ground freezing technique, as part of the BBT project.

The shaft was built by means of underpinning excavation, down to 27 m from ground level, in soil consolidated by jet grouting columns; consolidation was successfully performed even if some issues related to groundwater infiltration were seen during the excavation at depths greater than 20 m (with 13 m of hydrohead). It should be emphasized that the depth reached with these works cannot be considered as common practice and it is, quite probably, very close to the limit of technological feasibility. This limit is not strictly related to equipment capability in executing long jet grouting columns (in this case up to 38 m), but rather to the possibility of achieving the actual interpenetration of elements able to provide proper water-tightness to the structure during excavation. Drilling deviation has to be carefully considered at the design stage in situations in which the interpenetration of long columns must be achieved for water-tightness: the works described above were executed with jet grouting columns (d = 2000 mm) with a mesh of 1.65 m × 1.43 m; the average deviation of columns was recorded as approximately 1%, that is 20 cm in 20 m. In case of unfavorable distribution of drilling deviation after 20 m, the interpenetration of columns could not be guaranteed.

Tunnels and Underground Cities: Engineering and Innovation meet Archaeology, Architecture and Art, Volume 4: Ground improvement in underground constructions – Peila, Viggiani & Celestino (Eds)
© 2020 Taylor & Francis Group, London, ISBN 978-0-367-46868-2

Jet grouting application for soft soil tunnels in Norwegian clay

D. Gächter & O. Besler
Keller Grundbau GesmbH., Vienna, Austria

ABSTRACT: For the Folloline Project in Norway, running from Oslo to Ski, a double track railway tunnel is under construction. The sections of the project where the tunnels are entering from clay and quick clay into hard Norwegian rock are the geotechnical challenges due to the sensibility of such soils during construction. Keller is executing the main foundation and retaining works in Oslo and Ski using various geotechnical methods to find the best solution for the heterogeneous soil conditions. Especially in the Oslo Area a combination of Lime Cement Columns with Jet Grouting Columns ensured the flexibility and the required safety during execution to avoid movements and landslides in the Oslo City Centre. The dry soil mixing procedure is a standard process for soil improvement in the Scandinavian Countries. Jet grouting so far was not used as a standard method for the treatment of Clay and Quick Clay in relation to retaining structures, soil improvement and underpinning works. The combination of the above stated techniques with Keller´s leading position in the development of those methods were the key to the successful execution at site. To reduce the environmental impact of the ground improvement works, Keller used its own Water and Sludge Treatment plant and was able to decrease the Volume to 40%. A large number of Vibrating Wire Sensors were installed for a live time monitoring of the pore pressure and movements during the Jet Grouting works.

1 MOTIVATION

The use of jet grouting applications for soil stabilization and soil reinforcement is increasing around the world, due to many ongoing infrastructure projects. The projects are often located in highly populated areas with the requirement to execute underground works without influencing the surrounding buildings. The latest large railway projects in Norway are often a combination of typical hard rock tunnels and soft soil tunnels. The soft soil in Norway is often represented by clay and also quick clay. Jet grouting is commonly applied in soil conditions represented in the spectrum of sand to gravel but not as a standard treatment for clay with grain sizes smaller than 0.002mm. In the following paper the Jet Grouting process with its different applications will be described. Furthermore the recently executed stabilization and underpinning works in combination with soft soil tunnels in Norwegian Clay by using the Jet Grouting technique will be explained. As the jet grouting process is always related with back flow, a mixture of the treated soil with cement slurry, the positive aspect of using a filter press will be discussed.

2 SOILCRETE THE JET GROUTING PROCESS

The jet grouting process "Soilcrete®" is recognized as a cement soil stabilization. The name "Soilcrete®" derives from the concept "soil" to "concrete" a form of soil with a concrete consistence, a description that characterizes the type of soil stabilization. With the aid of high pressure cutting jets of water or cement suspension having a nozzle exit velocity ≥100m/sec eventually air-shrouded the soil around the borehole is eroded. The eroded soil is rearranged and mixed with the cement suspension. The soil-cement mix is partly flushed out to the top of

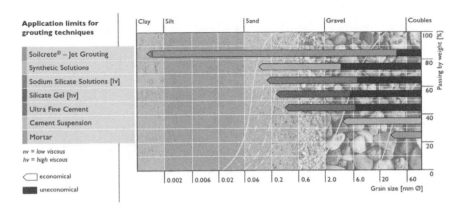

Figure 1. Application limits for grouting techniques.

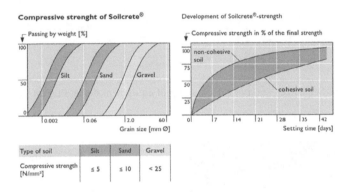

Figure 2. Compressive strength and Development of strength of different soil types depending on the applied jet grouting process.

the borehole through the annular space between the jet grouting rods and the borehole. The erosion distance of the jet varies according to the soil type to be treated, the kind of "Soilcrete®" - process and the jetting fluid being used, and may reach up to 5 meter. The technique is regulated by European Standard EN 12716.

In Contrast with the conventional ground stabilisation methods Soilcrete® may be used for stabilisation and sealing of all kind of soil ranging from loose sediments to clay. This applies also for non-homogeneous soil formations and changing soil layers, including organic material. Soft rock formations have also been treated by Keller – for example sandstone with weak grain texture.

Soilcrete® acts in the ground according to the specification either as a stabilization or as a sealing structure. A combination of both properties is increasingly required. The compressive strength of Soilcrete® ranges from 2 N/mm² to 25 N/mm² and is determined by the cement content and the remaining portion of the soil in the Soilcrete® mass. The sealing effect of Soilcrete® against water ingress is achieved by selecting a suitable grout suspension and if necessary by the addition of bentonite.

3 SITE INSTALLATION

Soilcrete® site installations consists of storage containers, silos and a compact mixing and pumping unit. Several high pressure hose connections and control cable lines connect the pumping unit with the Soilcrete® drilling rig at the installation point. The bore points are normally located in small trenches equipped with sludge pumps. From there the backflow material, water-cement-soil mixture is pumped to setting ponds or tanks. In the current project a

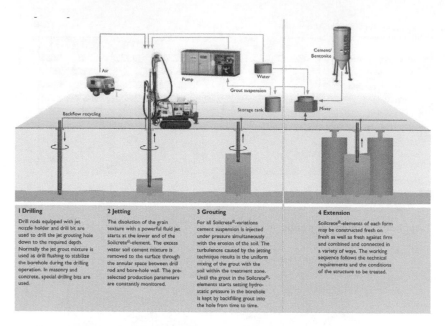

1 Drilling	2 Jetting	3 Grouting	4 Extension
Drill rods equipped with jet nozzle holder and drill bit are used to drill the jet grouting hole down to the required depth. Normally the jet grout mixture is used as drill flushing to stabilize the borehole during the drilling operation. In masonry and concrete, special drilling bits are used.	The disolution of the grain texture with a powerful fluid jet starts at the lower end of the Soilcrete®-element. The excess water soil cement mixture is removed to the surface through the annular space between drill rod and bore-hole wall. The pre-selected production parameters are constantly monitored.	For all Soilcrete®-variations cement suspension is injected under pressure simultaneously with the erosion of the soil. The turbulences caused by the jetting technique results in the uniform mixing of the grout with the soil within the treatment zone. Until the grout in the Soilcrete®-elements starts setting hydro-static pressure in the borehole is kept by backfilling grout into the hole from time to time.	Soilcrete®-elements of each form may be constructed fresh on fresh as well as fresh against firm and combined and connected in a variety of ways. The working sequence follows the technical requirements and the conditions of the structure to be treated.

Figure 3. Soilcrete® Sequences and typical site setup of equipment depending on the applied Jet Grouting Process.

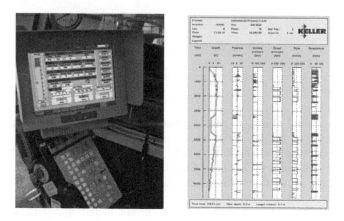

Figure 4. Soilcrete® Process Monitoring, left: live-time process parameters in machine, right: typical process data sheet for each single element executed.

backflow treatment plant as described later more in detail is used to reduce the amount of material transported to the landfill.

The whole Process is monitored over the total period of execution. Therefore up to 12 different parameters for the construction of the elements may be recorded and used by the engineer in charge of supervision and control.

4   QUALITY

The quality testing for the material, grout and the produced product is defined in the European standard EN 12716. Depending on the Jet Grouting application different test needs to

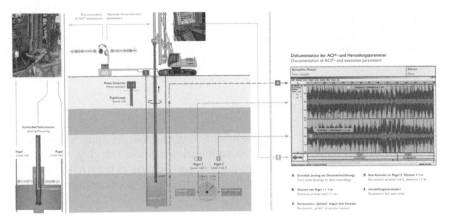

Figure 5. Acoustic Column Inspector – ACI, description of the measuring and execution process.

be performed. The tests are related to the material delivered to site, the produced grout, the backflow material during the grouting process and finally to the hardened product executed by any of the above described process variations.

With jet grouting apart from the control of the material strength the determination of the column diameter is essential. In layered soil conditions, columns have to be installed using varying parameters to achieve a uniform geometry. The ACI- Acoustic Column Inspector-method is used to an increasing extent, mainly where test columns cannot be excavated due to their great installation depth or confined space.

This system provides the opportunity not only to optimize and monitor the production parameters but also to actually prove the contact between jet grouting elements (full columns, half columns, lamellas) and for example bored piles or sheet pile walls. In the following Figure 5, the setup of the testing process is shown as well as the live time readings during the execution of the jet grouting element.

5   GEOTECHNICAL CONDITIONS IN NORWAY

The Soil where currently soft soil tunnels are under construction and going to be constructed in the near future in Norway is often characterised by different types of clay. In some parts the clay is soft and even characterized as quick clay, but in other areas the clay is described as stiff and medium in its sensitivity. Above the clay there is typically a layer of dry crust and fill that mostly varies in its thickness between 1m and 5m, locally maybe thicker and up to 10m. Underneath the clay there is often the presence of a moraine layer between 1m and 20m thickness before entering into the bed rock in form of Diabase and syenite porphyry. In the area of Oslo there is in the transition zone of moraine and bed rock, layers of alum shale. This is a black, shale like species of rock containing pyrites (sulphur ore) and pyrrhotite (magnetic pyrites).

5.1   *Geotechnical Definition*

Soft clay: Clay with undrained shear strength cu < 25kPa. More precisely it means soft and/or very soft clay according to definition of the NGF, Norwegian Geotechnical Society. Details can be found in the NGF melding nr. 2: Veiledning for symboler og definisjoner I geoteknikk. Identifisering og klassifisering av jord, Norsk Geoteknisk Forening, utgitt1982, revidert 2011.

Quick clay: Quick clay is a clay that in remolded state has a shear strength cur < 0,5 kPa. It means that the clay will be liquid in its remolded state. Quick clay is a soil with brittle fracture properties. Such properties mean that the clay will get a considerable reduction of strength due to strains larger than strains at maximum strength. As a consequence of this, the presence of quick clay may lead to the risk of having progressive landslides. A progressive landslide may, for example, start with a minor initial landslide that evolves in to a considerably larger landslide. Such landslides may lead to significant consequences for the affected area. It is known that quick clay may occur in Norway, Sweden, Finland, Russia, Canada and Alaska.

6 JET GROUTING APPLICATIONS

According to the above explained definition of different jet grouting applications two types are applied in the present project under the geotechnical conditions of clay and quick clay. In recent projects often a combination of Stabilisation and Sealing is required. Therefore the Jet Grouting with its various possibilities to adjust the process to local requirements is the ideal technique.

6.1 Joint Sealing– Strutting Slab

A common solution for a cut and cover construction in Scandinavian countries is the combination of a sheet pile wall with a strutting slab. As the ground conditions are mainly characterised by clay in its various forms the dry deep soil mixing technique (LCC – Lime Cement Column) is used to form a strutting slab in-between the sheet piles. To ensure a proper connection of the strutting slab and the sheet pile wall the jet grouting technique is used to seal the gap. The flexibility of the jet grouting process allows also to execute this type of strutting slab in areas where obstacles are present and the soil mixing technique is not applicable. The next figure shows the installation procedure of such a cut and cover excavation pit. Typically the sheet piles are installed until reaching the design depth, followed by the execution of the strutting slab. This activity is divided into 2 phases. First the dry deep soil mixing is performed before the jet grouting columns are placed to avoid any gaps within the slab and to ensure the connection with the sheet pile wall.

Figure 7 shows a design with a Dry Deep Soil mixing application as a rib pattern in combination with a block treatment and jet grouting connection to the sheet pile walls. The rib pattern was done with three rows of 800mm columns with an internal spacing of 700mm. For the connection to the sheet pile wall, Jet Grouting columns of 1000mm diameter every 900mm are executed.

Unexpected findings in the soil, such as boulders, timber piles, sheet piles or other parts of old foundations, makes the application of the dry deep soil mixing technique impossible. In such a case the Jet Grouting technique can realise the soil improvement due to its wide range

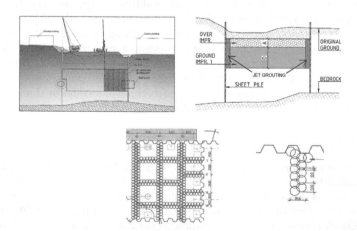

Figure 6.   top left and right: Joint Sealing for a strutting slab by using Dry Deep Soil Mixing (LCC) and Jet Grouting, Common execution procedure of a cut and cover excavation pit, bottom left: Planview of a grid and right: a rib pattern for a strutting slabs with 2 rows of columns.

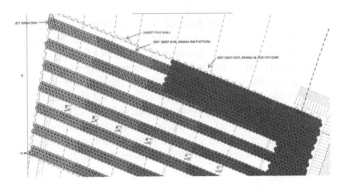

Figure 7.   Strutting Slab with Dry Deep Soil Mixing in rib pattern and block treatment with jet grouting to ensure a solid connection with the sheet pile wall.

of different drilling techniques. As there is the possibility to use a down the hole hammer, boulders and old concrete foundations are no obstacle for the jet grouting application. Moreover the hydraulic mixing process of cement slurry and soil allows to include such obstacles into the soil improvement pattern without removing them before starting the works. Figure 8 shows the as built of the above mentioned design. Due to the encountered obstacles, jet grouting was used to replace the dry deep soil mixing block. As with jet grouting a higher strength can be achieved than with the dry mixing technique, the amount of columns was reduced in the new design. The new design is highlighted below in purple, representing jet grouting columns with 1200mm diameter.

The execution process related to the jet grouting was performed after analysing the results of an extensive trial field at site. This included the investigation with the Acoustic Column inspector to ensure the connection of the jet grouting column and the sheet pile wall as well as to divine the most economical parameters to produce the required diameter. In this application, round columns were applied with depth up to 30m. To investigate the drilling deviations the drill string is measured by using a inclinometer.

For this kind of soil historically the Soilcrete® T – process would be the one used to get a good result in regards of diameter and strength of the element. Due to big developments in

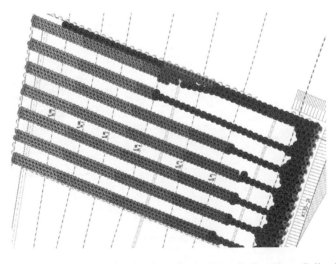

Figure 8.   as built of the strutting slab due to obstacles in the soil, Dry Deep Soil mixing is redesigned with Jet Grouting (highlighted in purple).

the equipment sector and as well in the area of Soilcrete® monitor and nozzles, Keller was able to develop a more efficient way of executing this columns. The decision, mainly influenced by the possibility of handling the process related backflow, was to use the Soilcrete® D – system. In this case an air shrouded grout jet is commonly used. Air in relation with soft clay and sensitive clay can result in large heaving's on the surface which is not permitted by the specification of the project. Instead of using air with all its negative side effects, Keller shrouded the grout jet with water to improve the workability of the backflow and maintain the positive effects by shrouding the jet. The influence of the additional water added into the process was tested during the trial field execution and the requirements regarding strength were proofed. This kind of application showed UCS values of 28 day core samples between 1.5MPa and 2.0MPa.

6.2 Underpinning – Sealed Box Structure

As part of the project an existing tunnel needs to be enlarged to comply with the changed requirements. The tunnel is situated directly under four crossing rail way lines with overburden soil of 2m to 3m. The cross-section is shown in the figure below. There is shown that one side of the tunnel is permanently in use and not changed at all and the second part will be enlarged, by lowering the bottom slab.

The existing tunnel was built in medium stiff clay without an additional foundation, such as Steel Core Piles. Due to some sand layers within the clay the groundwater table is maintained by pumps located in the middle of the existing tunnel. Therefore the purpose of the jet grouting works was divined as a combination of an typical underpinning application with a sealing slab. As this application is fundamentally different in relation of execution sequences, strength and tightness from the joint sealing works described above a trial field was performed prior to the execution of the main works. Core drillings in the concrete structure and the surrounding soil were the base for the design of the jet grouting tests. The soil was characterized as soft and medium stiff clay with some thin layers of sand in which ground water was encountered as well. The test field was similar to the one executed for the joint sealing in regards to the quality testing. To investigate the achievable strength of the jet grouting columns and the possibility to handle the backflow in the tunnel during the execution different methods were tested. For the main construction works the following execution process was defined to get the best result in respect to strength and water tightness. For the water tightness a triangular grid was designed

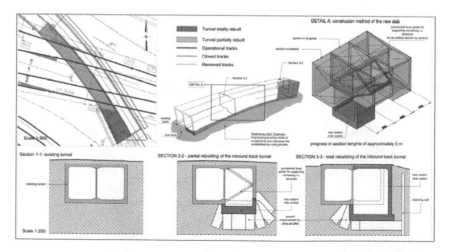

Figure 9.   Plan view of the tunnel and existing tracks, cross-section and 3D-Modell of the underpinning and sealed box structure executed by Soilcrete® application.

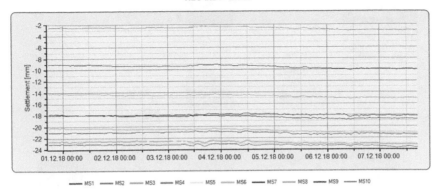

Figure 10.   Readings of the hydraulic level cells during jet grouting showing around 2mm settlement.

to ensure the necessary overlap of the columns in relation to depth and drilling deviation. The columns are executed by means of precutting using a water shrouded grout jet over the full length of the column. The suspension used for precutting has a w/c ration of 2.0 during precutting. Afterwards the column is grouted according to the procedure explained for the joint sealing application. This way of execution ensures the highest quality in respect of water tightness and strength for the underpinning box structure. During the trial field the UCS tests on 28 day core samples showed values of 6.5MPa to 8.0MPa depending on the applied water - cement ratio and lifting speed during execution. The existing tunnel is equipped with water level cells along the tunnel axis on all 3 walls and in addition tilt meter are positioned to alert in case of a rotation of the tunnel. Figure 10 shows an example of the settlement along the outer tunnel wall. The whole monitoring system is connected to the jet grouting machine so that the operator can react immediately to any changes in the live time readings.

7   BACKFLOW MANAGEMENT

The backflow management for jet grouting applications in clay is a very important task to ensure a continuous production. Depending on the project and its requirements different jet grouting procedures are used to execute the elements according to the specification. This indicates that also the back flow and more over the handling of the back flow is changing with the applied grouting process. The amount of back flow is also in direct relation to the process used. In the present project the behaviour of the back flow from the joint sealing was thick and just with additional treatment pump able. This is the reason why this kind of back flow was mainly treated by excavator and ponds. After the hardening time the material was sent to the landfill. Due to the pre-cutting process applied for the underpinning box structure the back flow was handled with common slurry hose pumps. As the produced amount of cement and soil mixture is a big impact on the environmental footprint of jet grouting, Keller has its own backflow treatment plant to reduce this impact significantly.

The backflow treatment plant is a combination of several machines to transport the backflow from the drilling rig to the desander and the filterpress where the actual treatment is taking place. Figure 11 shows the setup of such a plant and highlights the reduction in volume Keller achieves by using this technique. As the filtercake is a mixture of cement and clay mixed by the jet grouting process the amount of material is reduced to 40% of the original volume planned to be disposed on the landfill. This has a positive environmental effect as well as a positive economic impact on the overall project costs.

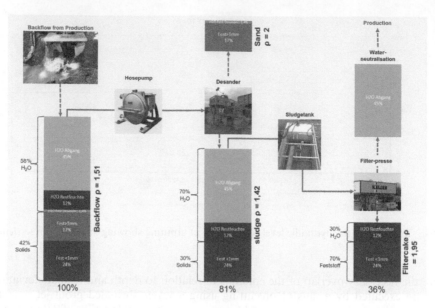

Figure 11. Back Flow cycle from Drilling rig to Filterpress, reduction from 100% at the drilling rig to 36% after treating with the Backflow Treatment Plant.

8 CONCLUSIONS

The successful application of jet grouting for Soft soil tunnels in the Norwegian clay shows that the process has a much wider range then commonly addressed also in respect to the economic aspect of jet grouting. The continuous development of new jet grouting monitors and nozzles in combination with new grout pumps and drilling rigs allows to widen the area of jet grouting applications for future projects in challenging soil conditions. As the environmental footprint is getting more and more in the focus of clients and owners the further development of back flow treatment systems will be an important part of the puzzle to keep reducing the environmental impact of construction sites to our nature. Keller is continuously working close together with universities around the world to develop this product further and to ensure that the environmental impact is decreasing by using state of the art techniques.

REFERENCES

Keller Grundbau The Soilcrete® – Jet Grouting Process. Brochure 67-03E, 0814 - 67-03E

NGF(2011). Norwegian Geotechnical Society, NGF melding nr. 2: Veiledning for symboler og definisjoner I geoteknikk. Identifisering og klassifisering av jord, Norsk Geoteknisk Forening, utgitt1982, revidert 2011.

Keller Grundbau Acoustic Column Inspector – ACI. Brochure 67-05D/E, 0215 – 67-04D/E

Folloline EPC Civil Oslo S – Contract Documents Appendix A – Scope of Works Area 123, October 2015

Keller Grundbau Filterpress – Koralmbahn BL 60.3 St. Kanzian, C. Deporta, 2015

Tunnels and Underground Cities: Engineering and Innovation meet Archaeology,
Architecture and Art, Volume 4: Ground improvement in
underground constructions – Peila, Viggiani & Celestino (Eds)
© 2020 Taylor & Francis Group, London, ISBN 978-0-367-46868-2

Tribunale Station of Naples Metro line 1. Design issues due to site subsoil conditions

S. Gobbi
Designer, I.G.T. srl - iDEAS GeoTechnics, Naples, Italy

F. Cavuoto
Project Manager Line 1 Metro Naples, Naples, Italy

ABSTRACT: The new Tribunale station of Naples Metro Line 1 does not require very deep excavations. Nevertheless, its design implied an accurate preliminary study and the resolution of relevant issues in consideration of the particular environment in which it is to be built. The site where the station is located, in fact, shows some peculiarities that strongly influenced the definition of the design solution. The selected area falls among buildings having direct shallow foundations, that in some cases are only few tens of centimeters apart from the earth retaining walls along the perimeter of the station. From the geotechnical point of view, the subsoil is characterized by a sequence of incoherent soils, mainly of pyroclastic origin, with interbedded peaty soil layers, the thickest of which is found few meters below the design bottom of excavations; groundwater table is close to the ground surface. Issues arising from the subsoil conditions are discussed, and the design solution is described. The results of a preliminary trial field test aimed to select the most appropriate injection parameters for jet grouting are also reported.

1 INTRODUCTION

Line 1 of the Naples Metro system is currently being extended from Garibaldi station (connection to the main railway station) and Capodichino (connection to the international airport). This new section represents a huge improvement in terms of overall urban mobility, since when it will be completed and operating all the main transportation systems, i.e. national and some local railway lines, harbor and airport, will be linked together into an integrated efficient network.

The new section is about 3.2 km long, part of which consists of artificial tunnels and part of natural tunnels having a maximum depth around 50 m. It includes three intermediate stations (Centro Direzionale, Tribunale and Poggioreale) before ending at Capodichino, in correspondence of the airport terminal.

2 SITE CONDITIONS

The new Tribunale station is located along a narrow strip of ground bounded by surrounding buildings on one side and the existing superficial tunnel of the Circumvesuviana railway on the other side. All the buildings close to the station range between 4–5 and 11 floors and have direct foundation, whose depth in some cases is less than 1.0 m below the ground surface. The total length of the station is around 190 m, and the maximum width is 25 m; the design depth of the bottom of the foundation slab is generally equal to 8.20 m with respect to the actual ground surface (average elevation ≈ +9.0 m), with the exception of the technical tunnel passing underneath the tracks that reaches 11.0 m depth.

Extensive geotechnical investigations have been carried out throughout the area of the station. They consisted of continuous coring drilled boreholes, static cone penetration tests using a

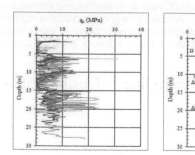

Figure 1. Results of the CTPU tests (tip resistance, q_c) and SPT tests (N-values) vs. depth.

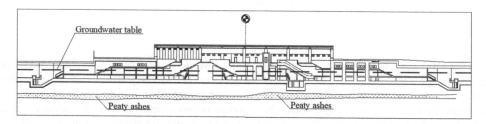

Figure 2. Longitudinal section of Tribunale station. Depth and thickness of the thickest layer of peaty ashes is also shown together with the short term elevation of the groundwater table.

piezocone (CPTU), dynamic Standard Penetration Tests (SPT), pressuremeter tests, a cross-hole test, MASW tests, monitoring of the elevation of the groundwater table, and laboratory testing. Since the knowledge of the coefficients of permeability of soils represents a matter of major concern in view of the design of the station, a preliminary pumping test and a subsequent accurate analysis of its results was also performed besides a series of on-site Lefranc tests.

The subsoil at the site is found to be constituted by a sequence of mainly sandy and silty soil layers of volcanic origin, with interbedded levels and/or thin layers of peaty ashes, the latters being more frequent among the first 10–15 m depth from the ground surface. More precisely, with reference also to the results of CPTU tests (tip resistance q_c vs. depth) and SPT tests (blows/foot vs. depth) shown in Figure 1, the peaty soils are randomly distributed as single thin levels along the depth above the design foundation level, while in the depth range between 12.0 m and 15.5 such soils are found at small/very small spacing giving rise to a kind of a unique thick layer. It follows that few meters below the design bottom of the excavation the presence of a layer around 2.5–3.0 m thick of weak and low permeability soil is to be considered. The position of this weak layer with respect to the longitudinal section of the station is show in Figure 2.

The bedrock, represented by the so called Neapolitan Yellow Tuff, was not reached during the drilling of boreholes until the maximum depth of 40 m.

The monitoring of the groundwater table extended over a period of more than 20 years, in accordance with some measurements taken at regional level, shows that the elevation is slowly rising up. The actual elevation is equal to +5.80 m, and a design elevation of 6.0 m has been set for short term analyses, as those performed to estimate the behavior of the earth retaining walls along the perimeter of the station as a consequence of the excavations to be made. The design short term elevation of the groundwater table is also shown in Figure 2.

3   GEOTECHNICAL MODEL OF THE SUBSOIL

On the basis of the results of all the investigations carried out the reference subsoil stratigraphy given in Table 1 has been established for the whole area of the Tribunale station.

Table 1. Reference subsoil stratigraphy for the Tribunale station area.

Layer	Soil description	Depth Top (m)	Bottom (m)
1	Made ground	0.00	4.00
2	Volcanic ashes	4.00	9.50
3	Volcanic ashes and pumices	9.50	12.00/13.00
4	Peaty ashes	12.00/13.00	14.00/15.00
5	Thin stratified sands	14.00/15.00	26.00
6	Pozzolana* A	26.00	31.00
7	Pozzolana* B	31.00	40.00

* Pozzolana is a local term for pyroclastic, well graded, mainly silty sands

Table 2. Characteristic values of the geotechnical properties of soils.

Layer	γ (kN/m^3)	γ_{sat} (kN/m^3)	c' (kPa)	φ' (°)	E_{sec} (MPa)	k (m/s)
1	17	19	0	32	6	1.0×10^{-4}
2	14	16.5	15	32	18	2.5×10^{-4}
3	12	15	15	34	20	1.5×10^{-3}
4	11	13	10	30	3	8.5×10^{-6}
5	15	17	15	34	40	2.5×10^{-4}
6	15	17	15	33	40	2.5×10^{-4}
7	15	17	15	33	25	1.0×10^{-4}

The characteristic values of the main geotechnical properties for each layer are summarized in Table 2.

4 DESIGN ISSUES ARISING FROM SITE CONDITIONS

Site conditions as those briefly explained in the previous paragraphs give rise to some issues in order to properly design the earth retaining walls along the perimeter of the excavation where the new Tribunale station is currently under construction.

The extreme closeness of the buildings along the eastern side of the station to the earth retaining walls, together with the fact that they have shallow foundations, require that the deformations immediately at the back of the face of the excavation shall be minimized all along the construction phases. This is of course a matter of primary concern, also because during the soil investigations some layers of weak soils were encountered, the thickest of which was found at shallow depth from the design excavation level, i.e. in a critical area for the control of the behavior of the earth retaining walls. Due to the same reason, no relevant variation of the elevation of the groundwater table can be considered acceptable.

At the same time, the lack in the nearby area of a suitable sewer where to discharge the water flowing into the excavation area from its bottom implies the need to limit to a maximum extent, or completely avoid, any water inflow.

5 DESIGN SOLUTION CHOSEN

The optimum design solution has been selected on the basis of extensive preliminary studies and analyses in such a way it meets either the technical requirements or the financial ones.

In this view, one of the best cost-benefit solution would be represented by a continuous slab of consolidated soil of small thickness and reduced permeability located just above the base of

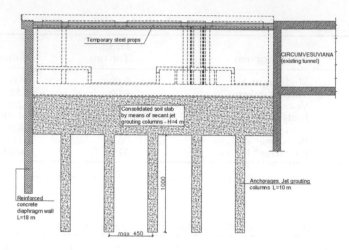

Figure 3.   Typical design cross section of Tribunale station. Temporary props and consolidated soil slab by jet grouting with anchorages are shown.

the diaphragm walls. This option would allow to overcome any possible mutual interaction between consolidation works and peaty soils, but it clearly implies a negligible effect in terms of control of the overall deformation of the earth retaining walls and the soil at their back. On the other hand, it is well recognized that the peaty soil layer around 4 m below the design bottom of excavations strongly affects the distribution of the water pressure in case of seepage and also the effectiveness of any consolidation work. Under these circumstances the usual solution consisting of a consolidated slab of adequate thickness capable of fully balancing the uplift pressures exerted by groundwater does not represent a viable solution. At the same time, it is worth to mention that a certain slab of conveniently consolidated soil located just under the bottom of the excavations plays a dramatic role in limiting the displacements of the diaphragm walls as a consequence of the excavations to be carried out. Besides this, some preliminary consolidation works are also necessary to ensure the stability of the bottom of excavations and lower the soil permeability in order to limit the water inflow into the excavated area, any drainage or temporary water pumping system being not possible because of missing discharge facility.

Taking into account the above, the chosen solution primarily consists of a series of adjacent 18 m long and 80 cm thick diaphragm walls with impervious joints. All diaphragms are connected on top by a reinforced concrete beam, which acts as a partition beam with respect to the temporary props to be installed before starting excavation works or when the excavation depth has reached 3 m below the ground level in function of the height of the nearby buildings. These props consist of hollow steel tubes, whose typical longitudinal spacing ranges between 4.5 m and 6.2 m so that they couple with the disposal of the main supporting beam of the reinforced concrete slab which will realize the roof covering the underground part of the station. Working from the current ground level, a continuous slab 4.0 m thick is realized by means of secant jet grouting columns D =180 cm right under the design excavation depth, so that it is entirely located above the main peaty soil layer. Since the weight of this slab cannot counteract the water uplift pressures, a series of vertical jet grouting columns of the same diameter acting as anchorages are realized in addition at spacing not larger than around 4.0 m. The total length of such anchorages is 14 m, the section crossing the peaty soils being completely neglected to calculate their tensile strength.

Figure 3 shows the typical design cross section of the station as described before.

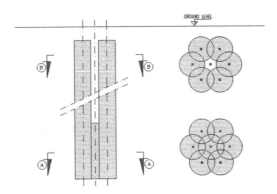

Figure 4.   Typical arrangement of multiple jet grouting columns aimed to preliminary field tests.

6   PRELIMINARY TRIAL FIELD TESTS

In order to assess the effectiveness of the proposed jet grouting consolidation works and to select the most appropriate injection parameters and methodology of execution, two preliminary trial field tests have been foreseen. Due to an incomplete availability of the construction area, at this time only one of them has been performed. It consisted of two series of columns having nominal diameter D = 180 cm, arranged as shown in Figure 4 and executed using two different sets of injection parameters, A and B, and no.4 single columns realized with the same parameters. A third series of columns with nominal diameter D = 220 cm was also arranged by the Contractor to check the option to reduce the total number of the columns and speed up the consolidation works. The main differences between the two sets of parameters pertain the cement content of the injection mixture, varying between 1272 daN/m and 1450 daN/m respectively, and the specific jetting energy, which is equal to 67 MJ/m in the first case and 78 MJ/m in the second one.

The portion of the columns above the groundwater table has been exposed by removing the surrounding soil four weeks after execution. From the direct observation and the results of a series of control drillings it has been noted that the actual average diameters of the columns are in good agreement with expectations; not surprisingly, at the same time, sudden relevant decreases in diameter result in correspondence of peaty soil layers. These elements confirmed the effectiveness of the jet grouting technology in order to consolidate the sandy and silty deposits, and, more in general, of the foreseen design solution.

The results of the laboratory tests carried out on a series of samples taken along the trial columns are shown in Figures 5–6. The effect in terms of unconfined compressive strength and secant modulus of deformation of the different cement content is clearly recognized. Nevertheless, taking into account the conservative design parameters assumed for calculation purposes, column diameter D = 180 cm and injection parameter set A have been selected for final execution.

7   WORK ADVANCE

At the time of delivering this paper (September 2018) construction works are still ongoing. In particular, with reference to Figure 7, all the diaphragm walls are now completed, as well as the consolidation works and the subsequent excavation in the area of the station facing Poggioreale. In the same area the concrete casting of the foundation slab has also been executed.

So far, no issue arose during construction, and the overall behavior of the strutted earth retaining walls and of the consolidated soils by means of jet grouting proved to be very satisfying. The measurements acquired through the monitoring system and their analyses will be the subject of a further paper once the station will be fully completed.

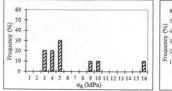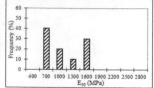

Figure 5. Parameter set A. Frequency distribution of unconfined compressive strength (σ_R) and secant modulus of deformation (E_{50}).

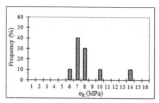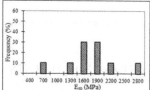

Figure 6. Parameter set B. Frequency distribution of unconfined compressive strength (σ_R) and secant modulus of deformation (E_{50}).

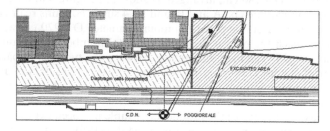

Figure 7. Plan view of Tribunale station. Work advance, September 2018.

8 CONCLUSIONS

Tribunale station is currently under construction as part of the prolongation of Naples Metro Line 1. Due to the combined presence of the groundwater table close to the ground surface and of a thick layer of peaty soils few meters below the ground surface an unusual solution has been developed to ensure the stability of the excavations and minimize the deformations of the earth retaining walls. On the basis of a preliminary trial field test, a kind of anchored slab made of consolidated soil by means of secant columns of jet grouting has been designed and realized. The measures achieved so far by the monitoring system prove the effectiveness of such solution.

ACKNOWLEDGMENTS

The design has been carried out under the supervision of Napoli Metro Engineering srl. Cooperation and support from eng. Sergio Spada and arch. Ernesto Amodeo are gratefully acknowledged.

Construction works are being executed by ICM Group. The kind availability of eng. Giancarlo Acchiappati and eng. Angela Di Nardo is highly appreciated.

REFERENCES

Croce, P., Flora, A. & Modoni, G. 2014. Jet Grouting. Technology, Design and Control. Boca Raton: CRC Press

Esposito, L. & Piscopo, V. 1997. Groundwater flow evolution in the CircumVesuvian plain, Italy. In John Chilton et al. (eds), *Groundwater in the Urban Environment; Proc. XXVII IAH Congress, Nottingham, 21–27 September 1997*. Rotterdam: Balkema

Plaxis 2D. Reference Manual 2018. Delft: Plaxis

Viparelli, M. 1978. Le acque sotterranee ad oriente di Napoli. Fondazione Politecnica per il Mezzogiorno d'Italia, 111. Napoli

Tunnels and Underground Cities: Engineering and Innovation meet Archaeology,
Architecture and Art, Volume 4: Ground improvement in
underground constructions – Peila, Viggiani & Celestino (Eds)
© 2020 Taylor & Francis Group, London, ISBN 978-0-367-46868-2

Brenner base tunnel & Isarco River underpass section: Several technical and operational solutions

P. Lunardi
Lunardi Geo-Engineering, Milan, Italy

G. Cassani, M. Gatti & L. Bellardo
Rocksoil S.p.A., Milan, Italy

ABSTRACT: The Brenner Base Tunnel is part of the TEN-T SCAN-MED corridor and allows the overcoming as a natural barrier of the Alpine ranges. The Isarco River Underpass represents the southernmost construction lot and links the Brenner Base Tunnel with the existing Brenner line and the railway station in Fortezza. The section will include civil works for the two main tunnels for a total length of roughly 4.3 km, as well as two interconnecting tunnels for a total length of 2.3 km that connect with the existing railway line. Construction will be extremely complex from a technical point of view: both the main tunnels and the interconnecting tunnels will pass under the Isarco River, the A22 motorway, the SS12 motorway and the Verona-Brenner railway line with a minimal leeway. The work is scheduled to be completed in 2023. The paper summarizes the design choices made for the construction of the underground works, with specific reference to the construction of the Isarco river underpass tunnels. The excavation will be performed with a very shallow cover, ranging between 5–8 m, very close to the river bed. It will start from shafts on the riverbanks, after ground improvement and soil freezing. The specific field test executed to define the grouting procedure will be focused on as well.

1 INTRODUCTION

The Trans-European Transport Network (TEN-T) aims to develop an integrated multimodal transport network allowing people and goods to move quickly and easily across the EU. This is intended to support the development of the internal market and reinforce economic and social cohesion. The Scandinavian-Mediterranean Corridor (n.5) is the longest of the TEN-T Core Network Corridors and is based in part on a series of former Priority Projects. It links the major urban centres in Germany and Italy to Scandinavia and the Mediterranean whilst crossing 7 different Member States: Finland, Sweden, Denmark, Germany, Austria, Italy and Malta. The Brenner Base Tunnel is part of the TEN-T SCAN-MED corridor and allows for overcoming the natural barrier of the Alpine ranges. For this reason, the EU is prioritising this tunnel among its infrastructural projects. Under the Brenner Pass, the longest railway link in the world is being built: the Brenner Base Tunnel (BBT). The Brenner Base Tunnel stretches for about 55 km between Innsbruck and Fortezza train stations. Together with the existing Innsbruck bypass, the tunnel will reach 64 km and reduce travel time for freight and passenger traffic significantly.

The "Isarco River Underpass" Section is the southern segment of the Brenner Base Tunnel. It is situated approximately one kilometre North of Fortezza, Prà di Sopra, Bolzano (Italy). The section will include civil works for the two main tunnels for a total length of roughly 4.3 km, as well as two interconnecting tunnels for a total length of 2.3 km with the existing railway line. Construction will be extremely complex from a technical point of view: both the main tunnels and the interconnecting tunnels will pass under the Isarco River, the A22 motorway, the SS12 motorway and the Verona-Brenner railway line with a minimal leeway

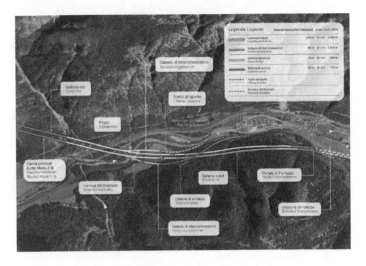

Figure 1. Isarco River Underpass overview map (www.bbt-se.com).

(Figure 1). Before the start of construction, several preliminary activities have been carried out on the surface, including the re-routing of the national road SS12, the construction of two bridges over the Isarco River and the Rio Bianco and the realization of the interconnecting area on the A22, which has been be required to facilitate the transport and supply of construction materials for the Project. The construction works are performing by the "Isarco" consortium, composed by Salini-Impregilo, Strabag, Consorzio Integra and Collini Lavori.

2   GEOGRAPHICAL AND GEOLOGICAL OVERVIEW

The project area is located in Val d'Isarco in the municipality of Fortezza, at an altitudes varying between 750 and 850 m above sea level. The geological sector extends from the Rio Bianco (at the north side) to Fortezza (at the south side). The corridor of the route is divided into two parts by the River Isarco with NW-SE direction. The valleys of Rio Bianco, Rio Vallaga and Rio Riol depart transversally from the direction of the main valley. The most prominent relief within the project area is Mount Riol (1547 m), with its steep side south and SE oriented known as "Hohe Wand". The main residential areas in the Isarco Valley between Fortezza and Rio Bianco are the villages of Pra di Sopra and Fortezza itself. In addition to these villages some farms and buildings are distributed in the project area too. From the geological point of view, this construction lot is ascribing to the Southern Alps sector (Fuoco, 2016).

The railway alignment enters into the South-Alpine crystalline basement, on the southern side of the Periadriatico Line, consisting of the Granite-Granodioritic Pluton of Bressanone, Gabbro del "Monte del Bersaglio" and the metamorphosis of encasing (Filladi, micascisti granitiferi) of the Fillade Quarzifera di Bressanone. The Bressanone Granite is the most widespread rock in the project area (Zurlo, 2013). The Isarco River Underpass passes through the alluvial deposit of the valley bottom and through the dejection conoids of the tributary rivers. These loose deposits, heterogeneous both in composition and in granulometry, consist of gravels and rounded sand, with frequent boulders and thick layers of sandy silt (Figures 2, 3).

The flanks of the valley are covered by coarse particle size material, composed by slope debris, alluvial sediments and weather material. The most voluminous rock fragments, up to 1 m, are made of granite. Except for the silty levels, which in any case constitute a minority, the loose deposits are characterized by friction angle value varying between 30° and 38°, nil cohesion (0–2 kPa) and elastic modulus in the range 40–60 MPa. Table 1 summarizes rock mass's geotechnical parameters. A large survey campaign has been carried out during each design stage.

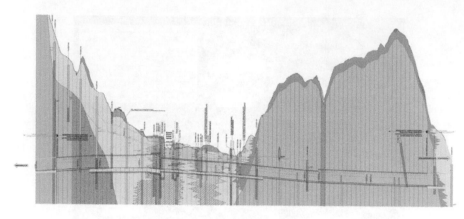

Figure 2.   Isarco River Underpass Geological Profile.

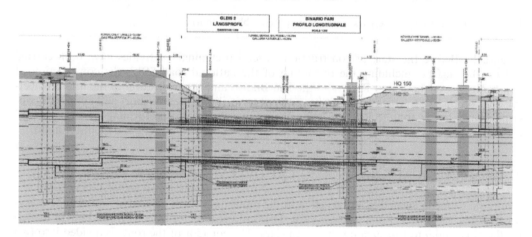

Figure 3.   Geological Profile: detail of Isarco River Underpass.

Table 1.   Rock mass geotechnical parameter.

	Rock Type	σ_{cm} [MN/m²]	γ [kN/m³]	ϕ [°]	c' [kN/m²]	E_s [MN/m]
GA-BG-01	Granite: Granodiorite large - medium fracturing	54 (39–69)	26,5	62	1900	15000
GA-BG-02	Granodiorite large - medium fracturing	22 (12–22)	26,5	58	1000	5000
GA-BG-03	Granite of Bressanone	1.4 (0,8–2,2)	26,7	30	2600	500
GB-G-GA 6	Granite of Bressanone	44	26,7	64	1800	19000
GB-G-GA 7	Granite of Bressanone Rio Bianco fault	7,4	26,7	52	300	1700

Final design required additional in situ and laboratory tests to check stratigraphy, ground's strength and deformability parameters. The water table flows along the same direction of the Isarco River while remaining independent from it. The water table is bounded at the sides and at the base of the rock surface by the Bressanone granite. Hydrogeological studies were carried out in the valley bottom sector, near future tunnels, with the aim of assessing the characteristics of the aquifer and in particular determining the maximum permeability and speed values of the water table, which are of considerable importance for the success of the consolidation intervention. Alluvial deposits have a maximum hydraulic conductivity equal to 1.9×10^{-3} m/s, with a maximum speed, estimated by tracer tests, equal to approximately 14–16 m/d.

3 PROJECT DESCRIPTION AND DESIGN CHOICE

With reference to Figure 4, the Isarco River Lot could be subdivided in three main sectors: 1) the tunnels north of the Isarco river, mainly interested by Granite of Bressanone (GB-G-GA6 and 7) with the special passage through the Rio Bianco fault; 2) the tunnels south of the Isarco River, interested by the Brixen Granite and granodiorite (BG- 01, 02) and 03) the Isarco river valley, where soft soil, mainly alluvial deposit of the valley bottom and the dejection conoids of the tributary rivers, are present (Lunardi, 2018).

The tunnels in rock will be excavated full-face, according to ADECO-RS Approach and by conventional excavation with D&B, with several confinement actions, such as steel bolts and ribs, shotcrete and, locally in fault zones, forepolings (Figure 5). The northern tunnels are double track tunnels and will be excavated starting from the shafts located near the Isarco's banks as the last sector of the Lot. The southern tunnels in rock are single and double track tunnels (double track tunnels from chainage 55+485 towards south) and they are being excavated by a mid-access, named NA4, which gives the possibility to work from four face, two toward south portal and two toward the Isarco river.

For soft soil tunnels, located into the Isarco River, several technical solutions will be applied depending on ground level interferences, overburdens and local geotechnical context. Where ground-level is clear, grouting from surface by jet-grouting technology has been used, by this way tunnels' construction process switches between excavation and final lining casting, without stops for grouting treatments (Figure 6). This solution speeds up tunnels' excavation. Where it was not possible to operate grouting from surface, it was planned to carry out the consolidation activities from the tunnel to guarantee the stability of the core-

Figure 4. Isarco River Underpass Lot.

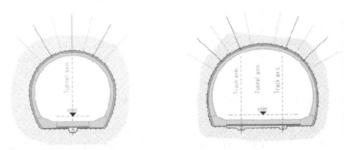

Figure 5.   Conventional excavation method section type for rock mass - Single track & dual track tunnels.

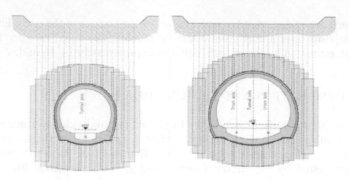

Figure 6. Section type for soft soil by vertical jet grouting - Single track & dual track tunnels.

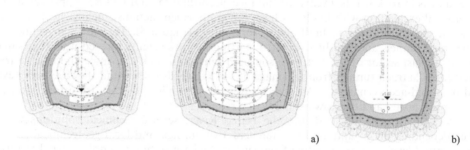

a)   b)

Figure 7. Section types for: a) soft soil with ground improvement by sub-horizontal jet grouting columns and injections - Single track & dual track tunnels and b) for river underpass by ground freezing.

face (Figure 7a), alternating excavation and consolidation to the final lining casting. This construction process will be applied in particular in the northern section for the underpass of the A22 motorway and for the underpass of S.S.22 state road. For these soft tunnels the ADECO-RS Approach has been adopted too. This delicate river underpass consists of 4 tunnels to be excavated, under the riverbed and the banks, without depressing the water table and without proceeding with any deviation of the watercourse. The underpass is safely performed starting from the 4 shafts previously excavated, thanks to the use of an eco-compatible technology that consists in freezing a crown of soil around the entire perimeter of excavation, after pre-treatment of the ground made by injection of cement mixtures and eco-compatible chemical supplements (Figure 7b).

The project choices foresee to totally avoid the lowering of the water table and to eliminate any interference with the flow of the Isarco River, thus reducing environmental impact. In addition the adopted technical solutions, on respect to a cut&cover solution with temporary deviation of the river, allows to: reduce the connection sections between different excavation technologies, thus reducing potential discontinuities and infiltrations, simplify the constructive process achieving faster and controlled progress and considerably reduce the volume of excavated ground and the transit of construction vehicles to depot.

The project guarantees the continuity of water flow both underground and at the surface level, both during construction and during operation. Finally, some sections, preliminary to the relocation of the historical line, will be constructed by cut&cover (Figure 8).

4   UNDERGROUND PROJECT SECTIONS - DETAILS

The project has a total length of 5629 m, 4842 m in conventional tunnel and 787 m in Cut&Cover. Tunnels will be excavated full-face; the variable geological context required different section for excavation. For good quality rock mass "A" sections have been

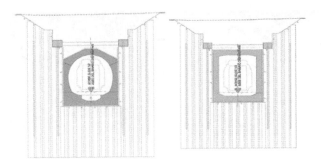

Figure 8. Cut & cover ground improvement by vertical jet grouting columns.

designed, "B" sections for fractured rock mass and "C" sections in case of debris flow and soft soils according to ADECO-RS Approach. The water table drawdown has been avoiding by grouting which provides waterproofing effect. Stability of the core-face is regulated by the intensity of the measures applied (Lunardi, 2015). Special guidelines allow to fine tuning the pre-confinements and confinements actions in order to control the face and cavity deformations and, consequently, the surface settlements. In the following, detailed description for the main project sections is reported (Lunardi, 2018).

4.1 *Rock tunnels*

The excavation, in good quality rock mass, will be performed by drill & blast, with rounds ranging between 3.0 m to 5.0 m, applying the "A" sections: A0, A1 and A2, with GSI values ranging between 50 and 70. The subsequent phases consider the installation of radial steel bolts, Swellex L=3.0–4.0 m capacity 200 KN with mesh approximately 1.20–1.5 m × 2.20–2.50 m, fibre reinforced shotcrete (thickness 5.0–10.0cm with wiremesh 610HD), the reinforced concrete invert (thickness 50.0 cm) and the concrete lining (thickness 36.5 cm for plain concrete and 41.5 cm for reinforced concrete). Very small diametric convergence, 1.0–3.0 cm, are expected considering an elastic behavior of the core-face.

The excavation, in fractured rock mass or in case of fault zones, will be performed by hydraulic hammer, applying the "B" sections: B0 for GSI in the range 35–50, B0V for GSI in the range 25–35 and B2V for GSI less than 25. The subsequent phase considers the installation of steel pipe umbrella (for the sections B0V and B2V), diameter 88.9/10 mm, with interaxis 40 cm and overlapping 5 m, steel arch (2IPN160 installation step 1.50m, 2IPN180 installation step 1.00m for sections B0V and B2V respectively), fibre reinforced shotcrete (thickness 25.0–30.0 cm), reinforced concrete invert (thickness 65.0–80.0 cm) and reinforced concrete lining (41.5–116.0 cm thickness). For the section type B2V the reinforcement of the core-face, by means of fiberglass elements, is also provided, to stabilize the face; a shotcrete layer, thickness 10–25 cm, is placed too. The expected deformations, in term of diametric convergence, are in the range 3.0–8.0 cm; basing on the deformation response is possible, according to defined Guidelines, calibrate the steel ribs step and the distance, from the face, to cast the final lining, invert and crown.

With this purpose a monitoring system was set up, to collect the geotechnical conditions of the face, by means of a continuous face-mapping, and the topographic measurements. Figure 9a shows a typical face condition: the excavation interests mainly granite in very good conditions, with GSI values ranging between 60 and 75. The good quality of the rock is confirmed by the measurement of the convergences that, up to now, have recorded maximum values around of 1.00cm. The mid-adit to excavate rock tunnels, named NA4, is placed in a hard rock slope. To prepare the starting face for the tunnels, hydraulic flat-jack have been used in order to avoid the vibrational disturbance linked to the drill&blast solutions (see Figure 9b).

Figure 9.   a) Tunel face in very good granite (GSI 60–70); b) Excavation with "super wedge".

4.2   *Tunnels in soils – Grouting executed in advance*

Excavation section C1 is applied in case of loose soils of quaternary origin. Particularly in the Northern sector, in correspondence of the Debris Flow with unstable behaviour, to underpass the motorway A22 and the SS12; the overburden is ranging between 25 to 30 m. It is applied in the Southern sector too, in correspondence of the detritus conoids where unstable core-face behaviour is expected. This section type considers pre-consolidation intervention around the cavity and at the core-face, aimed creating the conditions of stability and waterproofing of the natural soil. The design requires that excavation is performed without reducing the groundwater level, in hydrostatic conditions. For this reason it is necessary to create a grouted arch around the cavity and a plug at the face, so that the advance core is waterproofed. The interventions at the face are characterized by injections of cement mixture and waterproofing chemical mixture through PVC pipes with valves ("tube-a-manchette"), in the measure of 3 valves/meter. Half of these PVC pipes will also be reinforced by glass fibre structural elements, to increase the stability and the rigidity of the advance core, as well as providing an easier grip for the concrete sprayed on the excavation face. The stability of the tunnel's crow will be ensured by horizontal "monofluid" jet grouting interventions, $\varnothing=650$mm L=18.0 m, which will be performed after a preliminary intervention by cement mixtures injections through PVC pipes, to reduce the permeability of the soil and facilitate the realization of the Jet Grouting interventions (avoiding seepage effects).

In correspondence of the invert, only cementitious and waterproofing chemical mixture injections will be used. In presence of boulders local steel pipes will be placed in advance. The excavation sector will have a truncated cone shape with a length of 7.00 m, thus guaranteeing a double overlapping of the boundary interventions; the excavation will be performed for single step of 1.0 m, followed by the installation of steel ribs (HEA160/1.00 m) incorporated into shotcrete, 30 cm thickness (see Figure 10). To check the grouting benefits before starting the excavation, permeability tests will be performed in the pre-treated area by cement injections, looking for a reduced permeability: from 1×10^{-4} m/s up to 1×10^{-6} m/s. Additional interventions, or new injections, will be executed if the waterproofing effect is not achieved. Considering the heterogeneity of the soil, especially the presence of limes strata, the grouting activities should be calibrated during the construction process: injections up to 200 lt/valve are expected. The final lining will be casted in situ: the invert, 85 cm thickness, will be casted within 10–15 m from the face; the crown, with truncated cone shape, 66–150 cm, will be casted within 30–40 m from the face; these distances could be regulated during the construction according to the deformation response (predicted convergence value in the range 5.0–7.0 cm).

4.3   *Tunnels in soil – Grouting performed from surface*

An extension part of the tunnels in soils will be realised using jet-grouting treatments executed from the surface. The "double-fluid" system will be adopted to reach nominal diameter of 2000 mm; it should be really considered that very important grouted area will be executed so it possible to consider a "massive" behaviour of the grouted soil. Very important it is the grouting sequence: the drilling mesh will be realised in three step in order to gradually close

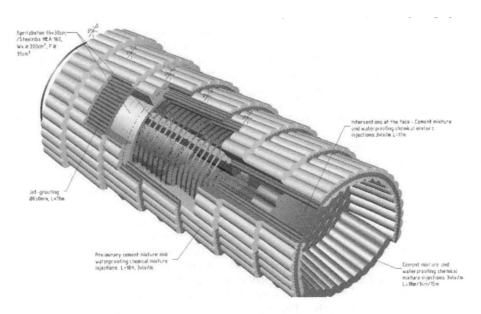

Figure 10.   Excavation section type C1.

the unconsolidated area and to obtain a treatment as widely as possible. The grouted soil must guarantee an increase of the mechanical strength of soil (> 5.0 MPa) and a reduced permeability ($<1\times10^{-7}$ m/s), so to proceed with the excavation in stable conditions and without water leakage. This technical solution will be applied in the odd track from km 54+465,00 to km 54+607,00 and in even track from km 54+440,00 to km 54+608,25 north; south between 54+700 and 54+968 for the even track, and between 54+711 and 54+889 for the odd track. The northern section required the application of complex excavation geometries related to the transition between the 4 single-track tunnels under the crossing of the Isarco river and the existing doubletrack tunnels at the mileage point km 54+608 approximately. The excavation sections, quite large in correspondence of the shaft, characterized by a diameter around 22.0 m, progressively reduce their dimensions to an excavation diameter of 13.50 m (single track). In this sector of the alignment, the overburden ranging from about 10.0 m up to 16.0–17.0 m, suitable for consolidation intervention from the ground level. The tunnel has been subdivided into sectors through the execution of treated soil partitions, by the jet-grouting technology too. The excavation is then performed by full section without depressing the water table; during the construction phase, the water pressure acts on the extrados of the treated soil partition, while in the long term it is expected that the final lining is able to fully support the piezometric head (-1 m from ground level). The jet-grouting technology is used to realize grouted soil of significant dimensions (thicknesses between 3.0 m and 4.5 m and develop of several hundred meters); a square mesh (1.65m x 1.43m) has been adopted also taking into account possible deviations of the perforation and some additional treatments is done after 3D measurements and control. The operational parameters for the execution of jet treatments have been defined after several tests (see 4.5). A cement/water ratio equal to 1 is used for injections at 400 bar with 10 bar air pressure, using 1 nozzle (7 mm in diameter) with stand-by 8.5–10.0 sec for 5 cm step and speed rotation 6/min; the specific energy is in the range 56–72 MJ/ml according to the deep of treatment To check the efficiency of the grouted performed, pumping tests will be executed for each excavation sector to verify the absence of defects in the treatments by estimating the residual leakage.

4.4   River underpass by ground freezing

The river underpass is with low overburden (5.0–8.0m) within a complex geotechnical and hydrogeological context, for this reason the underground excavation could take place only after ground improvement. Several ground improvement techniques have been designed for the execution of the shafts, at the river bank, and of the underpass tunnels. For the excavation of the shaft, elliptical in shape, depth up to 30 meters, "double-fluid" jet grouting is adopted with the same criteria of the chapter 4.3; three rows of jet-grouting columns will be placed all around the shaft in addition to the bottom plug.

The excavation of the shafts will be executed by single step of 2.5 m, followed by the casting of an annular reinforced concrete structure. For the tunnel excavation, ground improvement by means of cement injections and ground freezing will be adopted. The 4 tunnels (Figure 11a) have a length variable between 56 m and 63m and they will be excavated starting from the shafts. The adopted ground improvement techniques have been defined on the basis of a job site test results (see paragraph 4.5) and will be executed from the 4 shafts too with a central overlapping of 4–5 m. Preliminary soil improvement by cement mixture injections is required with a double scope: strengthening the soil mechanical properties and reduce the soil's permeability, to successfully freeze the soil. It's important to take into account that freezing will be performed very close to the river, so that it will be affected by the water flow. The pre-grouting intervention allows to minimize the water flow. The typical section of advancing is represented in Figure 11b: it considers 66 drill holes for the cement grout and 88 drill holes for the installation of freezing probes with length up to 35 meters. Drillings will be executed by "symmetrical" system (140 mm) to place PVC pipes for injections and using steel pipes (114 mm) for freezer probe. The preventer will be installed to avoid water drainage; topographic control will be performed for drillings to consider deviation (generally in the 1.0–1.5% range).

The cement grout sheath has a W/C ratio equal to 1.0, while cement mixture for waterproofing (silicate based) has a W/C ratio variable between 1.2 and 1.4 with a 48 hours compression resistance not lower than 3MPa; this cement mixture will be injected by PVC pipes equipped by valves (3 vlv/meter); the injection will be executed with pressure equal to 15–25 bar and residual pressure of 5 MPa, with flow about 5–8 lt/min and the goal to inject 180–200 lt/valve. Once soil improvement is reached, reducing the permeability value to 1×10^{-5} m/s, the freezing stage can start.

The primary purpose of artificial ground freezing is to draw heat from the ground until its temperature falls below the freezing point of the groundwater (freezing stage) and then to maintain the temperature level reached by appropriately regulating the flow of heat extracted until excavation and construction operations have been completed (maintenance stage). Freezing will be obtained thanks to circulation of "liquid nitrogen" inside the freezing probes (at temperature between -100°C and -60°C) equipped with two concentric pipes: an external AISI Inox piping, 76 mm, and an internal copper piping, 25 mm, till the temperature of -10° C will be reached in the soil at 0.50 m at a distance from the freezing probe, so to obtain a minimum freezing diameter of 1.0 m; the frozen soil must exhibit a compressive strength of 5 MPa. Freezing maintenance will be got by "brine".

Thermometric probes will be placed all around the tunnel cavity to verify the continuity of the freezing, moreover a drainage will be placed in the tunnel axis so to check the absence of water leakage in the core-face to be excavated. The excavation will be made for single step of 1.00 m, followed by the steel rib and shotcrete placing and finally by the final lining casting.

4.5 Testing activities

To develop in detail soil treatment technologies, several trial tests have been performed, starting from the spring 2015 up to now. The first field test was executed in the jobsite at the foundation shaft of the "Isarco Bridge", as showed in Figure 12, south on respect to the Isarco river. To define the best parameters for jet-grouting, test columns were executed with different combinations of extraction and rotation speed, pressure and nozzles' number and dimension.

The diameter columns, investigated by excavation, were ranging from 0.80–1.0 m for "mono-fluid" system to 1.80–2.20 m for "double-fluid" system. Other jet-grouting tests were

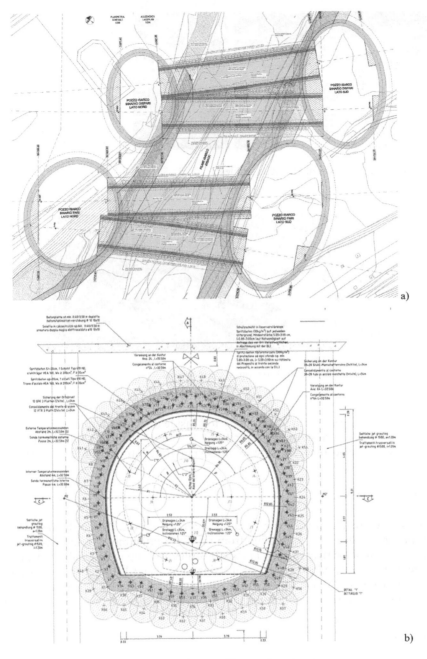

Figure 11.   a) Tunnels &Shafts of the river underpass; b) tunnel section type.

executed in the north part of the jobsite, where fine sand and silt are present to make more difficult the jet-grouting execution at depth (more than 25–30 m). In spring 2015 too, inside the shaft for the "Isarco bridge" foundation, a real test for freezing was performed, as showed in Figure 12b. Starting from the shaft wall, drillings, equipped by PVC pipes, were executed to improve soil by cement injections and check the residual permeability. After soil treatment, freezing procedures by liquid nitrogen were tested, experiencing the designed temperature in the frozen soil. An interesting trial test was executed in autumn 2016, performing a tunnel section by jet-grouting from surface, to check the efficiency of the "massive" jet-

Figure 12.   Job Site field tests and freezing real test.

Figure 13.   The underpass tunnels location and the four shafts.

grouting treatment by performing a pumping test inside the grouted tunnel. The experimental program provided also coring of the grouted soil, to control the RQD value and the compressive strength, and seismic analyses, able to investigate the continuity of the treatment.

5   STATE OF WORK

The current phase provides for the construction of the four shafts, the north ones under excavation, and for the execution of jet-grouting from surface. Furthermore, it includes the excavation of the tunnels south of the river, mainly in rock by drill&blast: up to 24th July 2018 the excavated tunnels' stretch is 2378 m, plus 640 m of connecting tunnels and 192 m of the access tunnel. As a preparatory measure for the relocation of the existing line, which will be carried out during the next construction phase, slope stabilization and earthmoving measures will be implemented.

REFERENCES

Fuoco S., Zurlo R., Marini D. & Pigorini A. 2016. Tunnel Excavation Solution in Highly Tectonized Zones, Excavation through the Contact between Two Continental Plates. *Proceedings of World Tunneling Congress 2016*. San Francisco.

Lunardi G., Cassani G., Gatti M., Bellardo L. & Palomba A. 2018. Brenner Base Tunnel & Isarco River Underpass Section: several technical and operational solutions. *Gallerie e Grandi Opere Sotterranee* (125): 33–44.

Lunardi P. 2015. Muir Wood Lecture 2015 – Extrusion Control of the Ground Core at the Tunnel Excavation Face as a Stabilisation Instrument for the Cavity. *Proceedings of World Tunnel Congress 2015*. Dubrovnik.

Zurlo R., Rea G. & Roccia M. 2013. Galleria di Base del Brennero. Descrizione dell'opera ed avanzamento attraverso la faglia Periadriatica. *Proceedings on the Italian Tunneling Society Congress: ExPo Tunnel 2013*:628–643. Bologna.

Tunnels and Underground Cities: Engineering and Innovation meet Archaeology,
Architecture and Art, Volume 4: Ground improvement in
underground constructions – Peila, Viggiani & Celestino (Eds)
© *2020 Taylor & Francis Group, London, ISBN 978-0-367-46868-2*

The St. Antonino tunnel reconstruction after the great collapse that had occurred on December 2007

G. Luongo, R. Lapenta, A. Pirrotta & U. De Luca
ANAS, Rome, Italy

P. Cosentino, A. Paravati & F. Crocetto
De Sanctis.S.p.A., Rome, Italy

A. Antiga
ÅF-Consult Italy S.r.l., Milan, Italy (previous: Soil S.r.l., Milan, Italy)

P. Coppola & M. Lorenzi
Soil S.r.l., Milan, Italy

ABSTRACT: On the evening of December 3, 2007, an impressive and sudden collapse affected the Reggio Calabria portal of the St. Antonino tunnel. The collapse, which mobilized about 500,000 cubic meters, led to the complete failure of the twin-tube tunnel that had reached about 180 meters. The geological model involves clayey rock in a particularly articulated tectonic structure. To this complex geological scenario it was possible to attribute the cause of the phenomenon. The recovery and completion of tunnels was highly demanding, due to the poor mechanical characteristics of the collapsed ground and to the detection of metal and concrete elements and construction machineries involved in the original tunnel collapse.

The upstream tunnel was completed the May 18, 2018.

1 "ST. BARBARA" COLLAPSE (03.12.2007)

The St. Antonino tunnel is part of the SS106 state road (in southern Italy). In the evening of the 03.12.2007, in coincidence with the preparation of the St. Barbara ceremony, during the excavation on the Reggio Calabria side, the slope at the tunnel entrance collapsed involving an impressive mass of ground, clearly delineated in the slope top part, which has overwhelmed and destroyed the tunnels under construction. The collapsed area (nearly 17,000 square meters) has a quadrangular shape and extends from the morphological saddle up to the entrance area. The upper limit (N-NW) is represented by a long traction fracture, probably generated by a main toppling mechanism, opens with width up to 250 cm. (Figures 1, 2).

The failure occurred in a complex and articulated geological and geomorphological context, mainly characterized by clayey-marly (Am) deposits, which rest on the so-called Argille Policrome (Polychrome Clay, Ap). The geological design model has shown that the twin-tube tunnel crosses the marly-sandy Miocene clays. The entrances on the Reggio Calabria side are set inside a tectonized marly clays, while those on the Taranto side are set in marly clays (Am) under a few meters thick detritic-colluvial blanket (Figure 7).

The stratification inside the clays has a position that plunges towards S-SE, inclined 20 °.

The excavation took place in a challenging geostructural condition, characterized by the presence of folds and traces of a marked lamination of the marlstone, such as to create a particularly severe initial stress state, as reported in the geomechanical survey of the tunnel face (Figure 3).

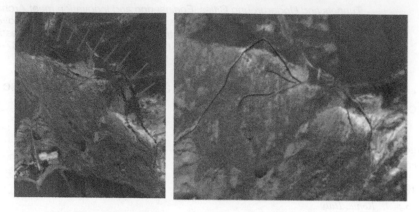

Figure 1. Aerial view of the collapsed area.

Figure 2. Traction fracture in the slope top part and view of the tunnel entrance.

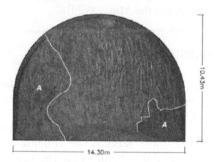

Figure 3. Geomechanical survey of the tunnel face.

At the time of collapse, the two tunnels (upstream and downstream) had reached a length of 171 m and 183 m respectively. The stabilization interventions (section type B2) was done with a provisional lining made with ribs and shotcrete; the inverted arch was 123 m away from tunnel face in the downstream tunnel, about 117 in the upstream tunnel (Figure 4).

From the news collected by the workers, immediately escaped following the first signs of collapse (breaking of the drilling rod during the improvement of the excavation core and large deformation of the steel ribs with splitting in the projected concrete), it appears that the phenomenon was sudden. It happened simultaneously on the two tunnel faces and without being preceded by any warning signs.

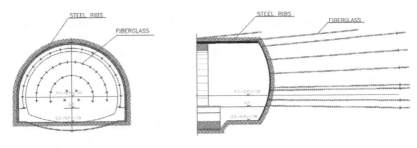

Figure 4. section type B2.

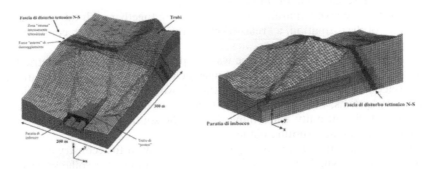

Figure 5. 3D Numerical analyses.

Until the moment of collapse, the excavation of the tunnels had advanced, without difficulties neither any signal of suffering of the provisional lining, in presence of centimetric convergences and of a stable tunnel face.

For the comprehension of the phenomenon and the reconstruction of the geological and geotechnical models, several surveying campaigns were carried out, based both on the geological-geomorphological surface survey and on the execution of geognostic bore holes, in situ tests and monitoring of inclinometers and piezometers. Several undisturbed samples were collected for laboratory tests. Moreover, deep geophysical prospections were performed with the tomographic method.

In order to analyze the possible causes of the collapse, by mean of tridimensional numerical analyzes the advancement of the excavations and the installation of the support works of both tunnels was simulated. (Figure 5)

2 THE NEW GEOLOGICAL AND GEOTECHNICAL MODEL

In the design review following the post-event analysis of 03.12.2007, the geology of the hill where is placed the entrance to the "St. Antonino "tunnel does not differ, in its general lines, from that identified in the project of 2006. It was confirmed a context, mainly characterized by clayey-marly (Am) deposits, which rest on the so-called "Argille Policrome" (Polychrome Clay, Ap).

From a structural point of view, the insights following the landslide event allowed to identify the presence of a faulty anti-form fold that was not visible on the surface. The axis of this fold runs along the ridge line of the geological structure to whose nucleus emerges, in correspondence of the morphological saddle, the formation of "Argille Policrome" (Polychrome Clays, Ap). With respect to the ridge line, therefore, the lay-out of the strata is with strata parallel to slope on both sectors. Based on the observations made on site, it was found that the landslide has characteristics compatible with those of a "complex type" sliding movement, in which the vertical component is predominant compared to the horizontal one. The set of site observations allows to associate a sudden translational kinematism to the landslide with geometry strongly influenced by the pre-existing discontinuities in the marly clays.

The phenomenon was activated along a composite biplanar sliding surface, which involved twin-tube "St. Antonino" tunnel causing the closure, from upstream to the downstream and a negligible translation to SW.

From the analyzes carried out, the collapse is due to a "crushing" which, consequently, has progressively involved the sections of tunnel excavated from the fronts up to about the entrances. The mass of land mobilized (about 500,000 cubic meters), once set in motion, has therefore been directed downstream by gravity towards the inlets on the Reggio Calabria side, as indeed was found, according to the direction of maximum slope.

The deep tomographic geophysical survey made it possible to represent the discontinuities intercepted during the excavation of tunnels (Figure 6).

The inclination of the layers (parallel to slope) induces, together with the tectonization and the clayey nature of the formation, a "relaxation" of the rock masses, with consequent triggering of numerous landslides at different scales.

Ultimately the main cause of the collapse was the presence of a pluri-decametric fault, with an N-S direction and an East dip, with a low-angle inclination towards the excavation.

This fault, once reached by the downstream tunnel, would have triggered the release of lateral discontinuity systems, longitudinal to the tunnels, which caused the failure. It is associated with zones of discontinuity with immersion to the south to medium-high angle which worked as rear retraction and rupture planes.

At the same time, the integrative investigations have allowed to reconstruct the inclination of the discontinuities to which the triggering cause of the impressive collapse was attributed (Figure 7).

The Sant'Antonino collapse is a well known case subject to judicial investigation and following prosecution of the main responsibles. As far as the writers are aware, legal disputes

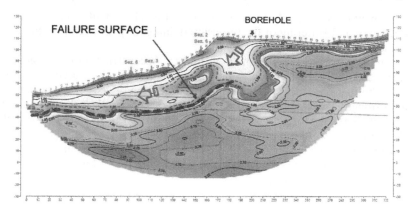

Figure 6. Tomographic geophysical survey.

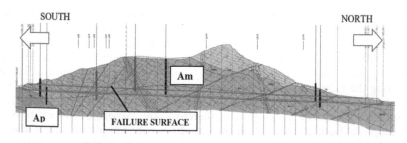

Figure 7. Longitudinal geological section of St. Antonino tunnel.

between the parties are currently under way and therefore we cannot express any evaluations regarding the responsibilities and possible concurrent causes of what happened. The authors believe that it is, in any case, certainly possible to exclude whatever cause related to the design and to highlight how the completion measures analyzed in this paper maintain the principles of the initial design substantially unchanged.

3 THE DEVELOPMENT OF THE WORKS CONTRACT (2015)

The collapse of 3 December 2007 led the court authorities to put under sequestration the construction site in February 2008. That event made the re-elaboration of the project necessary.

The landslide has also made the re-elaboration of the project necessary, adopting the necessary variations and additions, with the definition of both the stabilization of the collapsed slopes and the interventions for the reconstruction of the twin-tube tunnel. As a result of these vicissitudes, the termination of the Contract with the Contractor was reached.

The project of the entire lot, with upstream and downstream carriageways, involves the construction of a 3.7 km road, beginning with the connection with the cut and cover tunnel of the Bova Marina bypass and ending at the new junction of Palizzi on the SS106 "Jonica."

The road lot, in addition to the St. Antonino tunnel, is characterized by the presence of three other twin-tube tunnels and two viaducts with separate carriageways.

Considering the unavailability of the economic financing for the construction of the entire intervention, the Owner (Anas S.p.A.) proceeded to identify a first functional section, for which it was possible to start the procurement procedures. It consists in making accessible the single-carriageway upstream, to be organized in double sense of circulation.

In April 2015, Anas S.p.A. awarded the works of completion of the upstream carriageway to the company De Sanctis Costruzioni S.p.A. of Rome. The works started in August 2015.

4 THE RECOVERY OF WORKS

The tunnel was excavated advancing on both sides Reggio Calabria (south) and Taranto (north). The excavation from Reggio Calabria was carried out entirely within the collapsed section and therefore in a geomechanical and geomorphological context very peculiar and distorted by the landslide phenomenon. This area has been the subject of important morphological requalification and improvement of the geotechnical characteristics (Figure 8) and therefore no longer representative of the geomechanical and geomorphological conditions of the natural rock mass.

The excavation on the Taranto side, on the contrary, was built in an undisturbed rock mass that basically had the same characteristics of the rocky mass of the Reggio Calabria side as they appeared before the collapse.

4.1 *Reggio Calabria (south) entrance*

The landslide movement, which involved the already excavated sections of the St. Antonino tunnel, has destroyed the pre-existing morphological features of the slope. Therefore, the reconstruction of the twin-tube tunnel in the collapsed section, and its future maintenance, depends on the complete stabilization of the crossed slope.

This has led to the prerequisite to intervene on the slope through the remodeling, morphological and hydraulic, of the part of the slope crossed by the tunnel, directly or indirectly involved in the landslide movement (Figure 8). In particular, the remodeling of the slope provided for a substantial load reduction on the top and a contemporary backfilling in the area downstream of the tunnel, so to improve the stability of the slope with respect to the low resistance sliding surfaces identified.

The recovery design included in the first phase the earth movements that were carried out for the balancing of the collapsed slope, in order to obtain a first stabilization of this, sufficient to allow the safe development of the subsequent works.

Figure 8.   Reggio Calabria entrance: morphological and hydraulic remodeling.

In the subsequent phase it included the following design solutions, starting from the existing attack wall of the tunnel (Figure 9):

1. cut-and-cover tunnel section (L = 60 m) between large diameter piles.
2. bored tunnel section (L = 40 m) in which the ground improvement was carried out from the surface by cement grouting;
3. bored tunnel section (L = 70m) where the ground improvement was carried out from the tunnel face in progress.

During the drilling of the large diameter piles envisaged in the cut-and-cover tunnel section, operational difficulties were found to reach the depths of the project piles. This was because in the drilling phase the metal ribs of the collapsed tunnel have been encountered.

To overcome these difficulties an alternative design solution has been studied thet envisaged the displacement of the position of the entrance of the natural tunnel towards Reggio Calabria (section 15m long) and the construction, in this section, of a volume of improved ground by secant concrete plastic piles φ1500.

The excavation inside the cut-and-cover tunnel section was carried out from the existing entrance piles wall, once the first order of struts was completed. After having executed the second order of struts, the excavation was performed, up to the bottom of excavation, for segments (12m long) with contemporary construction of the inverted arch.

This procedure was adopted until the end of the cut-and-cover tunnel.

Subsequently, the reinforced concrete lining of the cut-and-cover tunnel and the subsequent backfill were completed.

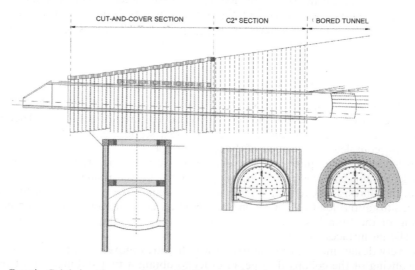

Figure 9.   Reggio Calabria entrance: design solutions crossing the collapsed section.

Figure 10.   Reggio Calabria entrance: excavation of the cut and cover tunnel.

During the excavation of the cut and cover tunnel (Figure 10), with (13÷20) m high excavations, maximum transversal displacements of the order of one centimeter were recorded. The concrete struts of the pile walls were monitored by means of strain gages in order to control the stresses during excavation. The monitoring data showed stresses maximum of 150MPa always in compression.

The subsequent stretch (L = 40m), to that in cut-and-cover, was bored by the section C2* (Figure 9). The design solution consisted in the improvement of a shell around the future excavation made by the external ground surface by grouted fiberglass tubes and the reinforcement of the tunnel core by mean of 55 (28 m long) fiberglass tubes with sleeves injected with cement grout.

Subsequently, for an extent of 87 m, the improvement interventions consisted in the reinforcement of the tunnel core made by mean of 55 (28 m long) fiberglass tubes with sleeves injected with cement grout and of a shell around the future excavation by mean of 108 (22 m long) fiberglass tubes with sleeves injected with cement grout.

The provisional lining for both sections consisted of steel ribs and shotcrete. The execution of the reinforced concrete inverted arch and of the final lining followed, at close distance, the tunnel face.

During the excavation crossing the previous collapsed section the ground was systematically affected by the presence of frequent collapsed steel ribs and other work material involved in the landslide event (Figure 11).

The presence of steel material inside the tunnel core made the implementation of the improvement scheme planned in the project quite difficult.

Given the random disposition of the findings and the consequent impossibility of predicting the conditions of the tunnel face that were progressively encountered, it was systematically necessary to redistribute the reinforcement fiberglass tubes according to the findings of collapsed ribs on the excavation face. In summary, for the reinforcement of the tunnel face, the standard design scheme was used as first attempt. This intervention was constantly adapted to the real context repositioning every single perforation according to the presence of drilling stops before reaching the designs lengths or directly visible obstacles on the excavation face.

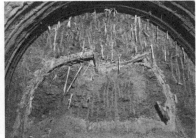

Figure 11.   Steel ribs support inside existing collapsed ground.

The most demanding situation was met once near the progressive reached by the collapsed tunnel where it was even encountered a shotcrete pump machine (Figures 12).

The Figure 13 summarizes the data collected by the monitoring system in terms of convergence and stress measured on the steel ribs. The results obtained show maximum values of radial convergence lower than 40 ÷ 50mm, perfectly matching with the forecasts of the design phase.

Higher convergence values were recorded starting from ~ 130m from the Reggio Calabria entrance at a fault zone (L = 30m). Also, the stresses measured on the steel ribs confirmed the forecasts of the design phase; only in the tunnel section between 120m and 130m from the entrance on the Reggio Calabria side it was necessary to reduce the distance between the ribs from 1.0m to 0.8m.

4.2 Taranto (north) entrance

The scheme of the ground improvement and support interventions used is reported in Figure 14. The ground improvement interventions consisted in the reinforcement of the tunnel core made by mean of 55 (28 m long) fiberglass tubes with sleeves injected with cement grout and of a shell around the future excavation by mean of 108 (22 m long) fiberglass tubes with sleeves injected with cement grout.

The provisional lining consists of steel ribs and shotcrete. The execution of the reinforced concrete inverted arch and of the final lining followed, at close distance, the tunnel face.

Through the monitoring data collected during excavation (convergence and extrusion measurements, strain gauges in the preliminary lining) it was possible to refine stabilization interventions and adapting them to the behavior encountered.

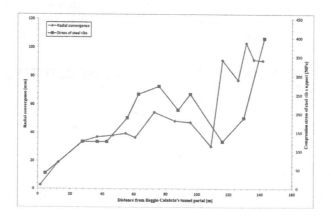

Figure 12.   Shotcrete pump machine involved in the collapsed ground.

Figure 13.   Reggio Calabria entrance: monitoring results.

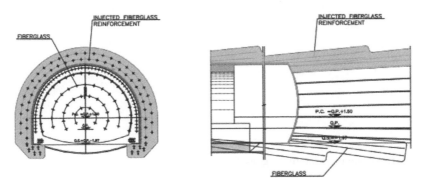

Figure 14.   Taranto entrance: ground improvement and support solution.

In various sections it was necessary to integrate the ground improvement interventions and proceed to the early closure of the preliminary lining with the reinforced concrete inverted arch because the measures of deformation were much higher of those forecasted in the design phase.

Particularly demanding was the section, of about 150 m, between kilometers 0+600 and 0 +750, where very high deformation phenomena, with instability of the shotcrete between the steel ribs and episodes of instability of the tunnel face, occurred (Figure 15).

As we can see from the data collected by the monitoring system (Figure 16), in this section high convergence values, with a maximum value of 21 cm, were reached (they were much higher than the design alarm threshold).

From the geomechanical survey of the tunnel face (Figure 15), the presence of a strongly tectonized clayey-marly rock mass showed a pervasive presence of cleavage and cracking of clays.

In this geomechanical context, the presence of a local accumulation of pre-existing tectonic stress was detected. This stress accumulation freed itself at the moment of the excavation giving rise to the development of significant deformation phenomena such as those observed.

In this section it was impossible to regulate the deformation phenomena only with the inverted arch (also reducing to a minimum the excavation fields of the inverted arch up to 3m). It was necessary to introduce the inverted strut to allow a further stiffening and strengthening of the provisional lining and at the same time a further limitation of the excavation field for the execution of the inverted arch.

Figure 16 summarizes the data collected by the monitoring system in terms of convergence and stress measured on steel ribs (stress also measured at the steel inverted strut).

As regard the provisional lining, the data collected indicate that the stresses in the steel ribs have always remained close to the limit values and sometimes even higher than the yield

Figure 15.   Taranto entrance: episodes of instability of the tunnel face.

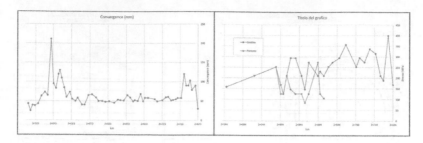

Figure 16.   Taranto entrance: monitoring results.

values to indicate how the designed solutions have been optimal in terms of utilization of the resistant section.

5   CONCLUSION AND CLOSING REMARKS

During the works, through the implemented monitoring system, it was continuously possible to check the stress and strain response of the rock mass. It was possible to adjust the distances between the inverted arch and the final lining from the tunnel face to maintain the deformations and the tensions below the design limit values.

In this way, it was possible to confirm the importance of a prompt closure of the temporary lining, to the aim of controlling the stress and strain behaviour subsequent to the tunnel excavation. This was obtained by pouring the reinforced concrete inverted arch at a short distance from the tunnel face and by limiting the excavation length of the inverted arch also using the steel inverted strut.

It was observed, in several situations, how the prompt installation of the inverted arch near the tunnel face permitted to block the rapid increase of the convergences and of the stresses in the preliminary lining.

The tunnel excavation, for the upstream tube, ended on 05/17/2018 (Figure 17).

The design of the downstream carriageway is currently in progress.

This will allow the contracting of works to complete the entire lot on two carriageways.

Figure 17.   05/17/2018: upstream tunnel is finished.

*Tunnels and Underground Cities: Engineering and Innovation meet Archaeology,
Architecture and Art, Volume 4: Ground improvement in
underground constructions – Peila, Viggiani & Celestino (Eds)
© 2020 Taylor & Francis Group, London, ISBN 978-0-367-46868-2*

Rock grouting and ground freezing for tunnelling at Duomo subway station in Naples

V. Manassero
Underground Consulting, Pavia, Italy, formerly Icotekne S.p.A., Naples, Italy

F. Cavuoto
Project Manager of Metro Line 1, Naples, Italy

A. Corbo
Geotechnical Engineer, formerly Studio Cavuoto, Naples, Italy

ABSTRACT: The paper deals with the application of grouting technology to improve soft rocks for tunnelling at Duomo station, within Naples subway Line 1 extension project. Located downtown in a deeply urbanized area, the station is composed of one access shaft, four platform tunnels and four inclined pedestrian access tunnels. The soil improvement was setup to allow boring the 8 station tunnels under a hydrostatic head of up to 30 m. An annular low permeability thick shell all around the tunnels to be excavated was achieved by penetration grouting. The treatment, entirely carried out from the already in place TBM tunnel, was fan-shaped. The tuff cracks were sealed by using at first high penetrability cement based grouts (fine and microfine cement) and then a silica-based chemical grout, through MPSP grouting pipes. In addition, one tunnel out of 8 was bored and constructed combining two soil improvement methods: grouting and artificial ground freezing.

1 INTRODUCTION

The Line 1 of Naples subway fits inside a complex railway transport system, where it will form, once completed, a closed ring around and under the city, characterized by a series of links with all the other important subway and regional railway lines (Figure 1). Eighteen stations are already operating, including four out of five new stations on the "low stretch" extension of the line, while one is still under construction.

The five new stations of the line extension are all located downtown, in a deeply urbanized area (Figure 2), and are composed of one access shaft, four horizontal platform tunnels 50 m long and four inclined pedestrian access tunnels 30 m long (Figure 3). Usually the access shaft is centred with the twin rail tunnels and all the station tunnels were excavated starting from this shaft.

To allow for safe tunnelling, soil improvement was carried out at all five stations. At four out of the five stations artificial ground freezing was applied as planned starting from the access shaft (Colombo et al. 2008). However, at Duomo station penetration grouting was chosen for soil improvement. This was performed from the twin TBM tunnels already in place and the platform tunnels were widened starting from these tunnels, while the access tunnels were excavated entirely within the improved soil.

This exceptional course of action at Duomo station was chosen due to the fact that during the excavation of the access shaft, located in the heart of the ancient greek-roman city, *Neapolis*, important archaeological findings were made (Figures 4 and 5). This significant event led to a considerable delay in the jobsite planning and to a substantial modification of the entire station design, which was initially based on the construction of the four platform tunnels first, before the TBM tunnelling. The delay in the activities and the need to avoid any stops, forced

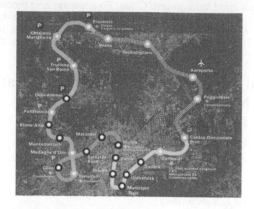

Figure 1. Line 1 of Naples subway: General Layout.

Figure 2. Aerial view of the "low stretch" extension of Line 1, with its five new stations Toledo, Municipio, Università, Duomo and Garibaldi.

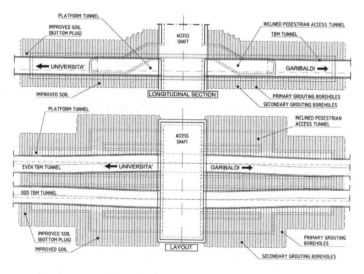

Figure 3. Duomo station: layout and longitudinal section.

the TBM to tunnel across the whole station in advance, thus allowing the platform tunnels to be later widened from the TBM tunnels.

The availability of the TBM tunnels, in advance to platform tunnelling, allowed the use of the tunnels themselves as a means to carry out the soil improvement, instead of the shaft. This

Figure 4. Duomo station: general view of the archaeological excavation within the access shaft.

Figure 5. Duomo station: detail of some archaeological findings within the access shaft.

considerably reduced the borehole length required to perform the soil improvement. The radial shape of the improved soil shell, instead of a cylindrical ahead of the face shape, is particularly suitable for permeation grouting, thus encouraging the choice of this technique as a soil improvement method.

Permeation/penetration grouting is a technique commonly used to improve soil/rock characteristics, in terms of both reduced permeability and increased strength and modulus. The soil improvement is achieved by filling the natural voids (pores or cracks) with suitable grouts, without any essential change to the original soil volume and structure.

Nevertheless, one of the four inclined pedestrian access tunnels, on the even track and Garibaldi side (see Figure 3), was bored and constructed combining two soil improvement methods: grouting and artificial ground freezing. Grouting was performed first from the TBM tunnel within the frame of the grouting treatment described above, while ground freezing was carried out later on from the central access shaft to form a protection made of a frozen soil canopy above the tunnel crown.

This exceptional combination was due to the fact that borehole drillings from the TBM tunnel resulted in the detection of an unexpected anomaly in the soil profile, namely a localized deepening of the roof of Neapolitan Yellow Tuff (NYT) bedrock. This entailed the tunnel crown being partially located within the overlying loose silty sand layer called *Pozzolana*. According to all the previous soil improvement experiences in Naples, saturated *Pozzolana* is too fine a material to be grouted by permeation, either with cement or with chemical grouts. For this reason an additional ground freezing treatment was conceived all around the crown section affected by the presence of *Pozzolana*, in order to support the soil and assure its waterproofing.

2 GEOLOGICAL AND GEOTECHNICAL ASPECTS

The geological history of the Naples soil formation was strongly influenced by two close volcanic systems: *Vesuvio* and *Campi Flegrei*. The subsoil is essentially composed of pyroclastic material (ash, lapillus, slag, pumice, pozzolana, and tuff), arranged sometimes in layers and sometimes chaotically (Viggiani & Rippa 1988). All these materials can be summarized into two main different layers: *Pozzolana*, a loose pyroclastic silty sand characterized by a particularly small grain size, located on the top, and the Neapolitan Yellow Tuff (NYT), a soft volcanic rock formed from the same material as above, but cemented, underneath. The typical geotechnical properties of these two main layers are shown in Table 1.

NYT, the formation mainly affected by soil improvement and tunnel excavation, is a soft rock with a widespread presence of cracks (Figure 6), both vertical (*scarpine*) and horizontal (*suoli*). These cracks, formed during the cooling process of the pyroclastic matrix, are the reason for the significant secondary permeability of the tuff, which may release a huge amount of water during excavation works. The water-table is located few meters below ground level, thus implying a maximum water head of 30 m.

Table 1. Typical geotechnical properties of *Pozzolana* and NYT.

Pozzolana		Neapolitan Yellow Tuff (N Y T)	
Property	Average	Property	Typical
Dry specific weight γ_d [kN/m^3]	11.5	Dry specific weight γ_d [kN/m^3]	11 - 12
Saturated specific weight γ_{sat} [kN/m^3]	17.0	Saturated specific weight γ_{sat} [kN/m^3]	16 - 17
Porosity	0.5	Porosity	0.55
Permeability k [m/s]	10^{-7}–10^{-5}	Permeability k [m/s]	10^{-7}–10^{-6}
Drained cohesion c' [MPa]	0	Unconfined compressive strength σ_{UCS} [MPa]	3.0
Angle of shear strength φ' [°]	35	Drained cohesion c' [MPa]	0.8 – 1.0
Poisson coefficient v' [kN/m^3]	0.3	Angle of shear strength φ' [°]	27 – 28
		Young Modulus E [MPa]	1500
		Poisson coefficient v' [kN/m^3]	0.3

Figure 6. Typical Neapolitan Yellow Tuff samples recovered by coring.

3 SOIL IMPROVEMENT FOR PLATFORM TUNNEL WIDENING AND ACCESS TUNNEL EXCAVATION

An annular thick shell around all the tunnels to be excavated was achieved by soil improvement with penetration grouting. The platform tunnels were widened from the initial TBM tunnel diameter of 5.85 m (ID) to the final dimension of 11.0 m, while the access tunnels, 6 m wide, were fully excavated through the improved soil. The treatment, entirely carried out from the existing TBM tunnels, was fan-shaped with 31 to 38 grouting pipes per fan and a fan spacing of 1 m: for each tunnel 26 primary fans were installed spaced 2.0 m apart and 25 secondary fans, with the same spacing, were installed in between.

The typical near cross-section (closer to the shaft), including both platform and access tunnels, is shown in Figure 7, while the typical distal cross-section, relevant only to the platform tunnels, is shown in Figure 8. The last five fans, at the end of the tunnel, were fully treated in order to ensure the function of bulkhead (bottom plug section).

The design of the grouting treatment took into account the results of a previous experience gained with the improvement by penetration grouting of the NYT (Manassero et al. 2008).

Grouting was carried out by using a selective procedure. Every grouting pipe injected only one grout mix; through the primary-fan pipes cement-based mixes were injected (high-penetrability grout made from either standard fine cement or microfine cement), whilst a silicate-based chemical mixture was injected through the secondary fan pipes. Within each fan a primary-secondary sequence was also applied; in particular, through the primary holes of the primary fans a standard fine cement-based mixture was grouted. The secondary holes of the same fans were grouted with a microfine cement-based mixture.

Grouting activity was monitored by a real-time data acquisition system, which automatically measured and recorded a set of parameters, allowing reliable analysis of the results. Furthermore, deformations and the groundwater level were constantly monitored, in order to

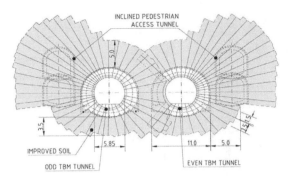

Figure 7.   Duomo station: near cross section, including platform and inclined access tunnels.

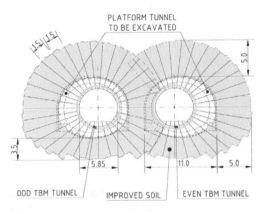

Figure 8.   Duomo station: distal cross section including only the platform tunnels.

Table 2.   Summary of the grouting parameters.

Type of grout	Volume V_{max} (l/m)	Grouting pressure P_{ref} (bar)
Cement-bentonite	100	20
Microfine cement -bentonite	150	20
Silacsol-S	250	20

control possible accidental induced effects on adjacent structures, buildings and utilities by the grouting activities and excavation works.

The grouting criteria ad opted were the following: the grouting of each valve was stopped when either the design volume or the refusal pressure was reached, whichever came first. The characteristic values of these grouting parameters are shown in Table 2.

The grouting flow rate was always kept within the range of 10 to 15 l/min. As far as the grouting pressure is concerned, they are intended as net pressures, i.e. after deducting the losses along the grouting line, packer and sleeved valve.

4   GROUTING OF ROCKS: THE MPSP METHOD

In general, different grouting methods may be applied (Manassero 1993). In rock: up-stage or down-stage methods or multi packer sleeved pipe method (MPSP); in soil: sleeved pipe method (i.e. tubes *à manchettes* method or TAM).

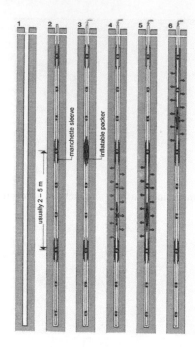

Figure 9. Typical MPSP operational sequence.

In rock, up-stage and down-stage methods are very efficient but also very expensive in terms of time and resources. On the other hand, the technique normally adopted for grouting loose soil is not applicable to rock because its stiffness makes very difficult, often impossible, to break the cement grouted sheath (necessary for injection through the sleeve).

For this reason, the Multi Packer Sleeved Pipe method (MPSP) was developed (Bruce & Gallavresi 1988); it is a hybrid system worked out to permit the utilization of sleeved pipes (that are very practical and less onerous) also for grouting rock. It is very suitable for improving grouting operations in those rock formations where:

– caving of drilled holes and pronounced weathering prevent the sealing off or collapse of sections of holes to be grouted at the design depth;
– where the down-stage method is not capable of achieving a sufficient consolidation and stabilization of the rock.

The system consists essentially of the installation inside a borehole of a plastic or steel pipe fitted at regular intervals with rubber grouting sleeves. Bag packers, fastened to the grouting pipe, seal off the holes at regular intervals (usually 2 - 5 m), between which the grouting is to be confined. The sealing action of the bag packers is obtained by expanding them against the hole walls by injection of grout into the bags. After the packer grout hardening period an upstage grouting by means of double packer is performed. The typical MPSP operational sequence is shown in Figure 9.

At Duomo station the MPSP pipes were equipped with bag packers, spaced at 2.5 m intervals; the first of them was placed at the hole collar, within the concrete segment.

5 THE CEMENT AND CHEMICAL GROUTS UTILIZED

5.1 High Penetrability Cement-Based Grouts

High penetrability cement-bentonite grouts were used within the project both with standard fine cement (Blaine specific surface over 5000 cm^2/g) and microfine cement. The background on high penetrability cement-based grouts has been explained in detail by, among others, Tornaghi et al. (1988), De Paoli et al. (1992a) and De Paoli et al. (1992b).

The use of a specific type of cement and bentonite together with a specific admixture as dispersing agent leads to a high penetrability grout with the following main properties:

- no bleeding;
- filtration rates much lower than those of conventional stable mixes, even at very low viscosity;
- low yield strength over an adjustable period of time;
- higher long term strength and lower permeability, as compared to conventional stable grouts with the same cement content.

As far as the composition chosen for the Duomo project is concerned, the C/W ratio adopted was 0.35, while B/W ratio was 3–4% and Admixture/W ratio 0.3–0.4%.

5.2 Silacsol-S Chemical Grout

Silacsol-S is a chemical grout composed of an activated silicate liquor and a calcium based inorganic reagent (Tornaghi et al. 1988).

In contrast to commercial alkaline sodium silicates, which are aqueous solutions of colloidal silicate particles, the liquor is a true silicate solution. The activated dissolved silicate, associated to the mineral reagent, produces calcium hydro-silicates with a crystalline structure quite similar to that obtained by the hydration and setting of cement. The resulting product is a complex of permanently stable crystals. Hence the reaction is no longer an evolutive gelation involving the formation of macromolecular aggregates and possible loss of silicized water (syneresis). On the contrary, it is a direct reaction on molecular scale.

This type of mix has the same groutability range as common silicate gels, allowing uniform treatment of medium to fine sands. Silacsol-S has a newtonian behaviour at fairly low viscosity, up to an effective groutability time chosen case by case (e.g. 60 to 100 min). Afterwards, the yield strength increases with time up to the final setting. Even in the case of larger voids or fissures created by hydro-fracturing, a permanent filling is ensured without any risk of syneresis. The activated silicate mix has the stability of cement grouts, preventing ground water pollution.

Other outstanding features of soil grouted with Silacsol-S are the far lower permeability and the better creep behaviour, in comparison with the same soil grouted by traditional silicate gel.

The main physical characteristics of the grout utilized are the following:
$\gamma = 1.34$–1.37 g/cm^3; viscosity = 10 mPaxs; setting time = 60–100 min.

6 GROUTING RESULTS

The results of the grouting treatment, in terms of grout take and pressure may be summarized as shown in Table 3.

The combination of the different grouting mixes solved the function which they were selected for, confirming the effectiveness of the soil improving method selected.

Table 3. Summary of the results of grouting.

Type of grout	Grout take		Actual average grout pressure (bar)
	Expected (l/m)	Actual (l/m)	
Cement-bentonite	25	25.4	13.4
Microfine cement - bentonite	75	52,2	12.6
Silacsol-S	150	234,3	14.0
Total	250	311.9	

Figure 10. Duomo station: widening excavation of one platform tunnel.

Figure 11. Duomo station: detail of the filling of the crack system by Silacsol-S.

During the tunnel excavation activities (Figure 10), a systematic actual filling of the widespread crack system was observed, sometimes by the cement mix, where crack width permitted, sometimes by the chemical mix, the latter able to permeate a large number of thinner cracks (Figure 11). The residual permeability of the improved tuff rock, reduced by two order of magnitude, allowed a safe, fast and almost dry tunnel excavation.

7 ARTIFICIAL GROUND FREEZING FOR ONE PEDESTRIAN ACCESS TUNNEL

When drilling grouting boreholes from the TBM tunnels we detected an unexpected localized deepening of the roof of NYT bedrock. This meant that the tunnel crown of one pedestrian access tunnel (even track, Garibaldi side) was partially located within the overlying *Pozzolana* layer (Figure 12).

According to all the previous experiences of soil improvement in Naples, the saturated *Pozzolana* is too fine a material to be grouted by permeation, either with cement or with chemical grouts. For this reason an additional ground freezing treatment was conceived all around the section of tunnel crown affected by the presence of *Pozzolana*, to support the soil and assure its waterproofing (Figure 13). Therefore this tunnel was bored and constructed combining two soil improvement methods: grouting and Artificial Ground Freezing (AGF). Grouting was performed first from the TBM tunnel as previously described, while ground freezing was carried out later on from the access shaft to form a protection made of a frozen soil canopy above the tunnel crown.

As is well known, AGF is a technology for temporary soil improvement in water-bearing soils, selected to create frozen ground bodies of appropriate thickness and characteristics to act as temporary soil support and/or waterproofing. The achieved frozen ground behaves like

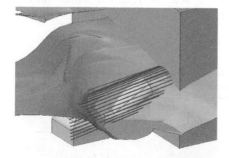

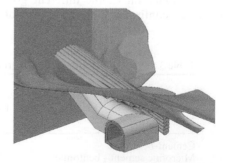

Figure 12. 3-D view from North side of the pedestrian access tunnel, NYT roof and freeze-pipes.

Figure 13. 3-D view from South side of the pedestrian tunnel, NYT roof and frozen soil canopy.

a conglomerate whose binder is the frozen water (ice) filling the soil voids; it is impermeable and its mechanical characteristics are remarkably improved compared to natural soil.

AGF is achieved by circulating a coolant fluid through a series of pipes (freeze-pipes) inserted into the ground, suitably disposed around the volume to be protected, to generate a shell cofferdam made of frozen soil.

The water-bearing soil is cooled by extracting heat until its temperature drops below the freezing point of the groundwater system (freezing phase). Each freeze-pipe forms a column of frozen soil; the columns grow and merge together with the adjacent ones, forming a resistant and impermeable retaining structure. Then, the achieved temperature level is maintained by adjusting the flux of heat extracted from the soil (maintenance phase) until the construction activities have been completed.

The AGF treatment was carried out from the central shaft. Both freeze and temperature pipes were installed by drilling through the shaft diaphragm wall and the loose soil behind up to penetrating the NYT downwards. 21 freeze-pipes were installed to achieve the frozen soil canopy and, within the soil near them, 3 thermometric pipes were installed to gather information on and control the growth of the frozen body both during freezing and maintenance phase (Figures 14 and 15).

Ground freezing was performed using the Liquid Nitrogen (LN) method, which is an open circuit process. The coolant medium is LN at -196°C delivered on site to vacuum-insulated storage tanks. The LN is circulated directly into the freeze-pipes through an insulated manifold system. On its way along the freeze-pipes the LN evaporates and extracts heat from the ground; the resultant gas is allowed to exhaust into the atmosphere at a temperature ranging from -100°C to -60°C.

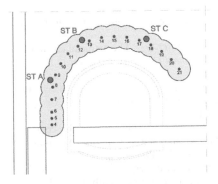

Figure 14. Cross section of the frozen soil canopy and arrangement of freeze and temperature pipes.

Figure 15. View of the front of freeze and temperature pipes.

Figure 16. View of the frozen crown from the central shaft during tunnelling.

Figure 17. View of the inclined pedestrian access tunnel during tunnelling.

The combined grouting/freezing treatment allowed safe, fast and almost dry tunnelling. Two views of the tunnel under excavation are shown in the Figures 16 and 17.

8 CONCLUSIONS

Penetration grouting was applied to allow the boring of 8 tunnels totalling 320 m at Duomo station for the Naples subway project: four horizontal platform tunnels 50 m long and four inclined pedestrian access tunnels 30 m long. A thick improved soil shell all around the tunnels to be excavated within the Neapolitan Yellow Tuff was created, using a fan-shaped grouting borehole arrangement drilled from the twin TBM tunnels. The MPSP grouting method was adopted and high-penetrability cement-based grout (both standard fine and microfine cement) and silicate-based grout (Silacsol-S) were applied. The results of the grouting treatment in terms of grout take and grouting pressures were in accordance with the design anticipated values. In terms of permeability, the reduction was of about two order of magnitude.

In addition, one of the eight tunnels was bored and constructed combining two soil improvement methods: grouting and artificial ground freezing. This was due to the fact that drilling grouting boreholes from TBM tunnel had detected an unexpected localized deepening of the roof of NYT bedrock, thus entailing that the tunnel crown of one pedestrian access tunnel was partially located within the overlying *Pozzolana* layer.

In all the cases the designed soil improvement treatments allowed safe, fast and almost dry tunnel excavation.

REFERENCES

Bruce, D.A. & Gallavresi, F. (1988). The MPSP system: a new method of grouting difficult rock formations. *ASCE Conference, Nashville, TN.*

Colombo, G., Lunardi, P., Cavagna, B., Cassani, G., Manassero, V. (2008). The artificial ground freezing technique. Application for the Naples underground. *World Tunnel Congress 2008. Agra (India), 19–25 September.* 910–921.

De Paoli, B., Bosco, B., Granata, R., Bruce, D.A. (1992a). Fundamental observations on cement based grouts (1): traditional materials. *Proc. Conference on Grouting. Soil Improvement and Geosynthetics, New Orleans.* Geothec. Spec. Pubbl. No. 30. 474-485. ASCE.

De Paoli, B., Bosco, B., Granata, R., Bruce, D.A. (1992b). Fundamental observations on cement based grouts (2): microfine cements and the Cemill process. *Proc. Conference on Grouting. Soil Improvement and Geosynthetics, New Orleans.* Geothec. Spec. Pubbl. No. 30. 486–499. ASCE.

Manassero, V. (1993). Different techniques for soil improvement and underpinning. *International Symposium on Novel Foundation Techniques.* EC Science Programme Symposium, Cambridge, UK.

Manassero, V., Di Salvo, G., Giannelli, F., Colombo, G. (2008). A combination of artificial ground freezing and grouting for the excavation of a large size tunnel below groundwater. *Proceedings of 6th International Conference in case Histories in Geotechnical Engineering. Arlington VA.*

Tornaghi, R., Bosco, B., De Paoli, B. (1988). Application of recently developed grouting procedures for tunnelling in the Milan urban area. *Fifth International Symposium on Tunnelling. London. UK.*

Viggiani, C. & Rippa, F. (1998). Linea 1 della Metropolitana di Napoli. Relazione geologica e geotecnica. Unpubl. work.

Tunnels and Underground Cities: Engineering and Innovation meet Archaeology,
Architecture and Art, Volume 4: Ground improvement in
underground constructions – Peila, Viggiani & Celestino (Eds)
© 2020 Taylor & Francis Group, London, ISBN 978-0-367-46868-2

Monitoring artificial ground freezing and relevant fundamental observations

V. Manassero
Underground Consulting, Pavia, Italy

ABSTRACT: The paper deals with some fundamental observations gathered by the author from monitoring a large number of Artificial Ground Freezing jobsites, at first as technical director at geotechnical contractors and then as independent expert advisor. Surveying the freezing boreholes and real-time monitoring of temperatures, groundwater pressure and displacements are of the utmost importance for a successful ground freezing job. For all of them some practical recommendations are given from the point of view of both design and construction. The observations deal mainly with three aspects: the influence of the position of thermometric gauges versus freeze-pipes, the estimation of the temperature distribution within and around the frozen body and the evaluation of the transient phase between construction and maintenance of the frozen body, when different coolant fluids are used, with practical reference to one freezing project in Rome and two in Paris.

1 INTRODUCTION

Artificial Ground Freezing (AGF) is a technology for temporary soil improvement in water-bearing soils, selected to create frozen ground bodies of appropriate thickness and characteristics to act as temporary soil support and/or waterproofing. Frozen ground behaves like a conglomerate where the binder function is accomplished by the frozen water (ice) filling the soil voids; it is impermeable and its mechanical characteristics are remarkably improved compared to the natural soil. When the scope of AGF is to allow a safe and almost dry excavation or tunnelling below the water-table, the design frozen body contributes, together with other elements (such as the possible existing structures, natural or artificial bottom plugs, etc.), to forming the boundary of a closed and impermeable chamber which will protect the excavation from collapse or intrusion of groundwater and loose soil. Excavation or tunnelling is then carried out within this closed chamber.

From the design point of view, although geotechnical design is the main aspect in AGF, it should be accompanied by a thermal analysis, in order to obtain the temperature distribution, elapsed freezing time and energy required to perform the job (Frivik 1980), and also by the technological design. From the construction point of view, further to excellent engineering and a very skilled and experienced geotechnical contractor, quality control and monitoring are key factors for a successful ground freezing job.

2 THE TECHNOLOGY OF ARTIFICIAL GROUND FREEZING

In principle, AGF technology entails the installation within the soil of a system of freeze-pipes suitably spaced around the perimeter of the improved soil body to be built.

The primary scope of AGF is to extract heat from the water-bearing soil until its temperature drops below the freezing point of the groundwater system (freezing phase). Each freeze-pipe forms a column of frozen soil; the columns grow and merge together with the adjacent ones, forming a resistant and impermeable retaining structure. Then, the achieved temperature level is

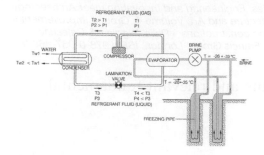 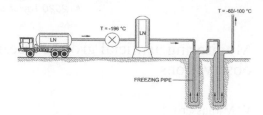

Figure 1. Indirect system - Brine cooled by a refrigerating plant, circulated into the freeze-pipes.

Figure 2. Direct system – Liquid Nitrogen circulated directly into the freeze-pipes.

maintained by adjusting the flux of heat extracted from the soil (maintenance phase) until after the construction activities have been completed.

Freezing is achieved by circulating a coolant fluid through freeze-lances, made of two concentric pipes: the outer one has a closed end while the inner one is open. Generally, the coolant fluid is pumped through the internal pipe down to its deepest point. On its way back through the annulus between the inner and outer pipes, the coolant fluid extracts the heat from the ground, thus decreasing its temperature. Three different construction methods of AGF are available.

The Brine method (Figure 1) is a closed-circuit process that requires the use of an industrial refrigerating plant, connected to a coolant system, which consists of a brine pump, surface manifolds and freeze-lances installed into the ground. The brine, usually a calcium chloride ($CaCl_2$) solution, is cooled by the refrigerating plant, typically at temperatures of -28 to -35°C, and pumped into the closed circuit. Alternative premixed coolant fluids are now available on the market and may also be used. The warmer fluid returning from the freeze-lances through the insulated surface manifold system is then re-cooled and re-circulated into the closed circuit.

The Liquid Nitrogen (LN) method (Figure 2) is an open circuit process. The coolant medium is LN at -196°C delivered on site to vacuum-insulated storage tanks. The LN is circulated directly into the freeze-pipes through an insulated manifold system. On its way along the freeze-pipes the LN evaporates and extracts heat from the ground; the resultant gas is allowed to exhaust into the atmosphere at a temperature ranging from -100°C to -60°C. The LN method is quicker than the brine method in achieving the frozen body and attains a higher strength on the frozen ground.

A third method consists of a combination of the two methods described above: the LN method may be applied to build-up the design frozen body (freezing phase) and the brine method to maintain the achieved level of soil temperature (maintenance phase).

The choice of the freezing method is a function of a number of design, construction and economic factors, such as ground and groundwater condition, speed of underground seepage, required design strength and elasticity modulus of the frozen ground, planning, duration of the maintenance period, logistics, costs, etc.

3 MONITORING ARTIFICIAL GROUND FREEZING

Monitoring is a crucial aspect of the AGF technology. Surveying the actual arrangement of freeze-pipes and temperature chains and monitoring of the soil temperature (within and around the volume to be frozen), coolant fluid temperature, groundwater pressures and displacements (on the existing structures and utilities nearby) are always to be implemented in AGF projects.

Monitoring should be carried out by means of an automatic reading, acquisition and recording system. The result of monitoring should be shared in real-time, through a specific web-platform, with all the key people involved in the project (owner, engineer, main and specialized contractor, consultant, etc.) in order to allow everybody to be up to date on the development of the frozen body and process as well as on possible groundwater overpressures and displacements, and allowing whoever may be concerned to react quickly if necessary.

3.1 *Surveying the actual arrangement of the net of freeze-pipes and thermometric chains.*

The on-going survey of the actual arrangement of the net of freeze-pipes and thermometric chains is essential to preliminarily detect any potential weak points within the future frozen body, so as to be able to supplement with additional freeze-pipes, and also for the understanding of temperature monitoring while freezing.

The actual arrangement of freeze-pipes and thermometric chains may result significantly different from the anticipated theoretical design arrangement. This is due to both possible errors during the initial positioning of the drilling rig mast and unavoidable borehole deviations from the theoretical design axis. The longer the drilling, the greater are these deviations.

For this reason, each freezing borehole is to be surveyed and its actual path measured by suitable methods in order to gather the actual picture of the 3-D arrangement of both freeze-pipes and thermometric chains. The recommended measuring methods are the inclinometric method for vertical or sub-vertical boreholes and the optical method for horizontal or sub-horizontal boreholes. The optical method is based on a direct reading of the actual global co-ordinates, X, Y and Z, of the casing pipe centre at pre-fixed depths. This measure is achieved by means of a theodolite installed on a fixed support at the near edge of the borehole casing pipe (Figure 3), and a special cylindrical device moving inside the pipe along its whole depth (Figure 4).

When necessary the optical measures may be assisted by alternative measuring methods (e.g. Maxibor or gyroscope), but these methods can give the actual path only by means of indirect measurements, therefore the author's advice is to exploit to the utmost the optical method.

The cylindrical device, equipped with suitable spacers with a slightly smaller diameter than the internal diameter of the pipe to be surveyed, is fitted with one LED at the centre of its nearest base. At the beginning of the survey the device is pushed in up to the maximum depth of the pipe and then withdrawn in steps of a predetermined length; at each step the readings of the actual global co-ordinates X, Y and Z of the LED are taken.

The actual paths are then included in an as-built 3-D model in order to allow calculation of any distance between adjacent freeze-pipes and between the thermometric chains and the closest freeze-pipe. The actual position of the freeze pipes and thermometric pipes may also be drawn on several representative as-built cross sections.

When the borehole path survey shows any local spacing between adjacent freeze pipes large enough to be potentially detrimental to the integrity of the frozen body to be built, additional boreholes are to be drilled and equipped with new freeze-pipes, in order to close these openings. The maximum allowed window opening is to be defined by the design technical specifications.

Figure 3. Theodolite installed on a fixed support at the near edge of the borehole casing.

Figure 4. Special cylindrical device equipped with LED inside the borehole casing.

These new boreholes are again to be surveyed for deviation in order to verify if the remedial target has been achieved. In case of failure in achieving the target, new additional boreholes are drilled and checked until around the whole surface of the body to be frozen there are no local zones with space larger than that fixed by the design technical specifications.

Furthermore, knowing the actual position of the thermometric pipes versus freeze-pipes allows calculation of the distance between each temperature gauge equipping the thermometric chains and the closest freeze-pipe (i.e. the closest source of chilling). This information is a key factor for the understanding of the freezing process development and for the estimation of the temperature distribution within and surrounding the frozen body, during both the construction and the maintenance of the frozen body.

3.2 Temperature monitoring.

Temperature monitoring during ground freezing entails the soil and the coolant fluid. Monitoring the soil is necessary to gather information on and control the growth of the frozen body, while monitoring the fluid allows us to control and guide the freezing process. For this reason, a specific temperature-gauge net is designed and installed within the soil nearby the freezing-pipes, within and around the design volume to be frozen, as well as at crucial points of the freezing circuit.

Suitable temperature gauges are installed inside purpose-drilled boreholes fitted with permanent casing-pipe. Chains of temperature gauges are assembled off-site, then inserted into the permanent casing-pipes and connected to a data logger for measurement, acquisition and recording of the temperature data. The longitudinal spacing of the gauges along the thermometric chains is specified by the design (usually 1 to 3 m, as a function of the total length). Particular care should be taken in the choice of the type of gauges: their characteristics must be appropriate to the anticipated soil temperature during the whole freezing process, mainly depending on the type of coolant fluid selected (brine or LN).

Further temperature gauges are installed at the outlet and inlet points of the refrigerating plant (s) and at the inlet and outlet points of each freeze-pipe (or each series of freeze-pipes if a parallel connection is adopted). These gauges allow the estimation of the temperature differences between the entry to and the exit from the whole freezing circuit and each freeze-pipe (or series of freeze-pipes), thus allowing us to appreciate the quantity of heat extracted by the net of freeze-pipes and by each freeze-pipe (or series of pipes), depending also on the fluid coolant flow-rate.

3.3 Water pressure monitoring.

When ground freezing is designed to form, together with other elements, a closed and impermeable chamber, monitoring of the water pressure inside the chamber is carried out.

Monitoring is usually performed by means of one or more open pipe piezometers installed within the unfrozen core to be excavated, fitted with valve and pressure gauge. Water pressure monitoring has a double purpose:

a) to detect the moment of the "closure" of the frozen shell by a sudden increase in the water pressure;
b) to detect any unsuitable overpressures on any of the structures forming the boundary of the closed chamber and, if necessary, release the water pressure by opening the valve and letting the piezometer pipe act as a drain pipe.

Furthermore, once the shell has achieved its "closure" and the design thickness, and before starting excavation or tunnelling, the piezometer pipe may be used as drain pipe, by just opening the valve, to let the water flow out, thus providing drainage for the unfrozen core to be excavated.

3.4 Displacement monitoring.

In displacement monitoring we are surveying the effects of soil freezing on the existing nearby structures and utilities. Displacements may occur both during freezing and during thawing (once the freezing has accomplished its scope). In the first case a frost heave or horizontal displacement may be expected, but if the movement is restrained, for example by a building load or by a retaining structure, a significant frost pressure may develop. Conversely, in the second case a thaw settlement may be expected.

The frost heave/horizontal displacement is a complex phenomenon related to the expansion of soil while freezing. When the temperature is lowered to the freezing point of the pore-water, water in soil pores becomes partly frozen and expands by 9% of its unfrozen volume. In fine grained, frost-susceptible soils the induced suction force causes additional water migration to and in the freezing zone. On freezing, the migrated water forms ice lenses and causes volumetric expansion in a direction perpendicular to the freezing front. In coarse grained, non-frost-susceptible soils expansion is supposed to be induced by closed system freezing only and on cases when the pore water can flow out of the freezing zone, no expansion is expected (ISGF Working Group 2 1991).

Thaw settlement is the generally uneven downward movement of the ground surface due to thaw consolidation (Harris 1995). It has different possible components, namely volume reduction due to phase change, self-weight of the soil and de-structuration of the soil while freezing.

Displacement monitoring may therefore be very useful and allow for suitable action to be taken to counteract frost heave/horizontal displacements (by properly controlling the freezing process both during construction and maintenance of the frozen body) and thaw settlement (by, for example, compensation grouting).

4 OBSERVATIONS ON SOIL TEMPERATURE MONITORING

4.1 *Influence of the position of the thermometric pipes versus freeze-pipes*

The interpretation of the soil temperatures during freezing cannot avoid correlating the measured temperatures with the distance between each temperature gauge and the closest freeze-pipe. In fact, the temperature measured by each gauge is obviously a function of its distance from the closest chilling source, following the rule that a closer chilling source entails a colder soil and a more distant chilling source entails a warmer soil.

As an example, we can see how this rule was observed during the freezing phase in a practical case where AGF was utilized to allow the safe and almost dry boring of a drift connecting a ventilation shaft with the TBM tunnel, for the Rome subway, Line B1. If we look at the graph in Figure 5, drawn for one specific thermometric chain, ST E, located on the right-hand side of the drift at the bottom, we can observe that after 6 freezing-days, the measured temperatures are following the above mentioned rule: the temperatures measured by the different gauges of the thermometric chain are variable in the range -80/+10°C, the distance being variable within the range of 17/109 cm. The variability of the temperature is completely congruent with the distance variability.

The same observation is even more evident on the graph Temperature vs. Depth (starting from the internal surface of the shaft) for the same thermometric chain ST E, shown in Figure 6. This

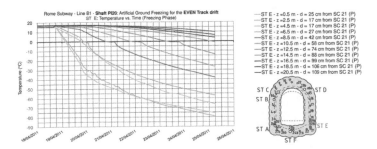

Figure 5. Example graph showing Temperature vs. Time: Rome Subway, Line B1, Shaft PI20, even track drift, thermometric chain ST E.

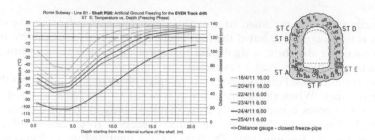

Figure 6. Example graph showing Temperature vs. Depth: Rome Subway, Line B1, Shaft PI20, even track drift, thermometric chain ST E.

graph shows six different moments of the freezing process: before freezing start, end of freezing and four intermediate moments; it gives the temperatures measured at each moment (color curves) as well as the distances between each gauge and the closest freeze-pipe (black curve, referred to the scale on the right). We can see that the shape of all the temperature curves is congruent with the shape of the distance curve according to the rule that a closer chilling source entails a colder soil and a more distant chilling source entails a warmer soil. This means both that the behavior of the soil when cooling is uniform along the whole thermometric chain and that the borehole path survey was performed correctly and the resulting calculated distances are likely to represent the actual distances between temperature gauge and closest chilling source.

4.2 Estimation of the temperature distribution within the frozen body

Knowing the actual position of the thermometric pipes vs. freeze pipes is furthermore of great help when estimating the temperature distribution within the frozen body during freezing (both freezing and maintenance phases) and thus for understanding the development of the freezing process.

Plotting the measured temperatures versus the distance between each gauge and the closest freeze pipe, we can find a correlation curve at any moment of the freezing process, thus allowing us to estimate at any time the position of the different isotherms as well as the evolution of the frozen front. As an example, in the graph in Figure 7, we can see Temperature vs. Distance between each gauge and the closest freeze-pipe for the same thermometric chain ST E we examined before and for the same six moments of the freezing process. The graph shows at each moment of the process the temperature distribution within the frozen body, calculated as a logarithmic regression of the plotted Temperature-Distance points. We can observe that the "end of freezing" regression fits very well with the Temperature-Distance points, confirmed by the very high regression coefficient $R^2 = 0.99$. The regression at the other freezing process moments fits also very well with the points, with a regression coefficient ranging between 0.95 and 0.99.

The same graph was plotted including the measurements at the gauges of all the six temperature chains and is shown in Figure 8. For clarity the plotted Temperature-Distance points are

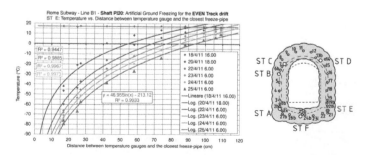

Figure 7. Example graph showing Temperature vs. Distance between each gauge and the closest freeze-pipe: Rome Subway, Line B1, Shaft PI20, even track drift, thermometric chain ST E.

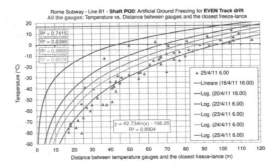

Figure 8. Example of graph Temperature vs. Distance between each gauge and the closest freeze-pipe: Rome Subway, Line B1, Shaft PI20, even track drift, all the thermometric chains.

shown only for the moment "end of freezing". The "end of freezing" regression still fits quite well with the points, with a regression coefficient of $R^2 = 0.89$. The regression at the other freezing process moments also fits quite well with the points, with a regression coefficient ranging between 0.74 and 0.88.

This analysis shows that the logarithmic curve represents the actual distribution of the temperatures within a frozen body very well and confirms the theoretical approach proposed by Sanger & Sayles (1979). This approach is founded on the following three basic assumptions:

a) Isotherms move so slowly they resemble those for steady state conditions (applicability of the Fourier equation for the steady state heat flow to a cylinder). This assumption may not be strictly true because the freezing front may not move in the same way as do the isotherms.
b) However, experience has shown that this principle is adequate for engineering design.
c) The radius of the unfrozen soil affected by the temperature of the freeze-pipe can be expressed as a multiple of the frozen soil radius prevailing at the same time (i.e. the soil surrounding the freeze-pipe is schematized by two concentric annular volumes, one frozen and the other unfrozen; the soil beyond these two annular volumes is not influenced any more by the temperature variation);
d) The total latent and sensible heat can be expressed as a specific energy which, when multiplied by the frozen volume, gives the same total as the two elements computed separately.

These three simple ideas have been developed into workable methods of design using formulae and nomograms.

The theoretical distribution of the temperature within the soil surrounding the freeze-pipe may be calculated by the following two formulae:

$$v_1 = v_s \frac{\ln (R / r_1)}{\ln (R / r_0)} \qquad \text{for } r_0 < r_1 < R \qquad (1)$$

And

$$v_2 = v_0 \frac{\ln (r_2/R)}{\ln (a_r)} \qquad \text{for } r_2 > R \qquad (2)$$

where:
r_0 = radius of the freeze-pipe
R = radius to the interface between the frozen and unfrozen soil (isotherm 0°C)
r_1 = generic radius within the frozen cylinder
r_2 = generic radius in the unfrozen region beyond the frozen cylinder
v_1 = temperature at the radius r_1 within the frozen cylinder
v_2 = temperature at the radius r_2 in the unfrozen region
v_s = difference between the temperature at the surface of the freeze-pipe and the freezing point of water

v_0 = difference between the original temperature of the ground and the freezing point of water

a_r = factor which when multiplied by R defines the radius of temperature influence of the freezing pipe, i.e. the radius of the zone of temperature influence is = a_r R.

The temperatures v are with respect to the freezing point of water and absolute values are used for convenience so that v_0 and v_s are always positive.

The typical temperature distribution around one freezing pipe according to the Sanger & Sayles (1979) approach is shown in Figure 9.

Several freezing jobsite experiences have shown that this theoretical approach has been confirmed by in situ temperature measurements and their interpretation during both construction and maintenance of the frozen body. For example, if we apply this approach as back-analysis to the Rome case study we obtain the graph in Figure 10. We can observe how well the theoretical curve fits the experimental curve within the frozen body, with the following hypothesis: r_0 = 0.04 m; R = 99 cm; v_s = -138°C (average between -196°C and -80°C); v_0 = 16°C; a_r = 2.

4.3 *Evaluation of the transient phase between construction and maintenance of the frozen body, when different coolant fluids are used.*

When the combined method is used, i.e. LN for the freezing phase and brine for the maintenance phase, there is a critical aspect to be considered, that is the transient phase. In fact, during the freezing phase the coolant fluid enters the net of freeze-pipes at -196°C and exits usually at -100/-60°C, while the brine enters at -28/-35°C and exits a few degrees higher. It is evident that the temperature difference between the two chilling sources, generated by different fluids flowing through the centre of the frozen body via the freeze-pipes, is quite significant and the distribution of the temperatures within and around the frozen body may not be the same in the two cases.

An example of estimation of the theoretical temperature distribution within and around the frozen body by the Sanger & Sayles (1979) approach, for LN and brine entering respectively at -196°C and at -35°C, is shown in Figure 11 (see the thick external curves). Furthermore, the graph in Figure 11a shows a theoretical transition simulation keeping constant the position of the isotherm -10°C, while the graph in Figure 11b shows the same simulation keeping constant the position of the isotherm 0°C (for both see the thinner intermediate curves).

From the two graphs we can observe that:

– the slope of the LN curve is always much steeper than that of brine;
– in LN freezing, to get the isotherm -10°C at 40 cm from the freeze-pipe axis the isotherm 0°C is located at 55 cm (+38%);
– in brine freezing case a), to get the isotherm -10°C at 40 cm the isotherm 0°C is located at 110 cm (+175%);
– in brine freezing case b), to keep the isotherm 0°C position constant at 55 cm, the isotherm -10°C regresses down to 24 cm (-40%).

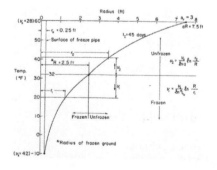

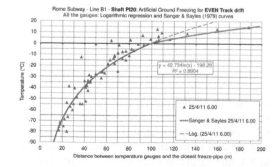

Figure 9. Typical temperature distribution around a freeze-pipe (from Sanger & Sayles 1979).

Figure 10. Graph showing the application of the Sanger & Sayles (1979) approach for a back-analysis to the Shaft PI20 Even track drift in Rome

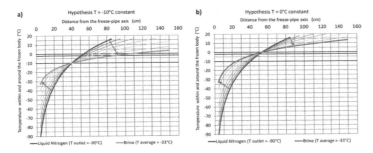

Figure 11. Example graphs showing the theoretical distribution of the temperatures within and around the frozen body during the transition phase between LN freezing and brine maintenance: a) T = -10°C constant; b) T = 0°C constant.

From these observations we can also understand which is the actual impact, when freezing is performed by brine, of choosing the design hypothesis of an active frozen shell included between the two -10°C isotherms only, and neglecting the portion of the frozen body with -10 < T ≤ 0°C (as sometime it is done). It means that the actual thickness of the frozen body (T < 0°C) is between 2.5 and 3 times the theoretical active thickness, being thermal energy and time necessary to achieve the target proportional to the actual thickness.

If we further analyse the transient phase graph, the temperature distribution curve, which is typical of LN freezing at the beginning, changes its slope becoming progressively less steep over time until it becomes a the typical brine freezing curve. This means that there is a redistribution of the temperatures within the frozen body leading to a re-equilibration of the temperatures. Closer to the freeze-pipe the temperature increases (compared with the beginning of maintenance) while farther away it decreases. This is due to a transfer of cold from the core of the frozen body to its periphery. In case a) the result of this re-equilibrium is also a progression of the frozen front (isotherm 0°C).

Taking into account the two limit simulations, the first one (a) may sometimes be a target – to keep the frozen body stable between the two isotherms -10°C – but it is not always possible to reach this target due to the lack of thermal power of the brine compared with LN. Therefore, sometimes the actual evolution of the temperature distribution during the transient phase is likely to be an intermediate between case a) and b). In a first phase we should observe both a regression of the isotherm -10°C and a progression of the isotherm 0°C; in a second phase the position of the two isotherms should become stable, thus showing that the steady state has been achieved.

As an example of this we may examine the actual redistribution of temperatures during the transient phase at the trial field of Aulnay-sous-Bois (Paris) within the frame of the new Line 16 of the Paris Subway. After freezing by LN, maintenance was performed by Temper (a premixed coolant fluid alternative to brine) entering the net of freeze-pipes at -28°C. Figure 12 shows a typical T-D graph specific to two thermometric chains (FT2 and FT3). We can observe that the transient phase is characterized during the first six days by a rotation of the curve around the point T = 0°C and D = 116 cm and then during the following 50 days around the point T = -9°C and D = 42 cm. The result is a regression of the isotherm -10°C during the first six days (from 80 to 35 cm) and then a further slight regression of isotherm -10°C and a significant progression of isotherm 0°C. From the graph we can also observe that if a colder (-35°C) coolant fluid had been used (see the dotted curves), the regression of the isotherm -10°C would have been less evident (from 80 to 45 cm during the first 6 days and to 67 cm after further 50 days).

Another example of the transient phase we are examining regards a freezing project at Clichy Saint-Ouen Station, within the frame of the extension of Paris Subway Line 14 toward the North. After LN freezing, a -35°C brine was circulated for the maintenance phase. Figure 13 shows a typical T-D graph specific to one thermometric chain (OT6). In this case we can observe that the use of a colder brine, compared with the previous case study, allowed us to control the regression of isotherm -10°C better, as well as to achieve the steady state sooner; in fact, we can see that after 12 days the re-equilibrium was practically completed.

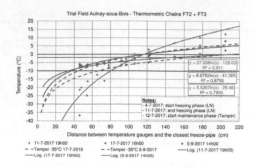

Figure 12.   Example of transient phase at Aulnay-sous-Bois trial field.

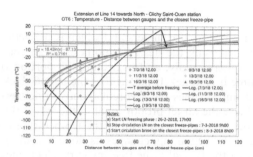

Figure 13.   Example of transient phase at Clichy Saint-Ouen station.

5   CONCLUSIONS

The lessons learned from a large number of AGF jobsites show that quality control and real-time monitoring are key factors for a successful ground freezing job, further to excellent engineering and a very skilled and experienced geotechnical contractor. Surveying of the freezing boreholes and real-time monitoring of temperatures, groundwater pressure and displacements are of the utmost importance to guide the freezing process well and to guarantee the best final result.

The interpretation of the soil temperatures during freezing cannot avoid correlating the measured temperatures with the distance between each temperature gauge and the closest freeze-pipe. In fact, the temperature measured by each gauge is obviously a function of its distance from the closest freeze pipe (chilling source), following the rule that a closer chilling source entails a colder soil and a more distant chilling source entails a warmer soil.

The actual temperature distribution within the frozen body during freezing is well represented by a logarithmic curve, as suggested by the theoretical approaches proposed by different authors and confirmed by several freezing jobsite experiences. The logarithmic approach allows also understanding the actual impact of design hypothesis, such as the choice of the isotherm defining the design boundary of the frozen body, and anticipating the transient phase from freezing to maintenance, when the combined method is used.

REFERENCES

Frivik, P.E. (1980). State of the art report – Ground freezing: thermal properties, modelling of processes and thermal design. *Proceedings of the 2nd International Symposium on Ground Freezing, Trondheim, Norway, 24–26 June.* 115–134. Rotterdam: Balkema.

Harris, J.S. (1995). *Ground freezing in practice.* London: Thomas Telford.

ISGF Working Group 2 (1991). Frozen ground structure – Basic principles of design. *Proceedings of the 6th International Symposium on Ground Freezing, Beijing, China, 10–12 September.* 503–513. Balkema.

Sanger, F. J. & Sayles, F. M. (1979). Thermal and rheological computation for artificially frozen ground construction. *Engineering Geology, Vol 13, Nos 1–4.* 311–337. Amsterdam: Elsevier.

Tunnels and Underground Cities: Engineering and Innovation meet Archaeology,
Architecture and Art, Volume 4: Ground improvement in
underground constructions – Peila, Viggiani & Celestino (Eds)
© 2020 Taylor & Francis Group, London, ISBN 978-0-367-46868-2

Ground improvement with accelerated micro-cement grout, Thessaloniki Metro, Greece

P.I. Maragiannis & P.G. Foufas
OMETE Edafostatiki S.A., Athens, Greece

ABSTRACT: Thessaloniki's METRO line in Greece foresees the construction of twelve pumping stations for draining purposes of the TBM tunnels. The stations are to be excavated from within the TBM tunnels by removing/demolishing part of the precast lining. Six of the stations are situated in Tertiary fluvial deposits consisting of alternations between soft silts and loose silty sands and gravels of permeabilities in excess of 10^{-1} cm/sec. Trial drillings conducted showed water ingress exceeding 60m³/h accompanied by severe flow of sand and gravels that instantly gave rise to surficial settlements, thus necessitating the implementation of ground improvement. High pressure injections (i.e. 24bar) with accelerated micro-cement grout using self-drilling anchors installed at a triangular grid, determined through finite difference analysis, were adopted. Non-accelerated grouting with normal cement was unsuccessful as the grout could not infiltrate into the silty sands and the groundwater velocity led to complete washout.

1 INTRODUCTION

Thessaloniki's Metro basic line comprises an ongoing project that includes the construction of thirteen (13) contemporary stations with central platforms, 7.70km of twin tunnels, approximately 6.00m in diameter each, constructed by two tunnel boring machines and about 1.80km of tunneling by means of cut and cover. The implementation of the project is done by a joint venture led by AKTOR S.A..

The excavation of three (3) sets (i.e. out of a total of six (6) sets) of pumping stations situated between Fleming and Voulgari Stations comprises part of the aforementioned project. Each set of pumping stations is to be excavated conventionally, alongside the axis of the twin TBM tunnels by demolishing part of the precast lining (Figure 1).

The aforementioned construction turned out to be rather challenging due to the physical and engineering properties of the ground encountered at the depth of excavation, and the diameter of the TBM tunnel, which limited the means that could be used for the excavation works.

In respect to the geotechnical regime, the poor strength (c_u<40kPa, N_{SPT}<4) and deformation (E<10MPa) properties of the encountered formations, which limited their stand-up time, along with their high coefficient of permeability (k>10^{-1} cm/sec) that led to severe water ingress followed by outflow of silt and sand, which in turn resulted to almost instant settlements at the surface of the ground, practically prevented the conventional excavation of the pumping stations without implementing first a ground improvement scheme.

Taking into account the limitation that any intervention in the direction of improving the ground conditions would have to take place from the TBM tunnels and not the ground's surface, all possible ground improvement scheme solutions (i.e. ground freezing, grouting etc) were evaluated in terms of cost and time effectiveness.

Finally, it was decided to proceed with a customized scheme involving the injection of accelerated micro-cement grout through self-drilling anchors installed at a properly determined

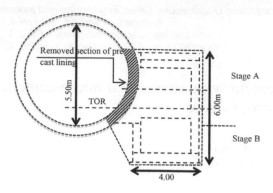

Figure 1.   Cross section illustrating the geometry of the excavation.

grid, in order to increase the stiffness and density of the soil formations while at the same time reducing their coefficient of permeability. The injection points (i.e. the installation grid of the anchors) were derived through finite difference analysis.

Upon full implementation of the above described ground improvement scheme, verification drillings were carried out in order to obtain undisturbed soil samples for laboratory testing and check the remaining, if any, water ingress. The results were beyond expectation as the water ingress measured was practically negligible whereas the major engineering properties of the subsoil material had risen to values similar to those of semi-rocks.

The aforementioned end result of the improvement scheme allowed for a "lighter", in terms of cost, redesign of the excavation methodology and temporary support of the pumping stations that was properly implemented, thus leading to the successful construction of all pumping stations.

The ultimate scope of this paper is to describe the steps that were followed in the design and implementation of a ground improvement scheme based on accelerated micro-cement grout, as well as to present the outcome of the implementation of the aforementioned scheme as documented during the subsequent excavation and through the readings obtained from the installed monitoring points.

The pumping stations to be excavated are approximately 2.50m wide, 6.00m high and 4.00m long (i.e. extending vertically out of plane compared to the axis of the TBM tunnels, refer to Figure 1).

The excavation was carried out in two major stages A (heading) and B (bench). The excavation step was initially set to 0.50m and hence eight (8) steps were required for the completion of each one of the two major excavation stages aforementioned.

2   GEOTECHNICAL REGIME

Fluvial deposits of poor engineering properties are encountered at the tunnel section between Fleming and Voulgari Stations. More specifically, the upper about 20.00m beneath the surface of the ground consist of soft, brown silt and clay of low plasticity with sporadic intercalations of loose silty sand and gravel. The cohesive sediments control the engineering behavior of this horizon, which is classified as Geotechnical Unit I. At the depth of interest and hence underlying Unit I, very loose sand and gravel, prone to liquefaction, extending at least up to the depth of 35.00m, are encountered. These non-plastic formations, which are classified as Geotechnical Unit II, comprise a highly permeable semi-confined to leaky aquifer, capable of transmitting significant quantities of groundwater at a flow velocity ranging between 2.00 and 2.50m/day (i.e. Darcy's flow).

The formations of Unit II were extensively investigated, both in the laboratory and the field. The pressuremeter tests that were carried out came back with elastic moduli values (E) as low as 3 to 5MPa, whereas the Standard Penetration Tests (SPT) did not exceed 4 blows in respect to $N_{(60)}$ values. Both test results practically point to the direction of poor material of practically negligible stand up time. Pumping tests were conducted in the aquifer of Geotechnical Unit II. The critical discharge rate under which the steady-state tests were conducted was derived to be about $160 \, m^3/h$ whereas transmissivity values in excess of $72 m^2/h$ were obtained.

Given the collapsing nature of the soil formations of Unit II, which would be further degraded as a result of the seepage forces that would be generated during excavation due to the extensive water ingress, any attempt to proceed with the excavation of the pumping stations, without first implementing a ground improvement scheme, would most definitely lead to extensive ground settlement, thus endangering the overlying structures.

3 EVALUATED GROUND IMPROVEMENT SCHEMES

The ground improvement techniques that were considered, bearing in mind all construction restrictions presented in §1 were *ground freezing, chemical grouting* and *cement grouting*.

The criteria, upon which the decision was based, were the time and cost effectiveness as well as the mobilization time of each method.

In respect to the *ground freezing* method it did not fulfill any of the criteria stated, as it required approximately 3 to 4 weeks for complete mobilization, since it would have to be implemented by a foreign company stationed overseas, another 3 to 4 weeks just for the freezing process and ultimately it was the most expensive, in terms of total implementation cost, compared to the other available solutions. Furthermore, there were also concerns regarding the effectiveness of the method as (i) the high groundwater flow velocity of the aquifer, which at places exceeds 2.50m/day (refer to §2), could leave openings – gaps in the developed frozen section (i.e. the frozen columns would not merge) and (ii) theoretical computations showed that during the ground freezing works heaving and settlements should be expected at the surface of the ground at a very sensitive area, where certain structures have already settled significantly.

Chemical grouting is most suitable for mediums with discontinuities (i.e. rockmass, semirocks), as it can permeate the network of discontinuities regardless of their aperture, as well as for very low permeability soil formations. Thus, in the case of high permeability, coarsegrained formations, it is common practice to commence the works with the implementations of a cement-based grouting scheme and if, at the end of the day, the volume of water ingress is still difficult to cope with, an additional, but of much lower extent and cost (i.e. the cost of a liter of polyurethane is about ten (10) times higher compared to that of a liter of a cementbased grout), chemical grouting program is carried out to further reduce the water inflow. Furthermore, another drawback, apart from the cost, of chemical grouting is that under high groundwater flow velocities (i.e. 2.50m/day generated in the aquifer system of geotechnical unit II), it can be washed-out before reacting regardless of the amount of catalyst added. The latter argument was verified through a field test during which about a thousand (1000) liters of polyurethane with 10% of catalyzer were injected in the soil unsuccessfully, as according to the observations done in the surrounding holes that were drilled, it was washed-out before it managed to react with the water.

Eventually, the only ground improvement method that could be quickly mobilized and implemented with the equipment available on site, thus being at the same time both cost and time effective was *cement grouting*. However, it would have to be customized – modified in order to prevent the washing-out of the injected grout, as initial field tests that were undertaken using CEM II/32.5R Portland cement, at a water to cement ratio of 1 to 1, were unsuccessful. More specifically, with the use of the aforementioned grout the injection process was ongoing without the slightest rise in pressure up to a total injected volume of 6000 liters, at which point the injection ceased. Trial drillings that were conducted the following day at the point of injection came back with nothing but soil, indicating complete washout of the injected grout.

Based on the aforementioned information it was decided to design a ground improvement scheme based on a cement grout "impervious" to water that would occupy and set at the soil to be improved instead of "travelling" far away from the area of interest.

4 ACCELERATED MICRO-CEMENT GROUTING METHOD

4.1 General

The accelerated grouting method is a well-established method for increasing the strength and deformation properties of a material, while at the same time reducing their coefficient of permeability (Bahadur, Holter, Pengelly, 2007). The main difference of this method compared to the conventional grouting techniques is the addition of an agent (i.e. accelerator), which decreases the gel and setting time of the grout, thus coping with high groundwater flow conditions as in this case (Garshol, 2007).

In the case of the granular sediments the increase of the engineering properties is mainly accomplished through densification whereas the fine-grained sediments exhibit a significant increase in stiffness basically due to hydrofracturing (i.e. claquage).

Although, it is a common approach when tunneling with Tunnel Boring Machines (TBM), as it is used for grouting the annular gap between the precast lining and the surrounding soil, or/and for sealing off leakages around tunnels (Pelizza & Peila, Sorge & Cignitti, 2012), there is lack of documentation regarding the use of this method in the direction of increasing the stiffness and reducing the coefficient of permeability of the ground, so that a conventional underground excavation can take place subsequently. Lack of documentation refers to the absence of engineering data regarding the behavior of the accelerated grout when injected into the ground under high hydraulic gradient and hence significant flow velocity. Without the aforementioned knowledge it becomes practically impossible to design an extensive ground improvement scheme based on accelerated grout that will be accurate and precise in terms of injection grid pattern, injection velocities, refusal criteria and grout mixtures.

The adoption of the aforementioned approach for improving the ground while at the same time coping with the unfavorable groundwater velocity, at the locations where the six pumping stations of Thessaloniki Metro were to be constructed, gave the opportunity to realize extensively the manner at which the accelerated grout reacts and permeates into the soil, thus revealing the above mentioned "missing" data.

4.2 Design approach

4.2.1 Injection grid

The first step in designing the selected ground improvement scheme was the determination of an optimum injection grid. Based on the geometry of the pumping stations (refer to Figure 1), grouting would have to be extended around the excavation by an adequate distance so that the improved ground can withstand the active earth pressure and the hydrostatic pressure. By assuming a 6.00m wide by 6.00m high by 6.00m long (i.e. in the out of plane direction) improved zone, an about 2.00m thick improved soil "plug" surrounding the excavation is created. Given that the crown of the excavation is situated about 20.00m below the surface of the ground and the ground water table sits at a respective depth of about 5.00m, the total force acting on the 2.00m thick improved soil is approximately 3000kN/m. If the shear strength of the improved soil plug is neglected the above approximated force must be carried by anchors.

Bearing in mind that the soil formations were collapsible and hence fiberglass anchors could not be used and that the water ingress and the accompanying soil flow once an injection point had been drilled were rather severe, the only solution was to adopt self-drilling anchors so that the drilling and injection procedure were completed in one step instead of two.

Given the above-approximated force that was going to be imposed on the improved zone of the soil, Φ32 self-drilling anchors with an internal diameter of Φ24 and yield strength equal to 200kN were selected.

By considering a factor of safety of 1.50 on the acting force of 3000kN/m and 80% of the yield strength of the self-drilling anchors it was deduced that 29 anchors would have to be installed in the vertical direction. Due to the fact that the required number of anchors corresponds to 1 anchor every 0.20m (i.e. too dense), it was decided to install 1 anchor every 0.50m in the horizontal direction, so that the acting force is reduced to 1500kN/m and hence the required number of anchors in the Y axis goes down to 15.

The outcome of the above-described procedure was the realization of a 0.50m x 0.40m grid, spaced in triangular manner, which corresponds to 180 injection-anchoring points.

4.2.2 Grout mixture

The next step in the design of the improvement scheme is to decide on the grout mixture and accelerator to be used. Given the grain size distribution of the silty sand and sandy silt of Geotechnical Unit II, it was deduced that micro-fine or ultra-fine cement would have to be used. The most suitable for the current ground conditions cement was found to be MICROCEM 8000 SR of Mapei S.p.A.

As mentioned before in order to deal with the increased groundwater flow velocity and the washout that it causes, the grout mixture had to be absolutely stable before injected into the ground. Hence it would have to reach gel time by the time it exits the self-drilling column. Therefore an accelerator had to be added right before the grout enters the self-drilling column, so that it has enough time to reach gel condition by the time it exits the column and enters the ground. In that manner, the injected grout would be completely stable, thus anti-washout, and controllable in terms of expansiveness. Since the total length of the self-drilling column would be 6.00m, the time required for the grout to travel through it, at an injection rate of 30liters/min, was estimated in the field at about 20sec. Therefore, the accelerator added would have to be able to achieve a gel time of no more than 20sec, while at the same time not affecting the ultimate strength (i.e. at 28 days) of the grout. An alkali-free accelerator labeled MAPEQUICK AF 1000, produced by Mapei S.p.A,, was found to be the most optimum, as when added in 1:1 (water: cement ratio) grout mixtures at 6–8% per volume of grout it allowed for a gel time of 15sec, while the loss in the ultimate strength of the grout did not exceed 5%.

Finally, in order to further enhance the stability of the accelerated grout mixture, a polymer liquid agent that increases the viscosity of the mixture while at the same time allowing it to retain adequate workability even at gel time as long as it flows at certain velocities, was added.

4.2.3 Refusal criteria

Upon deriving a proper injection grid and a suitable grout mixture for the soil conditions at hand, the following step is to establish the refusal criteria, thus the determination of the grout volume that needs to be injected in the soil and the pressure that needs to be raised and locked-in, so that an injection is considered successful.

In order to proceed with the above-mentioned determination, a two-dimensional simulation was designed, using finite difference flow analysis software, where the complete stratigraphy, physical, engineering and hydrogeological properties of the encountered formations was input properly. The injection points were simulated as recharge wells and the grout mixture as a contaminant bearing the viscosity of the actual end product to be injected.

The scope of the analysis was to get a first idea on the pressure under which the grout would have to be pumped into the soil as well as on the volume required in order to completely fill the area in between four surrounding injection points. It is stressed out that this computational approach was only conducted for indicative purposes, so that an estimate on the required quantity of micro-cement for fully implementing the grouting scheme is gained.

The analysis showed that due to the increased viscosity of the grout mixture significantly high pressures were required in order to be able to permeate the soil formations. More specifically, pressures exceeding 15bars were indicated by the computational analysis, whereas a total volume of about 375liters/meter (i.e. out of plane) needed to be pumped through the recharge well in order for an area of $1.2m^3$ to be completely filled with grout.

The exact volume required in order to conclude the ground improvement procedure was derived in the field by conducting actual injections and monitoring the expansiveness of the mixture. In more detail, one injection point and its nearest, diagonally, one were drilled, so that the latter is used for monitoring purposes. Injection of the grout commenced and once

Figure 2.   Grouted core of silty sand.

the mixture reached the monitoring point, the valve was closed and the injection process continued until the flow rate dropped to less than 9 liters/min. At that point, the injection valve was closed and the dissipation rate, if any, of the locked-in pressure was recorded to check whether the injection was successful or not. Many trial injections were conducted at different locations at the area to be improved, most of which came back with consistent results regarding the pressure at which the grout reached the monitoring point, the final pressure and the total injected volume. Almost all tests exhibited a "communication" pressure of 17bars, an ultimate pressure of 22bars and a total injected volume ranging between 1800 and 2500 liters.

Cores that were bored subsequently at the areas where the injections had taken place showed almost complete replacement of the silty sand with grout and severe hydrofracturing of the silty layers (Figure 2).

It is noted that for optimization purposes it was attempted to cease the injection process exactly at the pressure at which the grout reached the monitoring point, so as to reduce, if possible, the required quantity of micro-cement. However, not only were the cores obtained afterwards unsatisfactory in terms of improvement but also severe flow of water was still occurring. Hence, in order for an injection to be considered finalized a total volume of at least 1800liters would have to be injected and the flow rate should drop to less than 9liters/min at a minimum pressure of 22bars. Another criterion that was stated referred to the maximum injection pressure, which should not exceed 30bars so as to avoid any structural damage on the precast tunnel lining.

4.2.4 Injection velocity
The velocity at which an injection scheme based on accelerated grout of high viscosity shall be implemented comprises a very critical parameter. For instance, if the velocity is too low the grout mixture reaches gel time inside the injection column, thus leading to blockage and ceasing of the injection process, whereas if it is too high the distance at which the grout shall flow shall be considerably longer than required, because (i) the gel time of the mixture shall be delayed due to the interaction of the grout with the groundwater, (ii) seepage forces that shall be developed shall be greater leading to more intense internal erosion that will allow for the formation of more extensive flow channels for the grout to travel through and (iii) the adopted grout mixture is thixotropic and hence the higher the injection velocity the longer it retains its viscosity constant, thus the longer it takes to set and cause a rise in pressure that will create new-additional flow channels for permeation.

Bearing in mind that the injection velocity could not drop below 9liters/min, as this would cause blockage of the injection column, trial injections were conducted so as to determine the optimum velocity to be adopted. By injecting at one point and monitoring at different distances along the area to be improved, it was derived that with injection velocities between 25 and 35liters/min the grout's expansiveness is limited within the area of interest, while at velocities exceeding 40liters/min the expansiveness of the mixture exceeds 6.00m in linear distance.

Therefore, the injection process commenced at a velocity around 30–35liters/min. Once the pressure started building up the velocity was allowed to be decreased up to a minimum value of 25liters/min, at which point it was increased again at 35liters/min with an additional rise in pressure. The aforementioned procedure was followed until the injection met the failure criteria presented in §4.2.3.

4.2.5 Injection points

Given the collapsing nature of the soil formations, self-drilling anchors were employed (refer to §4.2.1) not just for supporting the face of the excavation but also to be used as injection columns. The anchors were equipped with $\Phi76$ drilling bits and they had a total length of 6.00m, consisting of four rods of 1.50m long each, which connected to each other with couplings. Once the drilling process was completed, a mechanical packer was installed at the face of the precast lining and the annular between the anchor rod and the packer was sealed off with a custom-made cylindrical two-part metal bolt.

Trial injections and subsequent sampling that were carried out showed that when the grout exited the drilling column from the drilling bit only, it did not fill the drilled hole uniformly but instead the grout bulb was "pear"-shaped, thus exhibiting significant expansiveness around the drilling bit and almost none as it approached the precast lining of the tunnel. Therefore, it was decided to place the couplings at 1.00m intervals (i.e. the 1.50m rods were cut down) and turn them into injection valves by drilling them properly, so that an almost uniform, in terms of expansiveness, grout bulb is formed around the whole length of the injection column.

It is pointed out that grouting with accelerated cement mixtures cannot be implemented in the form of tube a manchette because by the time adequate pressure is raised in the tube, in order for the valves to open up, the grout has already reached gel time and blocked them completely. For that reason, the couplings were turned into injection valves, with the first one placed at 3.00m and the second one at 4.00m to bear two $\Phi6$ holes each and the last one at 5.00m bearing four $\Phi8$ holes. In that way an almost uniform pressure was built up in the injection column allowing for the same volume of grout to exit each one of the couplings.

The reason for placing the first injection coupling 3.00m back from the precast lining and not any closer was to prevent a premature pressure build-up (i.e. at the first stages of the injection process) close to the segment that could cause it to crack. This argument was verified by conducting trial injections with pneumatic packers and monitoring their pressure as the injection process progressed. It was observed that at the very first stages of the injection process, the pressure of the pneumatic packer was increased as the injection pressure was also increased (i.e. if the packer had been inflated at 20bars and the injection pressure built up to 10bar, the pressure of the packer would rise to 30bars). After a few hundred liters of injected grout the pressure in the packer and hence at the back of the precast lining dropped again due to the setting of the grout at the area most proximal to the lining of the tunnel.

4.3 Equipment

Grouting with accelerated cement mixtures requires two hydraulic circuits. A *primary* one, driven by a master pump that accommodates the grout and a *secondary* one, assisted by a smaller slave pump that pushes the accelerator. Since the accelerated grout is very sensitive to any pulsations generated by a pump, because of the fact that even the slightest pulsation causes instant disruption to the flow rate, which in turn leads to blockages in the injection column, it is suggested to use screw pumps instead of piston pumps, which maintain constant injection pressure and hence flow rate, along the whole range of their operation. In the lines of implementing the current ground improvement scheme, a screw pump capable of delivering 40liters/min of grout at 30bars was employed. Regarding the slave pump that pushes the accelerator, it is best to use hose pumps, as their peristaltic operation allows for the "self-cleaning" of the drilling column. However, in the current situation such an option was not valid as the maximum pressure a peristaltic pump can reach is 16bars. Therefore, a progressive cavity pump, capable of delivering 3liters/min at 24bars was assigned to drive the accelerator.

The two aforementioned circuits are connected to a mixing device equipped with a small static mixer, which in turn is connected through a 0.50m long hose to the injection column. The accelerator is driven to the mixing device through a 0.5" hose, whereas the grout is pushed through a 1" hose. A very critical detail in the design of the mixing device, which again was custom-made for this project, is the exit point of the accelerator, which must be positioned exactly in the flow path of the grout so that it is drifted along by it.

The flow rate of the grout mixture is monitored through an automated read-out unit, while the flow rate of the accelerator, due to its small value, is read manually by a typical flow meter.

An important issue that needs to be mentioned is that the whole injection procedure cannot be automated by means of setting the master pump at a constant flow rate (i.e. 30liters/min) and the slave pump at another rate (i.e. 6% x 30liters/min) and wait for the pressure to reach the failure criterion of 22bars, because the constant flow rate of the master pump shall lead to premature blocking of the injection rods. This happens because once the grout mixture reaches a lower permeability zone and starts to set, the sudden increase in pressure introduced by the master pump, so that the flow rate is maintained to the set value, causes the grout to "congest" in the injection column, as there is not adequate time for new flow paths to the generated, thus ultimately leading to blockage of the rods. This is why once the pressure starts building up the flow rate is allowed to go down to a certain defined minimum value before it is brought up again to the designated flow rate value.

4.4 Verification

Forty-five (45) days were required for the full implementation of the above-described grouting scheme at each one of the six pumping stations to be excavated, with the average volume of grout injected being approximately equal to 250,000 liters (i.e. equal to 190 tons of micro-cement).

Upon finalization of the grouting procedure at each pumping station, sampling drilling was conducted in order to obtain samples for laboratory testing, as well as to check on whether or not water ingress has been halted. Regarding the latter, almost all locations that were drilled either exhibited "dry" conditions or practically negligible water inflow of the order of 3liters/min. As for the former, unconfined compressive strength tests and unconsolidated-undrained triaxial tests were conducted both on samples with obvious signs of improvement (i.e. with network of grout veins or cemented with grout etc) as well as on samples that did not show any signs of grout. The results that came back showed astonishingly higher values of strength compared to the initial conditions, as the ones obviously grouted exhibited average values of undrained shear strength around $c_u = 200kPa$, whereas the ones with no indications of improvement showed values around $c_u = 100kPa$ (i.e. 2.5 times their initial strength). Needless to mention that at the locations where the grout mixture had completely replaced the soil material the strength of the grout sample exhibited values exceeding 1MPa (i.e. after 24h). It is pointed out that the samples that did not exhibit any obvious signs of grouting and yet their strength came back double its initial value, also showed reduced values of moisture content and porosity-void ratio, indicating that their improvement came as a result of consolidation (i.e. sandy silt) or/and densification (i.e. silty sand).

Given the positive outcome of the post-grouting lab testing program that was carried out and since the initially severe water ingress had been halted, the excavation works could safely commence.

5 EXCAVATION - MONITORING

Based on the results of the laboratory tests and the field inspection that was carried out upon completion of the grouting scheme, the temporary support and excavation approach of the pumping stations was redesigned.

More specifically, the excavation step was increased from 0.50m to 1.00m, the shotcrete thickness was reduced from 0.25m to 0.10m and the HEB 160 beams were installed every 1.00m instead of every 0.50m. It is denoted that according to the computational analysis that was carried out there was no necessity for a steel frame installation. However, they were installed for contractual reasons solely.

Once the segments of the precast lining were removed and the excavation commenced useful conclusions were drawn from the view of the excavated face regarding the mechanism and behavior of accelerated grout:

- The fine-grained layers (i.e. sandy silt) were severely hydrofractured and reinforced with grout veins, thus significantly stiffer. Moreover, they exhibited signs of consolidation as a result of the pressure induced during the hydrofracturing. Vane tests that were conducted in situ showed values of undrained shear strength exceeding at places $c_u = 220$kPa (Figure 3).
- The coarse grained layers (i.e. silty sand) had either been cemented or completely replaced by grout. In both cases they showed significant densification resulting in refusal when pocket penetrometer tests were attempted to be carried out (Figure 4).

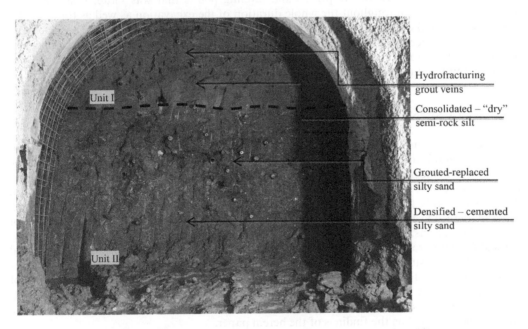

Figure 3.   Excavation face of Analipsi-Patirkios pumping station – Phase A – Step 2/4.

Figure 4.   Excavation face of Analipsi-Patrikios pumping station – Phase B – Step 1/2.

- All previous groundwater flow paths had been completely sealed off and could be mapped accurately based on the oxidizations that could be noticed on the grout that occupied these areas.
- Soil cavities that were formed during the drilling procedure prior to the injection had been completely filled with grout. It is characteristic that boulders of grout exceeding 0.50^3 in volume were identified.

The excavation proceeded in two phases (refer to Figure 1). In the first phase the upper 4.00m of the pumping station were excavated, whereas during the second phase the excavation was deepened by another 2.00m in order to form the reservoir of the pumping station.

The total duration of the first and second phase of excavation were 3 days and 2 days respectively; a rather quick time, considering that due to the limited space in the TBM tunnel most parts of the excavation were undertaken by hand, attributed to the enhanced engineering properties of the excavated material.

The settlements generated both at the surface of the ground as well as on the foundation level of the buildings, which are situated in the zone of influence of the excavation, were monitored through a network of 3D prisms and leveling points that was setup. Readings were obtained twice daily throughout the whole duration of the excavation as well as for at least a month following the construction of the permanent lining. The results that came back actually verified the enhanced, due to the grouting scheme that had been preceded, engineering properties of the subsoil formations as the total settlement recorded on the surface of the ground was less than 1mm and practically negligible on the buildings.

6 CONCLUSIONS

The successful construction of the underground pumping stations of Thessaloniki's Metro was an accomplishment, given the rather unfavorable hydrogeological regime along with the really poor engineering properties of the soil formations that had to be dealt with. The large-scale and extensive implementation of the accelerated micro-cement grouting scheme proved that it is probably the most cost effective solution for overcoming severe groundwater flows, while at the same time improving considerably the strength and deformations properties of the soil formations, as long as it is designed and implemented with precision and accuracy. Critical design parameters of this method, such as the velocity of injection, the stated refusal criteria, the viscosity of the end mixture or even the equipment used, are most of the times underestimated in the field, thus leading to unsuccessful applications and significant waste of grout mixture. The common perception that grouting schemes in general are unpredictable as there is no way of controlling the expansiveness of the mixtures in the ground is against the authors' belief based on the findings of the herein paper.

REFERENCES

Bahadur, A.K. & Holter, K. G. & Pengelly, A. 2007. Cost-effective pre-injection with rapid hardening microcement and colloidal silica for water ingress reduction and stabilization of adverse conditions in a headrace tunnel. In Bartak, Hrdina, Romancov & Zlamal (eds), *Underground space – the 4th dimension of Metropolises*. London: Taylor & Francis Group.

Garshol, K.F. 2007a. Pre-excavation grouting in tunneling. UGC International, Division of BASF Construction Chemicals (Switzerland) Ltd.

Garshol, K.F. 2007b. Using colloidal silica for ground stabilization and ground water control. Tunnel business magazine.

Pelizza, S., Peila, D., Sorge, R. & Cignitti, F. 2012. Back-fill grout with two components mix in EPB tunneling to minimize surface settlement: Rome Metro – Line C case history. In Viggiani (eds), *Geotechnical aspects of underground construction in soft ground*. London: Taylor & Francis Group.

Tunnels and Underground Cities: Engineering and Innovation meet Archaeology, Architecture and Art, Volume 4: Ground improvement in underground constructions – Peila, Viggiani & Celestino (Eds)
© 2020 Taylor & Francis Group, London, ISBN 978-0-367-46868-2

Thermal phenomena model for artificial ground freezing during a tunnel excavation for the Municipio Metro station in Naples, Italy

N. Massarotti, A. Mauro & G. Normino
Università degli Studi di Naples "Parthenope", Naples Italy

A. Di Luccio & G. Molisso
Ansaldo STS | A HITACHI GROUP COMPANY, Naples (Na), Italy

F. Cavuoto
Ingegneria delle Strutture, Infrastrutture e Geotecnica, Naples, Italy

ABSTRACT: Artificial ground freezing (AGF) in a horizontal direction has been employed in Naples (Italy) to ensure stability and waterproofing of the soil during excavation of tunnels connecting two lines of the Naples metro system. The artificial freezing technique consists in circulating a coolant fluid, with a temperature lower than that of the surrounding ground, inside probes positioned along the perimeter of the gallery. The water contained in the saturated soil solidifies and forms a block of frozen ground in the area surrounding the probes. To analyze the heat and mass transfer phenomena in the ground the authors have developed a 2D numerical model. This model taking into account the water phase change process and has been employed to analyse the phenomena occurring in five cross sections of the galleries. The aim of the work is to analyse the thermal phenomena occurring in the ground during the freezing stages and optimize the freezing process. In order to do that, the authors have taken into account the phases to realize the entire excavation of the two tunnels and the evolution of frozen wall during the working phases.

1 INTRODUCTION

Artificial Ground Freezing (AGF) is a consolidation technique used in geotechnical engineering when underground excavations must be executed in granular soils or below the groundwater level. The realization of relevant underground work in urban areas usually involves the management of delicate constructive problems that can also be related the presence of water in the excavation, especially if the soil has poor geo-mechanical quality. The AGF method consists in letting a refrigerant circulate inside probes located along the perimeter of the excavation, at a lower temperature much than the surrounding ground, so that the water contained in the soil passes from liquid to solid phase forming a block of frozen ground in the area surrounding the probes. The whole process can be modelled as a succession of different phases. Depending on the type of working fluid used, two different types of procedures can be identified: (i) Direct method, which is based on the use of liquid nitrogen entering the probes at a temperature of -197 °C and released in the atmosphere in gaseous phase at a temperature between -80 °C and -170 °C; (ii) Indirect method, which is based on the use of a solution of water and calcium chloride (brine) whose circulation temperature can vary between -25°C and -40 °C. The combination of the two previous methods is called the Mixed method. This process uses the direct method for the freezing phase and the indirect method for the maintenance phase, during excavation once the frozen wall reached the measurement thickness. The main advantages of this technique among the available ground consolidation and waterproofing technologies are: (i) security and compatibility with the environment since there is no injection and dispersion of products in the ground, the water naturally present in the ground is, in fact, freezed, using refrigerant fluids that will never be directly in contact with the ground and groundwater, avoiding contamination phenomena; (ii) applicability to any type of soil, from coarse to fine grain and rock.

Several numerical and experimental works analyzing the AGF technique are available in the literature. (Colombo 2010) invokes a well-known approximate approach for the a priori evaluation of the parameters that have an influence on the project, like the time required to reach the target temperatures or the needs in terms of power of the plants to be used. The results of this analysis, applied to the Neapolitan tuff, were then compared with those obtained from a series of numerical analyses conducted with the finite element method, and with the experimental data measured on site during the freezing operations carried out in the framework of the realization of the galleries for the stations of Piazza Dante and Piazza Garibaldi of metro Line 1 in Naples. (Papakonstantinou et al. 2010) first analyzed experimental data of monitored temperatures inside the ground during the freezing process and then developed a numerical interpretation through the FREEZE calculation code, a thermo-hydraulic software developed at the ETH Zurich. The thermal conductivity of the ground in the artificial ground freezing model can be accurately estimated by a latter numerical analysis when it is not known. From the analysis, the authors state that the thermal conductivity of the soil is an important parameter for the model and can be reasonably estimated by a posterior numerical analysis when not known. (Pimental et al. 2011) present and analyze the results of three applications of AGF in urban underground construction projects, by comparing experimental data with those obtained from a thermo-hydraulic coupled code model (FREEZE). The first case study concerns the construction of a tunnel for the underground in Fürth (Germany) in soft terrain with a significant infiltration flow. The second case study concerns a platform tunnel in a metropolitan station in Naples and aims to determine the relevant thermal parameters through a retrospective analysis and to compare the forecasts model with the on-site measures. In the third case, regarding a tunnel under the river Limmat in Zurich, numerical simulations are used to identify potential problems caused by geometrical irregularities in the well layout, in combination with infiltration flow. (Russo et al. 2015) analyze the experimental data collected during the excavation with the AGF technique and then develop a numerical model to evaluate the stress in the ground during the freezing and defrosting process of the frozen wall. This work describes the heat transfer analysis of on actual excavation of a tunnel with artificial ground freezing technique (AGF) to allow the safe digging of the service gallery, located half in the layer of the silty sand and half in the yellow tuff layer, below the groundwater. A phase of the tunnel construction was modelled and monitored with measure and control activities during the gallery excavation. Measurement collected during the construction process, in fact, allowed to control the subsidence due to the freezing-thawing process and the change in volume related to the excavation, providing useful information for future implementation of similar projects. The latter analysis in the test procedure was conducted using a complete 3d model implemented in the DFM Flac3D package.

The general agreement between measurements and calculations is satisfying. The authors in the present paper, have developed an efficient transient numerical model to effectively analyze heat transfer in the soil and at the same time, save computing resources. In general, the available literature proves that there is an increasing interest in numerical and experimental analysis for the artificial ground freezing based solutions. The proposed approach is based on the coupling of a heat transfer model between freezing probes and the surrounding ground with a heat transfer and passage phase of the soil. The model was used as a preliminary predictive analysis for the construction of two tunnels in Naples. The model has been verified against the experimental data of (Colombo 2010) and numerical data of (Ogoh et al. 2010). The numerical model has been then employed to perform a case study for two tunnels of the metro in Naples (Italy). The purpose of this work is to study the thermal and thermodynamic aspects related to the process of ground freezing for the realization of the tunnels between Line 1 and Line 6 of the Naples' underground. The analysis is carried out by a FEM model, using the Comsol Multiphysics computation program, for modeling the thermal interaction of freezing probes with the ground. In the next section, the authors describe the case study characteristics model the AGF technique, while in section three the numerical model developed is presented. The results of the analysis performed after model validation are reported in section four, while some conclusions are drawn in the last section.

2 DESCRIPTION OF AGF TECHNIQUE AND CASE STUDY

For the realization of two connecting tunnels between line 1 and line 6 of the Naples metro, the authors have used the mixed method that consists in using liquid nitrogen until the ice wall reaches the desired thickness and then brine to maintain the thickness of the frozen wall until the end of the construction activities. The mixed method process can be divided in different stages: (i) Phase 1-Nitrogen: use of nitrogen with an inlet temperature of –196°C, and expected outlet temperature of -110°C, for the time required for the formation of the desired thickness of the frozen wall (1.5 m); (ii) Phase 2-Waiting: end of nitrogen feeding, and expected increase of temperature within the probe to reach values adequate to the brine feeding; (iii) Phase 3-Brine: maintaining the ice thickness with brine fed inside the probes at a temperature of about -35°C.

Figure 1 shows the construction site and the cross section of the connecting galleries between line 1 and line 6 of the Metropolitan of Naples located near the 'Municipio' station, which is located in the historical center of Naples.

The soil affected by the excavation consists of a layer of pozzolana overlying a bench of tuff. As shown in Figure 2a, the horizontal distribution of the freezing probes is influenced by the actual development of the two tunnels, that have a slight curvature. The freezing probes,

(a)   (b)   (c)

Figure 1.   (a) place of activity; (b) view of the construction yard; (c) tunnel modelled in this work.

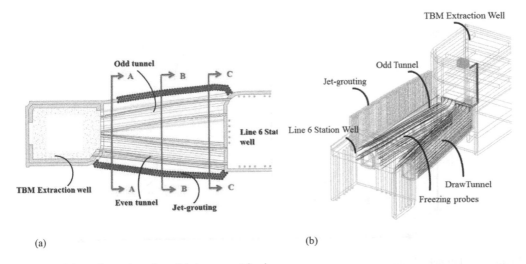

(a)   (b)

Figure 2.   (a) Horizontal section; (b) Axonometric view.

1411

for technological reasons, have a straight distribution along their axis. As reported in Figure 2a, the authors show the plan and the sections of the study. Section A-A is located 5.0 m from the Tunnel Boring Machine (TBM) extraction well, section B-B is in a central position with respect to the tunnels and section C-C is located 5.0m from Line 6 Station well. Figure 2b shows the axonometric of the case study and the position of the freezing probes.

3 NUMERICAL MODEL

The authors have developed an efficient transient numerical model to effectively analyze heat transfer in the soil during the work excavation procedure. The numerical model has been implemented within the commercial software Comsol Multiphysics. The ground subdomain has a depth of 20 m and a length of 35 m, as shown in Figure 3 and can be considered sufficiently large to avoid thermal interference with external environment and sufficiently deep to assume an undisturbed soil temperature. The study of the temperature field in the soil during at the time of artificial ground freezing, was implemented through a numerical two-dimensional model, representing a cross section of the tunnels located near the front, and solved using the Finite Element Method (FEM).

The cross section consists in a computational domain with a size of 20x35 m where the tunnels are to be dug; the boundary in between the two layers crossed by the tunnels (above Pozzolana and below tuff) and the freezing probes, arranged in such a way to allow the formation of the upper protective coating of the tunnels. The simplified assumptions underlying the model carried out in this work are the following: (i) homogeneous and isotropic materials in each layer of the computational domain; (ii) thermo-Physical proprieties of the soil varying with temperature, between the frozen and unfrozen phase; (iii) for the whole volume of soil, the transition phase takes place at a temperature between 0 °C and 1 °C; (iv) the temperature of the working fluid in the probes has a linear variation along the axis; (v) the heat exchange is purely conductive in the soil, due to the limited convective flow motion in the ground. Table 1 shows the thermophysical characteristics of the computing domains. In Table 2 reports the thermophysical properties of tuff and pozzolan and jet-grouting dependent on the frozen and unfrozen phase as thermal conductivity and heat capacity. The values used in this work phase have been derived from measured values in this area presented in the available literature (Rocca 2011).

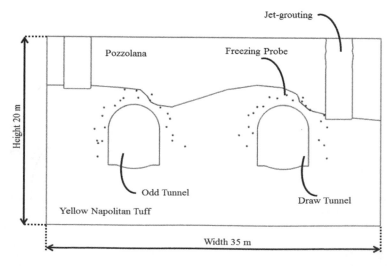

Figure 3.   Tunnel scheme.

Table 1. Domains characteristics.

Property	Porosity	Mineral density kg/m^3	Dry density kg/m^3	Wet density kg/m^3
Tuff	0.50	2713	1223	1723
Pozzolana	0.51	2392	1172	1733
Jet-grouting		3000		

Table 2. Thermal properties of the soil layers and jet-grouting.

Property	Thermal Conductivity W/mK	Heat capacity kJ/m^{3}K	Permeability m/s
Tuff	1.48/3.14*	3120/1990*	10^{-5}
Pozzolana	1.28/2.61*	3150/2790*	10^{-6}
Jet-grouting	1.40	900	

* Unfrozen/frozen ground.

3.1 Governing equations

The problem under investigation has been simulated by the authors by means of a dynamic model reproducing the 2D conductive heat transfer in the ground. The governing equations for transient conduction heat transfer are:

$$\rho_i C_{pi} \frac{\partial T}{\partial t} = \frac{\partial}{\partial x}\left(k_{x_i}\frac{\partial T}{\partial x}\right) + \frac{\partial}{\partial y}\left(k_{y_i}\frac{\partial T}{\partial y}\right) + Q \tag{1}$$

where ρ_i = density of the materials constituting the subdomain (kg/m^3), c_{cpi} = specific heat capacity (J/kg·K), k_i = thermal conductivity (W/m·K) and T = temperature distribution (K) of the and Q is the heat generated or absorbed within the control volume in the time unit. Through the latter term, the latent heat of solidification is modelled, i.e. heat absorbed or released by the volume units at constant temperature during the phase transition. In fact, the phase passage of the water causes a significant variation of the thermal diffusion coefficient and the specific heat of the saturated soil, in addition to the absorption of melting latent heat. As previously specified, a temperature decreases with a linear trend, between the refrigerant inlet and outlet along the freezing probes, during the experiments. For brevity in this paper, the authors do not report all the equations used for the phase changing, that can be found in the available literature (Ogoh et al. 2010). To evaluate the phase change the authors used an enthalpy method approach in the energy balance equation exactly when the material reaches its phase change temperature T_{pc}, it is assumed that the transformation occurs in a temperature interval between $T_{pc} - \Delta T/2$ and $T_{pc} + \Delta T/2$. In this interval, the material phase is modeled by a smoothed function, θ, representing the fraction of phase change before transition, which is equal to 1 before $T_{pc} - \Delta T/2$ and to 0 after $T_{pc} + \Delta T/2$.

3.2 Boundary Conditions

The boundary and initial conditions employed in the present model allow to describe the different phases of the procedure, dived to the domain in Figure 4. The temperature of the top, bottom and lateral surface of the soil has been assumed to be constant during the analysis, equal to the average yearly temperature of the region under investigation and equal to $T_0 = 16°C$.

1413

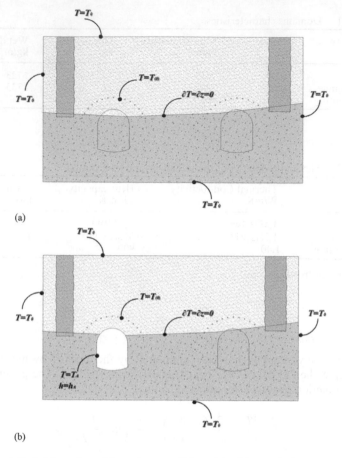

(a)

(b)

Figure 4.   Computational domain and boundary conditions (a) odd tunnel; (b) even tunnel.

The initial condition at each point of the domain is equal to:

$$T(x, y, 0) = T_0 = 16°C \qquad \forall x, y \in \Omega \tag{2}$$

where Ω = computational domain.

The following Dirichlet condition is imposed on the external surface of each probe for phase 1 and phase 3.

$$T(x, y, \vartheta) = T(\vartheta) \quad x, y \in \partial\Omega \tag{3}$$

The temperature boundary conditions applied to the external perimeter of each probe depend on the phase of the freezing process. In particular, during "Phase 1-Nitrogen", the temperature has linear variation along the axis, from -196 °C to -110 °C. For "Phase 2-Waiting", adiabatic conditions $\overline{\nabla}T \cdot n = 0$ have been imposed on the probes boundary. For "Phase 3-Brine" the temperature on the probe has been imposed equal to -33 °C. For the excavation of the tunnel (even), the same steps employed for the tunnel (odd) have been considered. During these phases, the excavated odd tunnel has been reproduced by removing the corresponding area of soil from the 2D computational domain and applying a proper boundary condition. This condition considers the presence of vehicles and air recirculation in the excavated tunnel, and is simulated by convective heat transfer on the walls of the odd tunnel:

$$-k\nabla T \cdot n = \overline{h}(T - T_A) \tag{4}$$

$$h(x, y, \vartheta) = h_A = 15 \, W/m^2 K \tag{5}$$

4 RESULTS AND DISCUSSION

4.1 *Model validation*

Given the importance of the validation procedure for numerical codes (Arpino et al. 2016, Massarotti et al. 2016), the mathematical model has been validated against experimental data (Colombo 2010) and numerical results (Ogoh et al. 2010) available in the literature. The graph in Figure 5 shows the conditions which determine the propagation velocity of the freezing front in the tuff.

This graph compares the data obtained from the FEM analysis carried out with Comsol Multiphysics by setting the conditions of physics and material as illustrated in the previous paragraphs and that reported in (Colombo 2010). To validate the calculation model exposed in § 3, concerning the heat transfer problem of the ground, with the phase transition effects caused by the freezing, it has been examined the treatment described by (Ogoh et al. 2010).

The study reports a simplified 2d model, which is solved analytically for the Stefan problem and compared with the results obtained by the FEM Analysis realized with the computational software Comsol Multiphysics. Figure 6 shows a comparison between the results of (Ogoh et al 2010) and those obtained from the model developed by the authors.

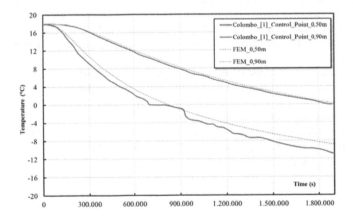

Figure 5.   Comparison with (Colombo 2010) with FEM analysis.

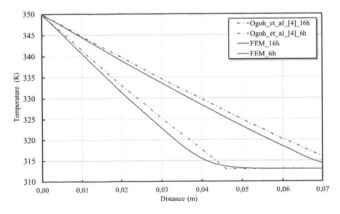

Figure 6.   Comparison of the results reported in (Ogoh et al. 2010) and FEM analysis.

4.2 Numerical and experimental analysis

The thermal characteristics considered for each calculation area are those showed in Table 1, while the boundary conditions have been illustrated in § 3.2 Boundary conditions. The analysis was developed considering the probes connected in parallel. Regarding the temperature of the probes contour, for Phase 1-Nitrogen, it was assumed a nitrogen inlet temperature of -196 °C and an output temperature of the exhausted, maintained equal to -110 °C. For Phase 2-Waiting an adiabatic condition has been imposed. While in Phase 3-Brine the temperature of the probes contour, for each freezing phase, assuming equal to -33 °C.

Figure 7 shows the section cuts considered to evaluate the thickness of the ice wall, reached after several days of activation of the artificial ground freezing with liquid nitrogen. The section in question is placed at 5 m to 6-line station pit diaphragm as shown in Figure 2. The authors have paid special attention to the study of this section because the excavation of the tunnel peers have started just by cutting the diaphragm 6 line station well. Therefore, for the safety of excavation, it was necessary to make sure that it had reached the minimum thickness of project and required by (1.5 m).

In order to assess the evolution of frozen wall that forms during the activation phase 1, Table 3 shows the results of the analysis through the thickness of the frozen wall, considering as a limit the thickness of 1.5 m, identified both through the -2° C isotherm at -4° C, the values of the thickness of the frozen wall for several days of activation of soil freezing with liquid nitrogen.

Table 3 shows that after 14 days of nitrogen activation, the wall would reach the project thickness at all points of the section. Figure 8 shows the temperatures measured at day 14 of activation of soil freezing, for the cutting section considered and shown in Figure 7. Figure 9a shows the ice arch as the isotherm contours below 0° C, considered as a limit between liquid and solid water present in the soil. Figure 9b instead represents the relative to -4° C isotherm.

Thermometric chains were installed in the pipeline that made it possible to monitor the trend of temperature and therefore to assess the evolution of the frozen wall during freezing. Finally, Figure 10 shows the comparison between the numerical predictions and the experimental results obtained during freezing process.

Numerical results are comparable with the experimental ones, for sensors E and N there is a difference, this may be due to a passage of water or to different characteristics of the soil in the location of where the probe was installed.

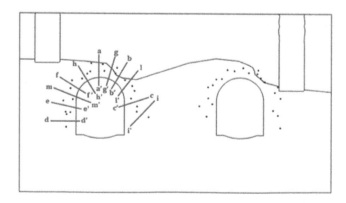

Figure 7. Linea where temperature of the frozen wall has been considered.

Table 3. Spessore frozen wall per isoterma -2 °C e -4 °C.

Cutting Line	8 days m	10 days m	12days m	14 days m
a-a'	1.9/1.8	2.1/2.0	2.3/2.2	2.6/2.5
b-b'	1.3/1.0	1.5/1.4	1.9/1.7	2.0/1.8
c-c'	1.3/1.2	1.6/1.5	1.9/1.8	2.1/2.0
d-d'	0.8/0.7	1.2/1.1	1.5/1.4	1.7/1.6
e-e'	1.7/1.6	2.1/2.0	2.4/2.3	2.5/2.4
f-f'	1.6/1.5	1.8/1.7	2.1/2.0	2.5/2.4
g-g'	1.4/1.3	1.8/1.7	2.0/1.9	2.2/2.1
h-h'	1.5/1.3	1.8/1.7	2.0/1.9	2.1/2.0
i-i'	0.6/0.5	1.0/0.9	1.4/1.3	1.6/1.5
l-l'	1.3/1.2	1.6/1.5	1.7/1.6	2.0/1.9
m-m'	1.4/1.2	1.7/1.6	2.0/1.9	2.2/2.1

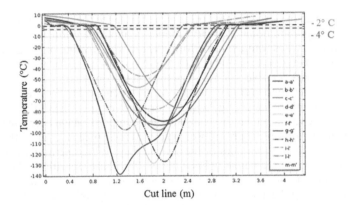

Figure 8. Trend of temperatures for the cut off and thick sections of the frozen wall.

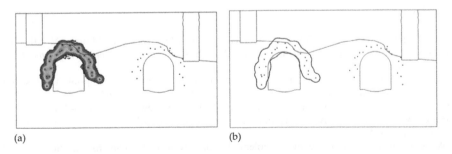

(a)   (b)

Figure 9. (a) Formation of the frozen wall capped at 0° C for the 14th day of freezing; (b) thickness of frozen wall identified with the isotherm at -4° C.

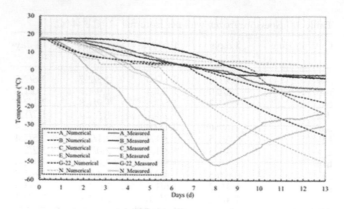

Figure 10. Comparison between experimental and numerical temperatures in the soil.

5 CONCLUSIONS

This work presents an analysis of the heat transfer performance for the artificial ground freezing, carried out by using a computationally efficient numerical model developed by the authors. In this study we describe the activities planned to perform the artificial freezing of the ground for the construction of the tunnels between Line 1 and Line 6 of Naples Metro. Through the realization of a 2D numerical model it was possible to reproduce the soil freezing phases and the evolution of the temperature field in the reference domain. The model has been used to illustrate the basic physical phenomena associated with phase change and the influence of latent heat, the influence of phase change and temperature variation. Several simplifying assumptions have been used to develop this model. In particular, (i) homogeneous and isotropic materials in each layer of the computational domain; (ii) Thermo-Physical proprieties of the soil varying with temperature, between the frozen and unfrozen phase; (iii) For the whole volume of soil, the phase transition takes place at a temperature of 0 °C within an interval of 1°C; (iv) The temperature of the working fluid in the probes has a linear variation along the axis; (v) Heat exchange in purely soil, due to the extent of the convective motion in the ground. More generally, the model does not consider the influence of mechanics on freezing. Despite these restrictions, the model appears pertinent if the aim is to be able to count on a reliable model to predict the performance of artificial ground freezing, according to the expected freezing times using this model if compared with the data reported in the literature.

REFERENCES

Arpino, F., Carotenuto, A., Ciccolella, M., Cortellessa, G., Massarotti, N., Mauro, A. 2016. Transient Natural Convection in Partially Porous Vertical Annuli. *International Journal of Heat and Technology*, (34), S512-S518.

Colombo, G. 2010. Il congelamento artificiale del terreno negli scavi della metropolitana di Naples: valutazioni teoriche e risultati sperimentali. *Rivista Italiana di Geotecnica*, 42–62.

Massarotti, N., Ciccolella, M., Cortellessa, G., Mauro, A. 2016. New benchmark solutions for transient natural convection in partially porous annuli. *International Journal of Numerical Methods for Heat & Fluid Flow*, (26): 1187–1225.

Ogoh, W. & Groulx, D. 2010. Stefan's Problem: Validation of a One-Dimensional Solid-Liquid Phase Change Heat Transfer Process. *Excerpt from the Proceedings of the COMSOL Conference 2010 Boston*.

Papakonstatinou, S. Pimental, E. & Anagnostou, G. 2010. Analysis of artificial ground freezing in the Pari-Duomo plataform tunnel of the Naples metro. *Numerical Methods in Geotechnical Engineering – Benz & Nordal (eds)*. 281–284.

Pimentel, E. Papakonstantinou, S. & Anagnostou, G. 2011. Numerical interpretation of temperature distributions from three ground freezing applications in urban tunnelling. *Tunn. Undergr. Sp. Technol.* (28): 57–69.

Rocca, O. 2011. Congelamento artificiale del terreno. *Hevelius, Argomenti di ingegneria geotecnica*.

Russo, G. Corbo, A. Cavuoto, F. & Autuori, S. 2015 Artificial Ground Freezing to excavate a tunnel in sandy soil. Measurements and back analysis. *Tunn. Undergr. Sp. Technol.* (50): 226–238.

Tunnels and Underground Cities: Engineering and Innovation meet Archaeology,
Architecture and Art, Volume 4: Ground improvement in
underground constructions – Peila, Viggiani & Celestino (Eds)
© 2020 Taylor & Francis Group, London, ISBN 978-0-367-46868-2

Design of an underground station within the Greater Paris metro line by conventional tunneling: Solutions and calculation methods

E. Misano, S. Minec & H. Sahnoun
Bouygues Travaux Publics, Guyancourt, France

A. Goulven & V. Dumoulin
Horizon, Vitry sur Seine, France

ABSTRACT: The "Vert de Maison" station belongs to the new Metro Line 15 – in the Municipality of Maison-Alfort in the southern suburbs of Paris. Due to the limited available space in a very dense urban context, the station is composed by an open-cut shaft and an underground cavern driven from the shaft beneath an historical residential building. The underground station has a complex geometry with lateral passages that come down from the shaft and join the main central cavern for an overall total width of about 35 m, an overall height of 20 m and a length of about 70 m. The presence of an operative railway and several historical buildings at the surface obligate the induced settlement field to be the slightest, the selected ground improvements include compensation grouting, ground freezing and mortar injections. The detailed construction design is presented, including construction methods, sequences and 3D FEM analysis for lining design and settlement estimation.

1 PROJECT AND SCOPE OF WORKS

The *Gare du Vert de Maisons* is located on the future Subway line 15 of the Grand Paris Express Project, on the premises of Maisons-Alfort and Alfortville towns, on a very con-strained construction site. The box station, of dimensions ~50m x ~40m, is built by means of diaphragm walls, following a top down construction process, the foundation slab is at a depth of about 40m.

The presence of an historical residential building at surface, which cannot be demolished, implies the necessity to complete the platform length by an underground cavern, excavated by conventional mining. The cavern is about 70 m long, 35m wide and 13m high, due to emer-gency issues in case of fire, lateral corridors are necessary to guarantee a correct evacuation to a different station level, therefore in the first part these corridors are inclined.

The geological stratigraphy implies mostly more or less altered limestones and marls, while along the crown and lateral passages there is a fine silty-sand, known as "Beauchamp Sand", which needs particular attention in conventional mining. The overburden of the cavern is about 30m, the excavation is driven under the "Seine" water table at a pressure of about 3,5 bars.

In this article, the means and methods to construction are presented as well as the calcula-tion methods used to design it and to foresee the induced settlements in a very sensitive urban context.

2 GEOLOGICAL CONTEXT

The geotechnical context is the typical one of the Paris alluvial basin, the first 10m are com-posed principally of urban fillings and loose modern alluvial deposits. Beneath that, the

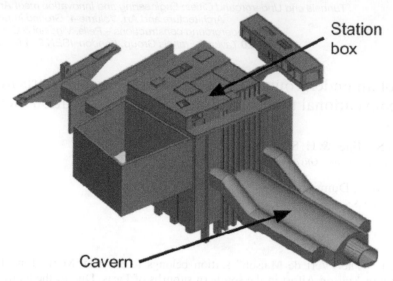

Figure 1. Vert de Maison Station 3D Model

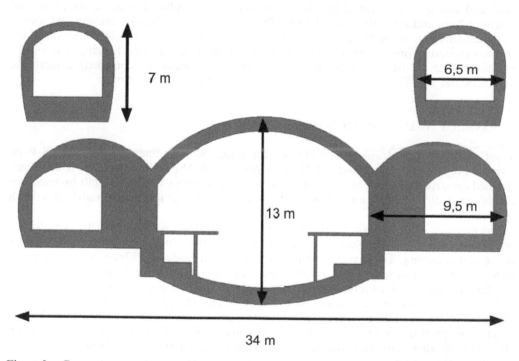

Figure 2. Cavern cross section

typical sequence of *Saint-Ouen Limestone, Beauchamp Sands, Marls and Pebbles* and *Hard Limestone* is found.

Saint-Ouen Limestone at this project location is composed of a more or less altered white limestone in a marl matrix, although it is not directly concerned by excavation, its mechanical behavior has major influence on induced settlements.

Beauchamp Sands is a compact sandy lime stratum, historically major problems of instability related to water pressure have been encountered within this layer, its excavation in conventional mining requires great precaution in waterproofing the material around the excavation in order not to incur in seepage accident with major solid transportation inside the tunnel.

Marls and Pebbles is the principal layer concerned by the excavation, is composed of coarse pebbles in an quite hard marl matrix, locally it can have a more altered facies, but it presents generally a good mechanical behavior.

The deepest excavated layer is the *Hard Limestone,* it is a locally very compact Limestone with high UCS values in its intact facies; historically this material, which is found beneath the whole Paris area, has been exploited for construction purposes, notably by the Baron Haussmann at the end of the XIX century in his great urbanistic renovation plan.

3 GROUND TREATMENTS

To build the cavern different ground treatments are foreseen to overcome the particular issues in this type of conventional mining:

- Ground Freezing: This treatment is provided in the *Beauchamp Sands* stratum to allow for conventional mining avoiding instabilities related to water pressure.
- Grouting injections: Massive grouting injections are foreseen over the whole volume of excavated soil, its aim is to provide a sufficient low permeability in order to allow for conventional mining
- Compensation grouting; Estimated settlement at surface exceed the maximum allowed for building safety, compensation grouting is provided to allow for uniform heave of building foundation during the cavern construction and avoid any issues with building damage.

3.1 *Ground freezing*

Ground freezing is used to reduce risks associated with construction of the main station cavern inside *Beauchamp Sands* stratum.

It is thus proposed to construct the initial high level sections of two access passages within the *Beauchamp Sands* stratum under the protection of an annulus of frozen ground. During

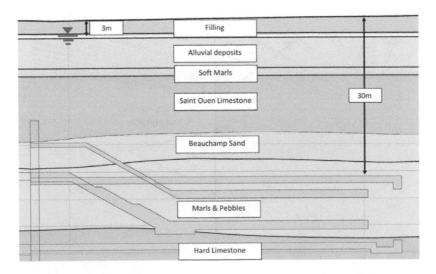

Figure 3. Geological profile.

excavation of the corridors, the annulus of frozen ground has low permeability and, together with a grouted soil body all around the cavern and corridors, provides a waterproofed soil body, allowing open-face excavation in almost dry conditions.

The frozen annulus will be formed by installation of freeze pipes drilled from the diaphragm wall station box both parallel to the inclined axis of the access passages and horizontally. A second application of ground freezing is to provide a frozen slab at the base of the *Sables de Beauchamp* over the entire footprint of the main station cavern. The purpose of the ground freezing is to ensure that there is no potential for construction issues associated with high water pressure instabilities within the *Beauchamp Sands*.

3.2 *Injection grouting*

The excavation is driven, except for the high level of lateral corridors where soil freezing is provided, in a soil with quite good geomechanical characteristics. Nevertheless the possibility of finding altered zones, mostly in the *Marls and Pebbles* layer requires to assure a sufficient low uniform permeability of the whole excavation volume plus a minimum of 3m in all directions to assure for some margin.

The treatment is driven from the surface with injection pipes, due to lack of space (building presence) all drilling are done from a few available spots. The grouting is composed of a cement-bentonite mix, injected with the IRS method. In the *Marls and Pebbles* stratum, the replacement ratio of the grouted volume over the whole treated volume is about 10–15%, which can be considered a satisfying result in this type of ground to allow for conventional mining.

The permeability target value of the soil treatment has been set to 10^{-6} m/s.

3.3 *Compensation grouting*

As the underground section of the station is located below a group of residential buildings from 20th century with very low settlements allowance, settlements control is a key issue.

Compensation grouting operations are foreseen in the first meters inside old alluvium layer below the surface in order to compensate the potential settlements that could occur after each major phase of the excavation works, based on observational measures at the surface and on the superstructures, in order to avoid any displacement of ground that could affect the buildings structural integrity. The threshold for vertical settlements beneath the building is fixed to 25 mm.

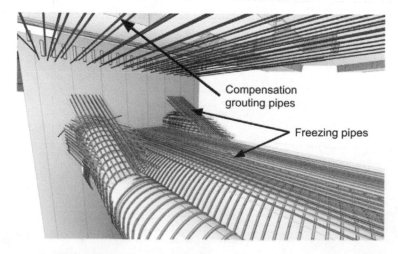

Figure 4. Ground freezing and compensation grouting pipes 3D layout.

4 CONSTRUCTION STAGES

Once injection grouting has been executed, and freezing and compensation grouting are ready to operate, the access corridors on both sides of the cavern will be excavated in first place. Beforehand the diaphragm wall of the station box will be cored and demolished, then the excavation will be carried out full section by roadheaders and shovel. This method allows excavation without excessive vibrations and an excellent control of the excavated profile for this kind of geology.

The excavation is supported by a pre-retaining system consisting in fiberglass bolts at tunnel face, forepoling above tunnel crown and a first lining composed of shotcrete and steel ribs. The advancement step will be adapted to the actual behavior of the ground, varying between 1 and 1.50 m.

Final lining of lateral corridors shall be concreted beforehand the excavation of the central cavern in order to give support to the foot ribs of the central crown.

Assumption is made that the tunnel boring machine will drive through the station box before the work begins in the cavern, thus the excavation of the cavern implies the demolition of the segmental lining tunnel (fiber reinforced).

Once the excavation of the central part of the cavern is completed, the inclined corridors are excavated from the lower part to the upper one; it is to note that the first part of the latral corrisors wich is just under the inclined one is a temporary excavation and it is backfilled with concrete prior to mezzanine corrdiors excavation.

Once alle the excavation operation are completed, the final lining are concreted, the structures are provided with a full round waterproofing coating.

In the following figures, the construction stages are resumed:

1) Lateral corridors are excavated and the supporting system is installed
2) Final lining of the lateral corridors are concreted in order to give support to the first lining of the central cavern
3) The upper part of the central cavern is excavated and first lining installed, the fiber reinforced segmental lining tunnel is demolished during excavation
4) All final lining are concreted

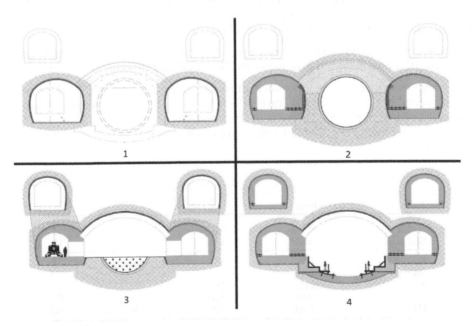

Figure 5.   Construction stages.

5 CALCULATION METHODS

The design of the underground works is based on a series of FEM analysis carried out with the Software Plaxis 2D and 3D.

The main scope of these analyses is to estimate the induced settlement trough at surface as well as dimensioning for temporary and definitive lining.

The excavation is driven in a grouted soil with reduced permeability, the analysis is performed in total stresses with soil parameters corresponding to the short-term behavior, nevertheless no improvement of geotechnical parameters due to the grout injection treatment has been accounted in the analysis.

The soil behavior is described by the Hardening soil constitutive model available in the Plaxis library, for all soil layers the parameter m has been set to zero, the other parameters are shown in Table1. It is to note that the value of E presented in Table 1 corresponds to the value of E_{50} and E_{oed} of the HS model, the elastic module E_{ur} has been set equal to 3 times the E value.

A transient thermal analysis has been performed with the software Ansys in order to estimate the necessary freezing time and power of the freezing system; for soil thermal proprieties and freezed soil mechanical characteristics an extensive laboratory test campaign has been performed.

5.1 3D FEM analysis

Due to the complex geometry and construction sequence, it has been stated to run the settlement analysis by the means of a tridimensional analysis, the advantage of such analysis is to reduce to minimum the overestimations on soil settlements coming from 2D model simplifications.

All structures and temporary lining are modeled with plate elements and construction stages are precisely reproduced accordingly to the works program, from the TBM passage until the end of the excavation of the cavern, totally the model accounts for about 200 calculation phases.

Boundary conditions are defined in order to prevent horizontal displacement on vertical borders and both vertical and horizontal on the bottom horizontal border, the construction of the station box is not considered in the analysis.

The main scopes of the 3D analysis are the following ones:

- Asses the building vulnerability according to a tridimensional settlement trough at surface
- Define the déconfinement ratios to use in the 2D models for structure dimensioning

Due to high inaccuracy on the calculated stresses field (compared to the displacement one), it has been stated not to estimate structural efforts by the tridimensional model and to use 2D models calibrated on displacement results of the 3D Model.

In the following figures the 3D model geometry results in terms of settlement trough at surface are presented.

Table 1. Soil parameters.

Soil layer	E MPa	c kPa	φ °	K_0
SO	124	45	20	0.5
BS	100	35	27	0.5
MPa	112	60	25	0.5
MP	600	120	30	0.5
HL	1380	100	40	0.5

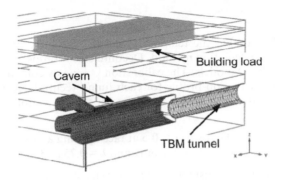

Figure 6.   FEM 3D Model.

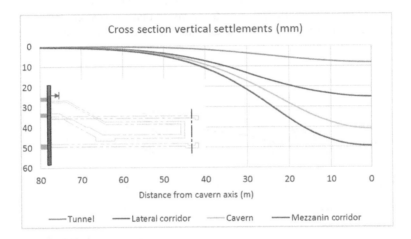

Figure 7.   Calculated settlement trough at surface on a transversal section for different stages.

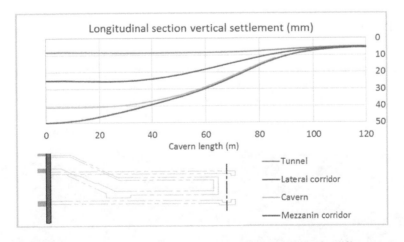

Figure 8.   Calculated settlement trough at surface on a longitudinal section for different stages.

5.2 2D FEM analysis

Plane models are used to design the temporary and the definitive lining accordingly to the convergence-confinement calculation method. Deconfinement ratios are defined by calibrating the 3D model results with 2D models. The calibration is done by comparing the displacements of several points on the excavation profile for each tunnel (central cavern, lateral and mezzanine corridors), the calibrated déconfinement ratios for the excavation profiles are the following ones:

– Lateral corridors and central cavern: $\lambda = 0,5$ at face opening and $\lambda = 0,8$ at 3 meters
– Mezzanine corridors: $\lambda = 0,65$ at face opening and $\lambda = 0,75$ at 3 meters

For first and definitive lining dimensioning no improvement of ground geotechnical characteristic has been taken into account for grout injection nor freezing, this choice is made in order to guarantee an adequate factor of safety if a failure of the freezing system occurs. For grout injections only a permeability criteria has been fixed in order not to have excessive water flow in trough the excavation. Nevertheless in order to verify that the freezed soil is capable to bear the efforts generated by the excavation a 2D analysis taking into account the freezed soil volume has been carried out.

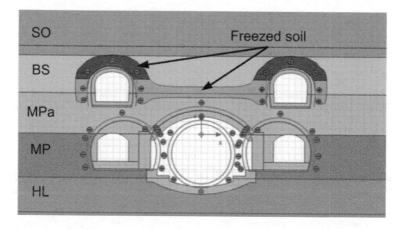

Figure 9.   FEM 2D Model with freezed ground volume.

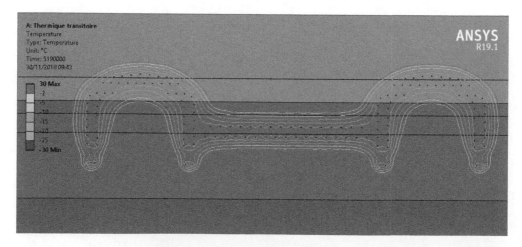

Figure 10.   Thermal analysis result on Ansys software after a freezing time of 2 months.

The geometry of the freezed volume is given by a transient thermal analysis carried out with the Software Ansys, the time is set in order to get an homogenous freezed volume around the excavation for the mezzanine corridors and a sufficient thickness of the slab over the crown to prevent the risk of high water pressure instability (about 2m).

The mechanical analysis is then carried out with Plaxis in order to verify the consistency of the efforts into the freezed volume. Mechanical properties of freezed ground have been assessed with specific test laboratory on freezed samples; cohesion of freezed soil vary between 2,5 and 5 MPa, Young Modules are about 100–200 times cohesion values in the range of temperature between -10°C and -20°C

The input for the thermal analysis are the freezing pipe pattern and the boundary thermal conditions, a constant temperature of -30° has been taken into account for freezing pipe (brine freezing system), the calculation shows that a time of 1,5–2 months is necessary to achieve the needed ground freezed volume.

6 CONCLUSIONS

Means and methods for the construction of a station cavern within the Greater Paris metro have been presented. Due to the complex geological and historically relevant urban context, different ground treatments are provided to allow the excavation in conventional mining.

The design of the cavern has been carried out by 2D-3D FEM analysis allowing for simple 2D dimensioning of first and definitive lining and for more complex 3D settlement calculation at surface.

Calculated settlements under the historical building are about 50mm (maximum value), the limit for building integrity are 25mm. The gap between the estimation and building limit will be filled at construction stage by the means of compensation grouting to assure that no building damage will occur.

A transient thermal analysis has been carried out to study design the ground freezing system, thermal and mechanical properties of freezing soils have been obtained by an extensive laboratory test campaign.

REFERENCES

AFTES 2001. GT7R6F1 – Recommandations relatives à la méthode convergence-confinement. Paris: Association nationale française des tunnels et de l'espace souterrain.

Carranza-Torres, C & Diederichs, M. 2009. Mechanical analysis of circular liners with particular reference to composite supports. For example, liners consisting of shotcrete and steel sets. *Tunnelling and Underground Space Technology*, volume 24 (number 5): 506–532.

Lunardi, Pietro 2008. *Design and construction of tunnels*. Milan: Springer.

Panet, Marc 1995. *Calcul des tunnels par la méthode convergence-confinement*. Paris: Presses de l'Ecole nationale des ponts et chaussées.

Tunnels and Underground Cities: Engineering and Innovation meet Archaeology,
Architecture and Art, Volume 4: Ground improvement in
underground constructions – Peila, Viggiani & Celestino (Eds)
© 2020 Taylor & Francis Group, London, ISBN 978-0-367-46868-2

Ground improvement carried out via an existing railway tunnel for the construction of a new underlying tunnel

A. Pauri

Italferr S.p.A., Rome, Italy

ABSTRACT: The new "Induno" railway tunnel is located in the municipality of Induno-Olona (Varese, North Italy) approximately on the axis of the historical line; it develops under the single track of the historical railway line built at the end of the 19th century.

Geologically, the excavation of the natural tunnel involved fluvioglacial and glacial deposits mainly characterised by silt-sand-gravel lithology. Pebbles are generally heterogeneous, highly rounded and poorly sorted (from fine gravels to boulders), sometimes well cemented.

To improve ground characteristics, the project is based on the consolidation of loose soil through injection treatment extended also to the area of contact between the ground and the underlying rock formation.

The new railway tunnel has an excavation length of 478 m approximately; the excavation section is about 115 m^2 and the maximum coverage is approximately 45 m.

1 INTRODUCTION

The project named "Nuovo collegamento Arcisate–Stabio" (New Arcisate-Stabio Link) is a part of a larger cross-border railway link program dedicated to the transport of passengers between Lugano–Mendrisio (Switzerland) and Varese–Gallarate Aeroporto di Malpensa (Italy) as specified in the Framework Agreement entered into by the Canton Ticino (Switzerland) and the Lombardy Region (Italy) on 25 July 2000.

The link allows to connect Varese with Canton Ticino (Mendrisio – Lugano) and Varese with Como as well as to directly connect the Malpensa Airport with the cities of Southern/Central Switzerland and with the Sempione traffic axis (Lausanne, Geneva, Bern) and the Gottard traffic axis (Bellinzona and Lugano) via interchange at the Gallarate station (Figure 1).

The new line, about 9 km long, will run for about 5.5 km along the existing route of the Varese-Porto Ceresio railroad and for about 3.5 km along a new route.

Among the main works that characterise the new railway line there is also the new "Induno" tunnel. It is located mainly along the same axis of the existing line and develops underneath the current (single track) tunnel. The distance between the invert of the existing tunnel and the extrados of the crown of the new one varies from a minimum of 0.5 m to a maximum of 2.5 m.

The new bored tunnel has an overall excavation length L = 478 m; maximum cover amounts to about 45 m.

The double track railway section constructed has approx. 80 m^2 of usable surface area and about 115 m^2 of excavation cross-section.

Geologically, the excavation of the tunnel involved fluvioglacial and glacial deposits mainly characterised by silt-sand-gravel lithology. Pebbles are generally heterogeneous, highly rounded and poorly sorted (from fine gravels to boulders), sometimes well cemented.

Figure 1. Railway schematic map.

In order to improve the characteristics of the soil to be excavated, the project envisaged actions for consolidating the loose formations via injection treatment, partially extended within the contact between soil (over) and underlying rock formation (under) as well as in the degraded rock portion. In detail, the project included the execution of consolidation via glass-fibre structural elements, injected with cement-based and chemical (silicate) compounds.

2 THE PROJECT

The tunnel was excavated, for the first 250 meters, in a mixed geology section which has involved glacial till and fluvioglacial sediment and the base substrate consists of a calcarenite rock formation with variable share between the crown and the tunnel invert; for the remaining part the tunnel excavation was conducted entirely inside the basic rock formation.

The main design issues connected to the performance of the work regarded the following aspects:

- In the initial portion of the tunnel, the crossing of soils with poor geotechnical properties that imply the instability of the excavation face and of the bore, with respect to the size of the excavation sections;
- In the central portion of the tunnel, the crossing of highly permeable soils, together with the presence of a water table with free surface at crown level, that causes the risk of water inflow;
- Along the entire tunnel, the construction underneath and along the axis of the existing tunnel, that acts as a disturbing element in excavating the new tunnel as a consequence of the small distance between the invert of the existing tunnel and the extrados of the crown of the new one.

In view of the possible design issues described above, it proved necessary to conduct preventive soil consolidation operations capable of preventing the onset of mechanisms that could cause instability of the face and of the tunnel walls. The consolidation works were not conducted progressively with the development of the excavation but preventively, exploiting the presence of the existing railway tunnel.

After strengthening via steel ribs and shotcrete the existing tunnel (old railway line was not in operation), the project envisaged the consolidation phase via repeated and selective low/medium-pressure permeation grouting and/or plug cementing using specific compounds.

These operations, performed from within the existing tunnel (Figure 2) provided benefits and advantages in the construction phase because the low pressures of the compound induced negligible coactions on the structure of the existing tunnel and, moreover, the technology applied, featuring the absence of discharges during the injection phase, made it possible to operate in very safe and clean conditions.

The technology applied was also highly suitable for the consolidation of heterogeneous soils such as those present at the site.

The reinforcement and consolidation treatment planned was subjected to a preventive field test.

This was necessary in order to verify the performance in terms of mechanical characteristics of the consolidated ground (the minimum required values resulting from the calculations were: effective cohesion amounts to 50 KPa; shear modulus drained soil is equivalent to at least twice that of natural soil).

2.1 Geometries

By way of general method, perforations were planimetrically located at the vertices of a triangular mesh so as to obtain a centre distance between the holes of the same pair of rays not exceeding 0.9 m.

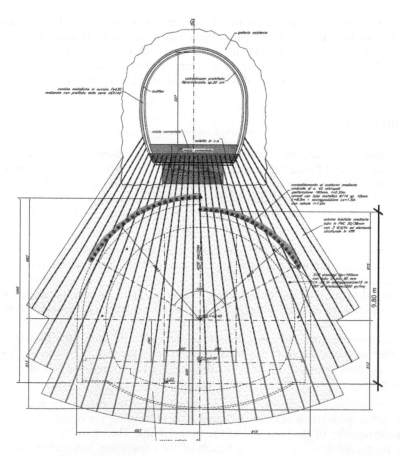

Figure 2. Cross consolidation section.

As regards the typical sections that foresee consolidation operations from the existing tunnel or from the existing railway trench, the vertical sections have been laid out in a fan pattern in a position and with inclinations such as to include the significant volume involved by the excavation and to maintain a centre distance (between the holes of the same pair of rays) not exceeding approx. 0.9 m at the end of the hole (Figure 2).

The holes were equipped with structural elements made of fibreglass (improved adherence flat bars, 40 mm x 6 mm, in pairs); structural element was fitted with a PVC injection hose $\varnothing \geq 1"1/8$ having manchette valves placed every 30 cm, to which it is fastened using spacers.

In the case of loose soils, the holes were extended all the way to the invert on the entire section of the face in order to reinforce the excavation face itself.

2.2 Executive phases

In loose soils rotary drilling was performed, with temporary casing or use of polymer or cement drilling fluids.

Once drilling was completed, into each hole a structural element fitted with the PVC injection hose was installed. In presence of loose soil alone, the sheathing mixture was immediately put in place in order to solidify the soil around the injection hose. The sheathing mixture was injected at low pressure (1-2 bar) starting from the deepest valve until the annular cavity between sleeved pipe and sides of the hole was completely filled. Once the sheathing mixture was in place, the injection hose was cleaned out thoroughly.

After completing the plug cementing operation, and after allowing for adequate time, injection under pressure was performed. In the loose soil, through all of the valves, the mixtures were injected using a double packer piston pump, starting from the deepest valve and proceeding from the holes on the side walls to the keystone.

The injections were performed on a controlled volume and pressure basis: the injection of each single valve was considered done only once the volume limit Vmax was reached or when the pressure limit Pmax was reached and maintained for at least 1 minute.

At the end of each injection phase, the sleeved pipe was washed out before a new injection phase was started.

Two types of mixture were injected into the loose soil in two different phases using the same sleeved pipe, using the cement-based mixture first and then the chemical mixture.

In function of the average porosity of the rock mass (estimated to be 25÷40%), the project was to inject a volume of mixture per each hole meter equivalent to:

- Cement-based mixture: 90 litres (VM)
- Silicatic mixture: 77 litres (VS)

The injection of each section was conducted, without interruptions, until the project volume (VM, VS) was reached. The injection was anyhow deemed completed once the maximum pressure (Pmax = 10÷15 bar) was reached even when the volume injected was lower than the project volume (VM, VS).

3 FIELD TESTING

The purpose of field testing is to verify the effectiveness of the consolidation system envisaged by the design, as well as to optimise the injection execution parameters in terms of volume and pressure values. To this end, two rosettes, identified by A and B (Figure 3), each consisting of 10 injection columns and featuring a different triangular treatment layout were performed, as shown:

The cement-based mixtures were prepared using the following components:

- water;
- cement: CEM I 52.5 R (for injection mixture), CEM II/B-LL 32.5 R (for sheathing mixture);
- bentonite: Sipag Bentogel PL;
- dispersing additive: Lamberti Lamsperse HS.

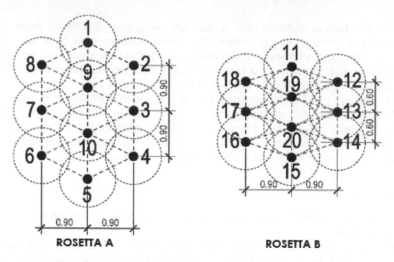

Figure 3. Test's rosettes.

The chemical mixture was prepared using the following components:

- water;
- sodium silicate solution;
- alkaline reagent;
- suspension of inorganic calcium salts.

Once the field tests were ready, the final tests were performed for each rosette, consisting of:

- 2 boreholes;
- Lefranc permeability tests at various depths and Lugeon test in the underlying rocks;
- pressiometric tests at various depths;
- cross-hole seismic testing.

Analysis of the results showed similar results for both rosettes, and therefore Rosette A was selected; in detail:

- the permeability tests performed inside the rosettes showed an average reduction in permeability by one or two orders of magnitude in the incoherent strata (fluvioglacial and glacial deposits) while permeability decreased practically to zero in the rock strata;
- The pressiometric modulus of the consolidated soil increased on average by 4% in the higher stratum (gravel) and by 40% in the sand/silt stratum.
- The seismic tests showed an increase in the individual parameters Vp (P-wave velocity), Vs (S-wave velocity), G (shear modulus), E (elastic modulus of the soil):
- The material extracted by coring was subjected to visual inspection that revealed the presence of the cement-based and chemical mixtures.

4 SITE ACTIVITIES

The boreholes/injections were carried out over a period ranging from August 2011 to June 2012.

The boreholes were drilled using 5 pieces of equipment: 3 smaller ones were used inside the existing tunnel (over a length of about 350 m) and 2 larger ones were used outside (over the remaining approx. 130 m). Overall, about 10,500 boreholes were carried out over a total length of about 97,500 m (Figures 4 and 5).

Figure 4. Injection phase.

Figure 5. Drilling phase.

The work was conducted without stopping, based on 3 daily shifts using 24 workers/shift (10 on perforation, 10 on injection and 4 as assistance).

Considering the small size of the existing tunnel and the encumbrance of the injection pipes that emerged out of the floor by 30 cm, to guarantee worker safety an escape route was created and kept constantly cleared consisting of a metal runway 70 cm wide that was extended in step with the progress of the work (Figure 6):

Figure 6.   Metal runway.

The following parameters were recorded for each hole:

- Number and type of hole
- Injection phase
- Mixture used
- Date and time of injection start/finish
- Section/valve depth
- Location (depth) of the piston pump
- Design mixture volume and actually injected mixture volume
- Design maximum pressure and pressure reached during injection.

5   POST-CONSOLIDATION TREATMENT SURVEYS

Prior to the performance of consolidation, 8 seismic profilings had been conducted, each using 24 geophones placed at a distance of 5 m one from the other.

The post-treatment seismic surveys were performed about 1 month following treatment completion; they consisted in 4 test arrays, each consisting of 24 geophones placed at a distance of 5 m one from the other.

Also performed were 3 couples of boreholes down to a maximum depth of 18 m from the work plane inside the existing tunnel, to be used for the performance of cross-hole tests.

The results of the tests performed after consolidation treatment showed an overall increase in seismic wave velocity, both for P-waves and for S-waves.

As can be seen in the colour mapping distributions, this increase appears to be drastic and homogeneously distributed at a level with the loose granular deposits where the rigidity contribution of the VTR reinforcements and of injections is decisive (Figures 7 and 8).

Vice versa, at rock level, the effect of the treatment was to bond the local disarticulated portions and to fill in the cracks with secondary effects on the mass's dynamic rigidity.

Due to the treatment, the velocities Vs of the granular soils within the context of the volumes involved by the excavation (the first 20 m measured from the work plane), went from $1000 \div 1500$ m/s to values in the $1500 \div 2500$ m/s range.

The direct measurements performed using the cross-hole tests confirmed these results.

Since the dynamic modules are proportional to the square of the velocity ($Go = \rho \cdot Vs^2$), the stiffness values of the consolidated soil are $2 \div 3$ times higher than those of natural soil.

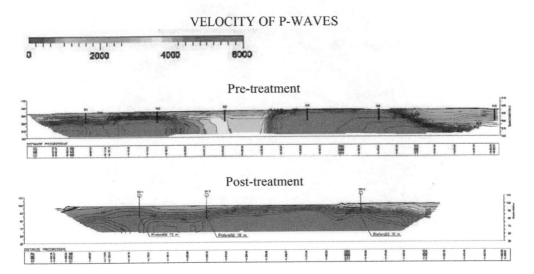

Figure 7. Velocity of P-waves.

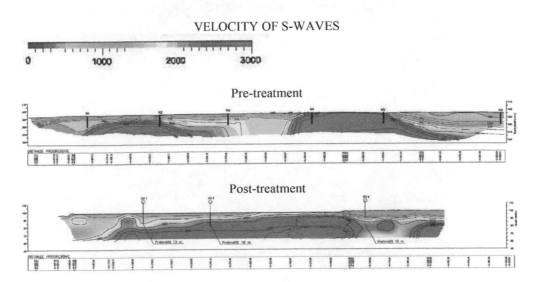

Figure 8. Velocity of S-waves.

This confirms the design assumption that the modulus of the consolidated soil should be twice as high as that of natural soil.

6 FINDINGS DURING THE EXCAVATION PHASES

Throughout the excavation phases, the face remained stable and dry.

No detachments were observed either from the face or from the walls of the excavated tunnel, as can be seen in the following Figures 9 and 10:

Figure 9.   Excavation face.

Figure 10.   Excavation face.

7 CONCLUSIONS

Preventive soil consolidation, although expensive and time consuming, has allowed the excavation of the new tunnel in complete safety without any problem during the construction phase of the tunnel.

Tunnels and Underground Cities: Engineering and Innovation meet Archaeology,
Architecture and Art, Volume 4: Ground improvement in
underground constructions – Peila, Viggiani & Celestino (Eds)
© 2020 Taylor & Francis Group, London, ISBN 978-0-367-46868-2

Optimization of the tunnel face bolt reinforcement

P. Perazzelli
Pini Swiss Engineers, Zurich, Switzerland

G. Anagnostou
ETH, Zurich, Switzerland

ABSTRACT: In conventional tunnelling ground reinforcement using bolts is a very popular measure for stabilizing the face because of its relatively low cost, its great flexibility and its adaptability to local geological conditions. The face reinforcement, as almost all geotechnical auxiliary measures in tunnelling, is carried out ahead of the tunnel face intermittently with the excavation, thus slowing down construction progress considerably. In view of its implications for construction time and cost, face reinforcement needs a particularly careful design. In addition to experience and engineering judgment, calculations are indispensable. The paper presents a computational tool for the design of bolt reinforcement that can be applied to heterogeneous, layered ground as well as to reinforcement layouts that are arbitrary in terms of the spacing, length, longitudinal overlapping and installation sequence of the bolts. Several examples are presented in order to discuss possible optimizations of the bolt reinforcement.

1 INTRODUCTION

In tunnelling through soft soils one of the main hazard scenarios is the collapse of the excavation face (Figure 1). Ground reinforcement using bolts is a very efficient effective measure for stabilizing the face in conventional tunnelling (Figure 1). The bolts are carried out ahead of the tunnel face intermittently with the excavation. Their installation slows down construction progress considerably, then affects significantly construction time and cost; for this reason face reinforcement needs a particularly careful design.

Several computational methods exist for the stability analysis of a tunnel face. Literature reviews can be found in Anagnostou and Perazzelli (2015) and in Perazzelli and Anagnostou (2017). The majority of existing works in the literature looking at tunnel face stability are concerned with estimating the support pressure required for stabilizing the face, while relatively few works investigate the effect of face reinforcement.

The stabilizing effect of face reinforcement is tackled either by so-called homogenization approaches, i.e. by smearing the effect of the bolts and considering an equivalent higher strength ground (e.g. Indraratna and Kaiser 1990, Grasso et al. 1991) or by considering an equivalent face support pressure (e.g. Bischof and Smart 1975, Peila 1994) or by taking account of the individual bolts (e.g. Peila 1994).

Anagnostou and Perazzelli (2015) and Perazzelli and Anagnostou (2017) suggested a computational method for cohesive frictional soils and purely cohesive soils, respectively, which considers the failure mechanism of Anagnostou and Kovári (1994) as well as the support forces exerted by the individual bolts (see Section 2). The method can be applied to heterogeneous, layered ground as well as to reinforcement layouts that are arbitrary in terms of the spacing, length, longitudinal overlapping and installation sequence of the bolts. Under the simplifying assumptions of homogeneous ground and uniform bolt distribution, closed form solutions and design charts are provided by Anagnostou and Perazzelli (2015) and Perazzelli and Anagnostou

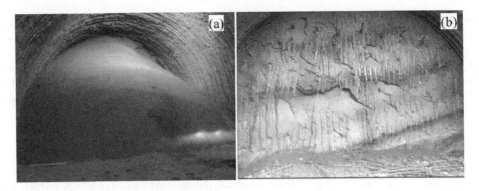

Figure 1.   (a) Collapsed tunnel face, (b) stable tunnel face with bolt reinforcement.

(2017). In the most general case of layered ground and arbitrary bolt distribution, the computation of the minimum required number of bolts consists in a multivariable optimization problem that can be solved only numerically (Anagnostou and Perazzelli 2015). A standalone computer application with graphical user interface was developed for this purpose and is presented in this paper (Section 3). Several application examples in cohesive frictional soils are investigated and possible optimizations of the bolt reinforcement are discussed (Section 4).

2   COMPUTATIONAL METHOD

2.1   Overview

The considered mechanisms consists of a wedge and a prism (Figure 2). Failure will occur if the trapdoor load exerted by the ground upon the wedge, i.e. the minimum vertical force needed to stabilize the ground upon the wedge, exceeds the bearing capacity of the latter, i.e. the maximum vertical force that can be sustained by the wedge at its upper boundary. At the limit equilibrium the trapdoor load is equal to the bearing capacity of the wedge.

For cohesive frictional soils (i.e., the ones considered in the present paper), the trapdoor lood is calculated on the basis of the silo theory, while the bearing capacity of the wedge depends on the support force offered by the face bolts (Section 2.2) and is determined by applying the method of slices. Basic equations can be found in Anagnostou and Perazzelli (2015) and in Perazzelli et al. (2017) for dry und drained conditions respectively.

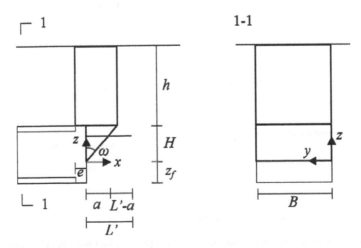

Figure 2.   Failure mechanism (Anagnostou and Perazzelli 2015).

Both the trapdoor load exerted and the bearing capacity of the wedge depend on the specific geometric parameters of the failure mechanism, i.e. on the inclination ω of the slip plane and on the elevation z_f of the wedge foot (Figure 2). These parameters are unknown at the start. Their critical values, i.e. the values that maximize the support requirements, depend on the reinforcement layout, on the shear strength and on the stratigraphy of the ground, and they will be determined iteratively (only under simple assumptions of homogeneous ground and bolt distribution) or by means numerical optimization methods (Section 2.3).

2.2 *Support pressure given by the bolts*

As the tunnel advances, the transmission of force between the bolts and the ground occurs solely via the shear stress acting at their interface, because any anchor plates (which might have been installed during an excavation standstill) are removed during excavation. The supporting effect of the face reinforcement thus relies solely on the strength of the bond between grout and soil. In general, the support force offered by each bolt depends on its tensile strength F_t; on the diameters d and d_b of the grouted borehole and the bolt, respectively; on the bond strength t_m of the soil - grout interface; on the bond strength t_g of the grout - bolt interface; on the anchorage length a inside the potentially unstable wedge under consideration; and on the anchorage length (L'- a) in the ground ahead of the inclined shear plane, where L' denotes the bolt length at the considered excavation phase (Figure 2).

Note that the bolt lengths (and thus also the effectiveness of the reinforcement if the bolts are short) decrease with every excavation round (Figure 3). The most favourable situation comes immediately after the installation of new bolts, while the most unfavourable situation comes just before the installation of a new group of bolts. Between these two stages bolt lengths decrease from their initial (or "installed") length L to a minimum length L'. The minimum bolt length L' depends on the installed bolt length L and on the installation interval l, i.e. on the gap distance between the locations of the successive installation stages (Figure 3). Depending on these two parameters, bolts of equal or different lengths may be present at the face in the most unfavourable situation (Figure 3). If the successive bolt groups do not overlap too much (more specifically, if the installation intervals l are greater than $L/2$), then all bolts will have the same length L' in the most unfavourable situation. L' is in this case equal to over-lapping length L-l (Figure 3a). If the installation intervals l are shorter than $L/2$, then bolts of different lengths will be present in the most unfavourable situation (Figure 3b).

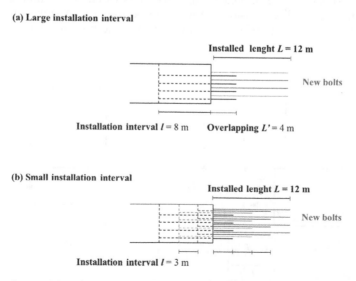

Figure 3. Dependency of the minimum bolt length L' on the installed bolt length L and on the installation interval l.

In general, the support pressure s offered by the face reinforcement reads as follows:

$$s = n \min \left[F_t, \; \max(\min(dt_m, d_b t_g)\pi a, F_p), \; \min(dt_m, d_b t_g)\pi a(L' - a) \right]. \tag{1}$$

This equation takes into account the following failure mechanisms: tensile failure of the bolt (1st term), failure of the anchor plates (2nd term) and shear failure at the bolt–grout–soil interfaces inside (2nd term) and ahead of the wedge (2nd and 3rd terms).

As the anchorage lengths a and $(L'\text{-}a)$ vary over the height of the wedge and, moreover, also depend also on the specific mechanism (i.e. on the angle ω and on the elevation of the wedge foot z_f; Figure 2), the support pressure offered by the bolts varies over the height of the wedge (even if the bolts are uniformly distributed) and changes with the geometry of the wedge. The impossibility of defining a constant support pressure can be illustrated by the examples of Figure 4a, which shows the distributions of the support pressure given by uniformly distributed bolts (sufficiently strong bolts and grout, no anchor plates) for several failure mechanisms. If the wedge foot is at $z_f = 0$ and the wedge angle $\omega = 35°$ (solid line), then the support pressure initially increases linearly with the elevation z, reaches a maximum value at $z_1 = 0.5L'/\tan\omega$, and then decreases linearly up to $z_2 = L'/\tan\omega$, where it becomes equal to zero. At the upper part of the face ($z > z_2$) the support pressure is zero. Between the elevations 0 and z_1 the support pressure is a linear function of the anchorage length a, which increases linearly with elevation z. Between the elevations z_1 and z_2 the support pressure is a linear function of the anchorage length $(L'\text{-}a)$, which decreases linearly with the elevation z. The support pressure becomes equal to zero when the anchorage length $(L'\text{-}a)$ is null. Similar remarks apply to the other wedges of Figure 4a (dashed lines). In conclusion, the support pressure offered by the reinforcement will be non-uniform even if the reinforcement density is constant (Figure 4a).

As explained above, depending on the initial lengths of the bolts and the longitudinal overlapping of the successive installation stages, bolts of different lengths may be present at the same time (Figure 4b). The effect of the installation sequence on the distribution of the support pressure is evident in the examples of Figures 4a and 4b: In both cases, the bolts are 12 m long. Figure 4a assumes the installation of 1 bolt/m^2 face every 9 m of advance. In the case of Figure 4c, the installation stages are more frequent (every 4.5 m of advance) but only half of the bolts of Figure 4a are installed at every stage (0.5 bolt/m^2). So the total quantity of reinforcement (expressed as linear meters of bolts installed per linear meters of tunnel) is exactly the same for Figures 4a and 4b. It is interesting to note, however, that the support pressures are completely different. The number of bolts in the ground ahead of the face is equal in the two cases (bolts of a single installation stage in Figure 4a and those of two successive installation stages in Figure 4b), but the average anchorage length of the bolts in Figure 4c is greater than the uniform bolt length in Figure 4b. As a consequence, the support pressure offered by the bolts in Figure 4b is higher. The installation sequence has a relevant effect on the stabilizing effect; a staggered application is more favourable for stability but necessitates interruptions of the excavation and support works more frequently.

The support pressure exerted by the bolts is constant over the face only in the case of uniform reinforcement density and extremely high bond strength (or in the presence of anchor

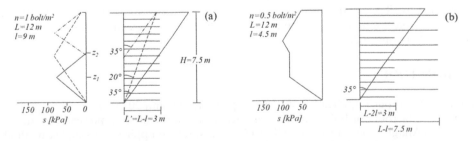

Figure 4. Support pressure distribution for different reinforcement densities and installation sequences (Anagnostou and Perazzelli 2015).

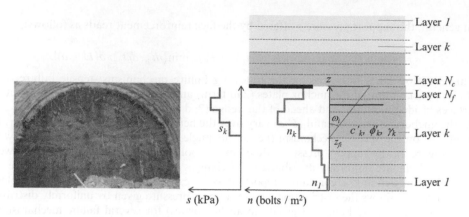

Figure 5. Division of the ground into layers.

plates and very long bolts). In these (rather theoretical) cases, failure is due to the tensile bearing capacity F_t of the bolts and the support pressure $s = n\, F_t$.

2.3 Minimum required number of bolts

In the general case (heterogeneous ground, arbitrary reinforcement distribution), the ground upon the wedge and the wedge are subdivided into an arbitrary number of layers of finite thickness (N_f for the wedge and N_c for the overlaying soil in Figure 5). Each layer k is considered as being homogeneous (constant values of c_k, ϕ_k and γ_k) and supported by a uniform pressure s_k. The latter is calculated from Eq. (1), taking account of the reinforcement density n_k (or equivalently the number of bolts N_b) of the specific layer and the average anchorage lengths (i.e., the anchorage lengths at mid-height of the layer).

The minimum required number of bolts $N_{b,cr}$ is such that the trapdoor load V_{Nc} exerted by the overlaying ground is equal or bigger than the bearing capacity of the wedge V_{Nf} and that the vertical load at the upper boundary of each layer V_k is equal or bigger than zero for all the considered failure mechanisms, i.e., for all ω_i, z_{fi} (a negative V_k means that the contact force is tensile and, consequently, the wedge must be "hung" in order to remain stable; as this is impossible, a negative V_k indicates face collapse up to layer k):

$$N_{b,cr} = \ \min \sum n_k B \Delta z_k \tag{2}$$

$$V_{Nf}(w_i, z_{fi}) \ \geq \ V_{Nc}(w_i, z_{fi}) \tag{3}$$

$$V_k(w_i, z_{fi}) \geq 0 \tag{4}$$

The problem stated above is a linear multivariable optimization problem that can be solved only numerically. The linearity is due to the fact that the trapdoor load V_{Nc}, the bearing capacity of the wedge V_{Nf} and the vertical load inside the wedge V_k are linear combinations of the variable n_k (Ackermann 2015).

3 DESIGN TOOL

A standalone computer application was developed (Ackermann 2015, Gheri 2017) for dimensioning or verifications of bolt reinforcement according to the method presented in the previous section (Anagnostou and Perazzelli 2015). The software applies the simplex method for solving the linear multivariable optimization problem. It contains a GUI (graphical user interface) for all input and output data and an automated report generator (Figure 6).

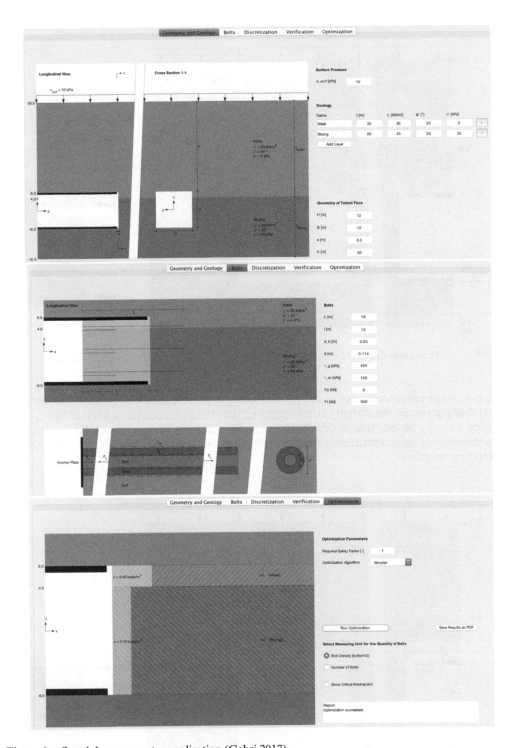

Figure 6.   Standalone computer application (Gehri 2017).

Figure 7 shows an example of verification problem. This consists in the computation of the safety factor of the tunnel face for a given bolting layout (i.e., given type of bolt, total length, overlapping length, number and spatial distribution of the bolts). The safety factor corresponds

1443

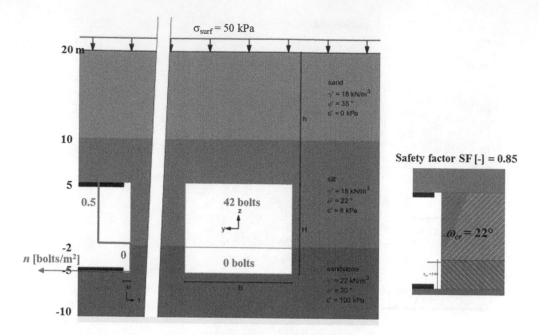

Figure 7. Verification problem ($H = B = 10$ m, $L = 18$ m, $l = 12$ m, $t_m = 70$ kPa, $d = 114$ mm).

to the minimum reduction factor of the ground strength parameters (cohesion and friction coefficient) that guaranties the stability conditions, i.e., that fulfils the equations (3) and (4).

Figure 8 shows an example of design problem. This consists in the computation of the minimum number of bolts and their spatial distribution for fixed type of bolt, total length and overlapping length.

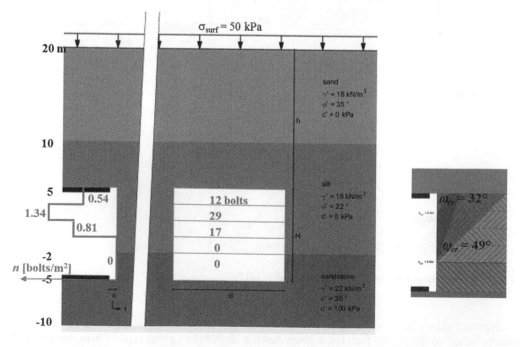

Figure 8. Design problem ($H = B = 10$ m, $l = 12$ m, $t_m = 70$ kPa, $d = 114$ mm, safety factor SF = 1).

4 APPLICATION EXAMPLES

4.1 *Effect of tunnel shape*

Figure 9 shows the required reinforcement density n_{cr} as a function of the soil cohesion c for three tunnel shapes, which concern three different excavation sequences: full face excavation (cross-section A), top heading bench excavation (cross section B) and side drifts excavation (cross-section C).

The effect of the tunnel shape becomes evident from the comparison between lines A, B and C of Figure 9. Cross-section C has the same height as A, but is narrower and therefore needs a considerably lower reinforcement density than cross-section A. The reason is that the stabilizing effect of the shear resistance at the vertical shear planes of the wedge is more pronounced if the face is narrow. A comparison between the reinforcement required in the cases of tunnel cross-sections B and C, which have the same area, shows that the favourable effect of the smaller width in case C outweighs the unfavourable effect of its greater height: Case C needs about the same quantity of bolts as case B, in spite of the greater height of the cross-section.

For a specific set of geotechnical parameters ($c' = 5$ kPa) the same figure shows the total number of bolts required in cases of full face excavation, top heading bench excavation and side drifts excavation. Top heading and bench excavation necessitates the smallest quantity of reinforcement.

4.2 *Effect of the installation interval*

The installation interval l determines the frequency of bolt installations (Figure 3) and is decisive for the distribution of the support pressure (Figure 4). Figure 10 shows the required reinforcement density n_{cr} as a function of the soil cohesion c for bolt installation every $l = 9$ m (solid line) or 4.5 m (dashed line). The total quantity of bolts (meters of bolts installed per linear meter of tunnel) is given by $n_{cr}BHL/l$ and amounts up to 102 m/m' in the first case, and up to 68 m/m' in the case of more frequent installation. This considerable saving is due to the more favourable pressure distribution (compare Figure 4a with 4c). Frequent installation of bolts interferes, nevertheless, with the excavation work and may therefore be advantageous only if the reinforcement quantities are large.

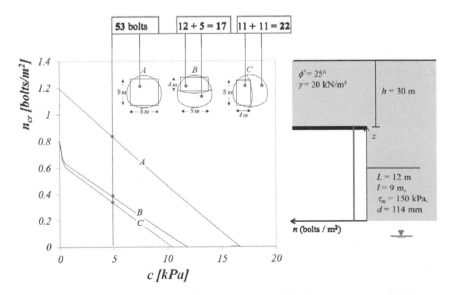

Figure 9.   Required reinforcement density n_{cr} as a function of cohesion c for different shapes of the tunnel cross-section. ·

4.3 Effect of the reinforcement distribution

For a homogeneous ground Figure 11 shows the required number of bolts for uniform distribution (Figure 11d) and multi-parametric reinforcement distributions (Figure 11a to 11c). The latter consist in two, three and four portions of the excavation face with different bolting density. As expected the number of bolts decreases with increasing number of face portions with different bolting density. The number of bolts needed in case of uniform distribution is 20% bigger than the number of bolts needed in case of optimum distribution (Figure 11d). The latter is characterized by the absence of bolts in the lower part of the face.

4.4 Heterogeneous ground

Figure 12 considers a tunnel intercepting two horizontal layers. A strong layer is encountered at three different heights over the face. Bolting density is assumed uniform inside each layer. As expected, in all cases bolts are required only in the weak layer. The most favourable condition for the stability (indicated by the smallest quantity of bolts) is the one with a the strong layer in the upper part of the face; the strong layer in the upper part of the face prevents the transfer of the load from the overburden to the underlying weak layer.

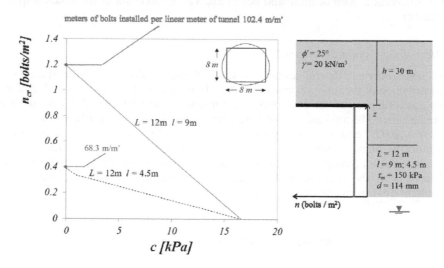

Figure 10.   Required reinforcement density ncr as a function of cohesion c for two installation intervals l (Anagnostou and Perazzelli 2015).

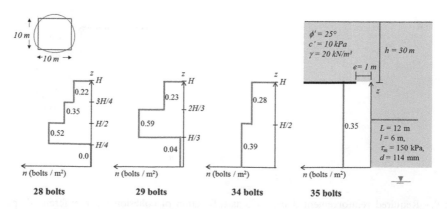

Figure 11.   Required number of bolts for different reinforcement distributions.

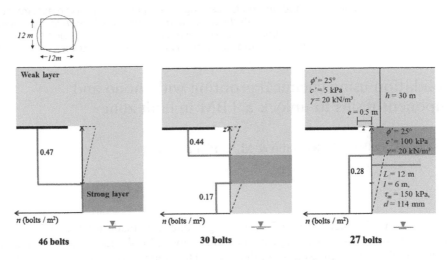

Figure 12. Required number of bolts for different reinforcement distributions.

5 CONCLUSIONS

Experience and predictions prove that ground reinforcement using bolts is a very effective measure for stabilizing the tunnel face. The computational examples reported in the present paper show how the excavation sequence and the installation interval of the bolts affect the required quantity of bolts. Top heading and bench excavation and small installation intervals allow to reduce significantly the quantity of bolts to install during advance. The uniform distribution is not the optimum one. A standalone computer application, implementing the analyses method of Anagnostou and Perazzelli 2015, was developed for the computation of the optimal distribution and it is presented in the paper.

REFERENCES

Ackermann, T. 2015. Optimization of tunnel face reinforcement with fiberglass bolts. *Master thesis, ETH Zurich*.

Anagnostou, G., Kovári, K. 1994. The face stability of slurry-shield driven tunnels. *Tunnelling and Underground Space Technology*, 9 (2), pp. 165–174.

Anagnostou, G., Perazzelli, P. 2015. Analysis method and design charts for bolt reinforcement of the tunnel face in cohesive-frictional soils. *Tunnelling and Underground Space Technology*, 47, 162–181.

Bischof, J.A., Smart, J.D. 1975. A method of computing a rock reinforcement system which is structurally equivalent to an internal support system, in *Proceedings of the 16th Symposium on Rock Mechanics*, University of Minesota, pp. 179–184.

Gehri, N. 2017. Graphical user interface - Verification and design of tunnel face reinforcement with fiberglass bolts. *Master Project Work*, ETH Zurich.

Grasso, P., Mahtab, A., Ferrero, A.M., Pelizza, S. 1991. The role of cable bolting in ground reinforcement. In: *Soil and Rock Improvement in Underground Works*, ATTI, vol. 1, Milano, pp. 127–138.

Indraratna, B., Kaiser, P.K. 1990. Analytical model for the design of grouted rock bolts. *International Journal of Numerical and Analytical Methods in Geomechanics* 14, 227–251.

Peila, D. 1994. A theoretical study of reinforcement influence on the stability of a tunnel face. *Geotechnical and Geological Engineering*, Vol. 12, No.3, pp.145–168.

Perazzelli, P., Anagnostou, G. 2017. Analysis method and design charts for bolt reinforcement of the tunnel face in purely cohesive soils. *Journal of geotechnical and geoenvironmental engineering*, 143 (9), American Society of Civil Engineers.

Tunnels and Underground Cities: Engineering and Innovation meet Archaeology, Architecture and Art, Volume 4: Ground improvement in underground constructions – Peila, Viggiani & Celestino (Eds)
© 2020 Taylor & Francis Group, London, ISBN 978-0-367-46868-2

Consolidation using chemical grouting with mono and bicomponent resins to unlock a TBM in fault zone

P. Petrocelli, A. Bellone, F. Rossano & M.A. Piangatelli
CIPA SpA, Rome, Italy

ABSTRACT: A 12% inclined and 2.8 km long tunnel was made for the execution of the new TERNA S.p.A. 380kV power line connecting Sicily and Calabria. It finished in a vertical 300 m deep shaft with a 7 m excavation diameter. The tunnel was made with a double-shield 4.1 m diameter TBM. It passed through a large part of the Calabro Arch's crystalline-meta-morphic base. This rock mass presents variations and fault areas with extremely high grades of broken rock, with very low GSI (Geological Strength Index) values. It also presents a granulometry similar to a sand-consistency, sometimes similar to a clay-consistency, with a significant quantity of water. These fault areas gave huge problems during the tunnel excavation, with very low progress areas and three "geological events" which caused TBM stops. The three TBM interruptions required the improving of the rock mass property by the use of mono and bi-component resins. These interventions allow the excavation of small tunnels to "unlock" the TBM.

1   INTRODUCTION

The work is constituted of a tunnel that is 2842 meters long with a 12% slope and by a 300 meter deep shaft (Figure 1) . The TBM passed through heterogeneous areas with strong water presence and faults.

These "fault" zones blocked the TBM and required unlocking interventions. Four major geological events required heavy soil treatments (Figure 2).

2   THE GEOLOGICAL EVENT AT THE KM 2+100 ON 12TH DECEMBER 2014

The geological event at Km 2+100 presented complex geological situations. All other cases are attributable to this, but have shown operational difficulties and conditions that were less prominent.

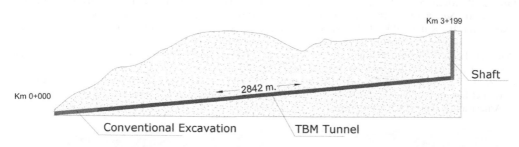

Figure 1.   The tunnel and the shaft.

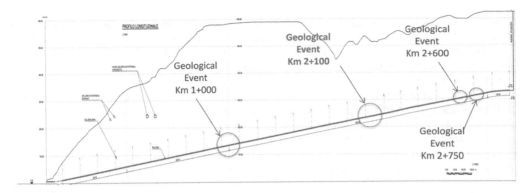

Figure 2 . Geological event occurrences.

2.1 *Block mechanisms*

Three kind of block mechanisms occurred during the geological events:

– Rotating head block
– Excavation chamber filling
– Front shield filling

The TBM went through zones of highly altered rocks and with a lot of water where the three mechanisms occurred in rapid succession, triggering a loop process that required additional interventions to restart the TBM.

The TBM crossed a clay diaphragm a few meters thick, entering in a zone that in the geomechanical profile was referred to as the main fault. After the TBM head of passed this diaphragm, the considerable presence of pressurized water caused a siphoning phenomenon.

A significant amount of material was carried into the excavation chamber, the front shield and the rear shield of the TBM.

In the end the material completely filled the machine, causing an instant block.

The situation differed from the first event at km 1+000:

– The material that flowed into the TBM was predominantly a clay matrix
– The presence of water was considerably higher than the first case
– Contrary to the first case, the phenomenon of the material inflow continued spontaneously and cyclically, without interruptions.

The phenomenon triggered by the diaphragm breakage was of considerable size.

After several days, the phenomenon did not seem to finish and after the attempts at TBM cleaning, it resumed vigorously, causing serious worries to the technicians. The zone of altered rocks was more than 20m thick. The water flow changed cyclically, from 50 l/s to 80 l/s and water pressure at the TBM head was about 13 bar. Immediately after the event, the material entering into the TBM had temporarily blocked the phenomenon (Figure 3), so that it seemed to have stopped, but the water flowing into the TBM was always cloudy, indicating a transport of material, so it was assumed that the phenomenon could not be considered exhausted.

The attempts to clean the shields and the excavation chamber caused the spontaneous return of material into the TBM, with force and intensity.

The TBM, before encountering the fault, had met a clay diaphragm that had created a waterproof barrier that allowed the water to reach high pressure.

Prior to starting with the consolidation actions, an attempt was made to find out the fault thickness and to reduce the hydraulic pressures by draining the area to be treated.

The TBM shield is provided with n.8 holes Ø89mm to make destructive drillings. A "Beretta T43" drill rig (Figure 3) was mounted on the erector slide. The drill rig had been modified and adapted to work in narrow spaces and it could do every kind of intervention (destructive drillings, drainages with steel casings, bolts and injections, etc.). This machine proved very effective.

Some destructive perforations were made, recording the push pressure, the rotations and the time needed for the penetration, in order to determine the thickness of the fault. All these values have been recorded and interpreted by engineers and geologists to rebuild the most plausible scenario (Figure 5 and Figure 6).

With constant drainage it was possible to reduce the pressure of the water, but performing continuous and constant drainage was not a simple task, because the clay layer often behaved as a plug and therefore the drained water decreased over time and pressure increased again.

This issue was resolved adopting self-drilling drainage pipes, with big water-soluble valves (more than a centimetre in diameter) and therefore hard to be obstructed by fine materials (Sidra Drain, Figure 7).

Figure 3. The material flowed into the TBM.

Figure 4. The material flowed into the TBM.

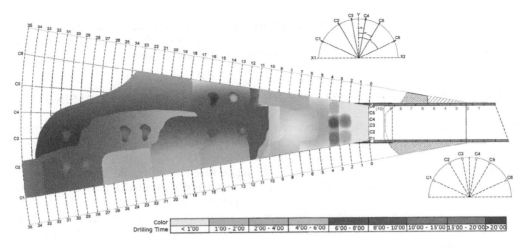

Figure 5. Chromatic representation of the time needed to the drillings to go through the soil.

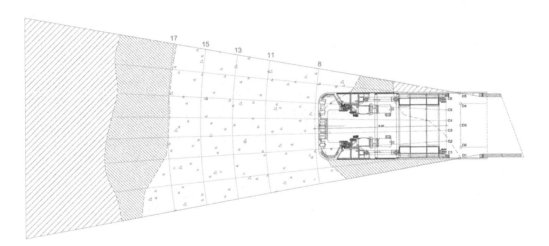

Figure 6. Reconstruction of the fault geometry.

Figure 7. Continuous drainage with self-drilling drainage pipes, with water-soluble valves.

With this method, the area to be treated with resins was well drained of the water that would have reduced the effectiveness of the resin injection. Also, the pressure decreased and the phenomenon of the cycling water flush and material inflow ended.

3 INJECTIONS WITH MONO-COMPONENT AND BI-COMPONENT RESINS

The resin injections had been hypothesized by using PVC pipes with "manchettes" valves. This would have allowed injections of each type of resin to the various set distances, by the use of packers to isolate the different injection portions.

Unfortunately, all attempts to install these pipes failed because the presence of water allowed the material to flow into the casing. This material prevented the pipe from being installed, or dragged it out during the casing extraction to expose the valves.

After several attempts, the adopted solution was the direct injection into self-drilling bolts. After the injections, the bolts were extracted for a subsequent re-drilling that allowed a new injection to a new distance.

The general criterion followed was:

– Try to consolidate areas whose time for drilling had been less than 10 minutes;
– Try to differentiate the injections according to the following guide lines:
  o Filling injections to protect TBM shield with the mono component resin "Basf MRoc MP 355 (MS) 1K DW";
  o Consolidation injections with the bi-component resin "Basf MRoc MP 355 (MS) PTA + PTB".

The conditions round the TBM were unknown and the behaviour of the resin in those particular conditions was not foreseeable, so the technicians implemented a scale model of the area to be treated to simulate the behaviour of the resin in the presence of large amounts of water (moreover in drainage condition, Figure 8).

In the box where the tunnel excavation material was accumulated, an experimental study was conducted by simulating multiple bi-component resin simulations combining two different factors:

– Amount of water;
– Different mix-designs involving resin, accelerator and water quantities.

Figure 8.   Test field for mix-designs.

Figure 9.   Some tests for mix design.

The various mix-designs were injected into the soil with the same technique and the same lengths as for the real case, and then results were analysed after dissecting the treated areas (Figure 1). It was possible to determine the right mix-design to achieve two results:

– Good mechanical strength of the combination rock + resin in order to create an arc effect at the passage of the TBM;
– Good set speed to allow the resin to travel in the tube and expand around a significant injection area without getting dragged by flowing water.

Once the mix design, modes and timing had been defined, the real intervention started. In order to perform an effective intervention, the volume to be injected was shared into a two-dimensional grid (Figure 10 and Figure 11), also taking into account the forced inclination of the holes in the gripper shield.

The two-dimensional grid was designed with a 1m x 1m mesh according to the following items:

– The injections should be carried out by the injection of self-drilling bars R32 with three-hole drilling head.

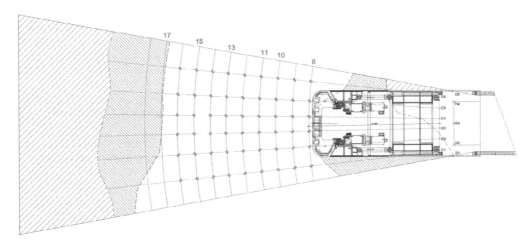

Figure 10.   Some tests for mix design.

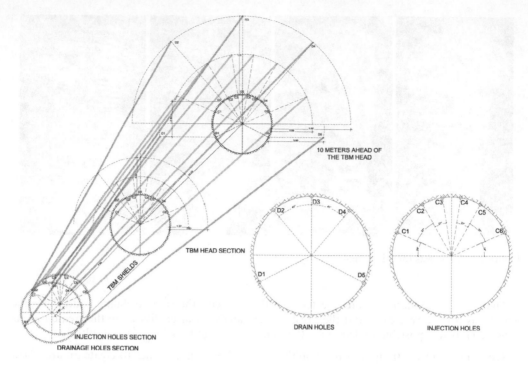

Figure 11.   3D projection of drains and injections.

–   Each bar could be injected only from the bottom and only one injection could be carried out per each bar.
–   Range of action. The test field showed that the resin solidified in a diameter of about 1–1.5 m.

Intervention sequence:

1.   Insertion of the bar at the set depth;
2.   Injection of the resin up to saturation or up to a predetermined volume;
3.   Pull back of the bar used for the injection;
4.   New drilling at a lower depth

For each injection hole, identified with C1÷C6, at least 8–9 perforations had to be made. Due to the injection time, it was decided to make the first consolidations up to 19 m, then to evaluate the result by using a local tomography that returned the physical characteristics of the combination soil + resin. At the end of this step, the scheme of the injections is shown in Figure 12.

Before the TBM restarting, a series of tomography and seismic tests were carried out within the drainage pipes to determine the resulting physical characteristics (Figure 13 and Figure 14).

The results obtained were encouraging, so another series of injections were made on the left side, that resulted less consolidated. After this second step, new tests were carried out (Figure 15).

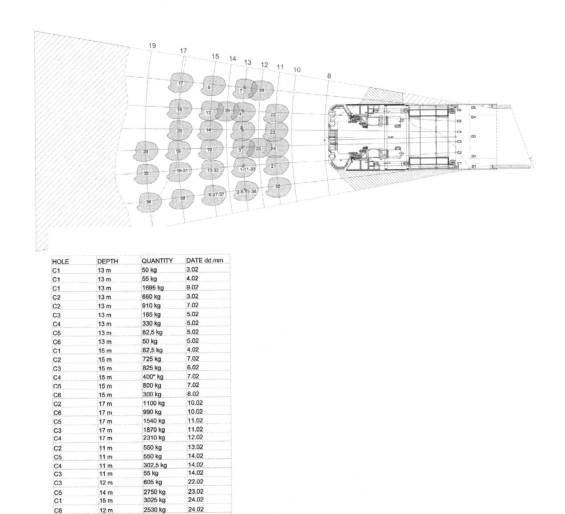

HOLE	DEPTH	QUANTITY	DATE dd.mm
C1	13 m	50 kg	3.02
C1	13 m	55 kg	4.02
C1	13 m	1695 kg	9.02
C2	13 m	660 kg	3.02
C2	13 m	910 kg	7.02
C3	13 m	165 kg	5.02
C4	13 m	330 kg	5.02
C5	13 m	82,5 kg	5.02
C6	13 m	50 kg	5.02
C1	15 m	82,5 kg	4.02
C2	15 m	725 kg	7.02
C3	15 m	825 kg	6.02
C4	15 m	400* kg	7.02
C5	15 m	800 kg	7.02
C6	15 m	300 kg	6.02
C2	17 m	1100 kg	10.02
C6	17 m	990 kg	10.02
C5	17 m	1540 kg	11.02
C3	17 m	1870 kg	11.02
C4	17 m	2310 kg	12.02
C2	11 m	550 kg	13.02
C5	11 m	550 kg	14.02
C4	11 m	302,5 kg	14.02
C3	11 m	55 kg	14.02
C3	12 m	605 kg	22.02
C5	14 m	2750 kg	23.02
C1	15 m	3025 kg	24.02
C6	12 m	2530 kg	24.02
C3	19 m	2640 kg	13.03
C2	19 m	1980 kg	14.03
C2	17 m	1320 kg	15.03
C2	15 m	1430 kg	16.03
C2	13 m	2255 kg	16.03
C1	19 m	950 kg	17.03
C1	11 m	2750 kg	19.03
C1	13 m	1925 kg	19.03
C1	15 m	1595 kg	20.03
C1	17 m	2975 kg	20.03

Figure 12. The injection grid – The injection distribution at the end of the first step.

4 CONCLUSION

A new power-line tunnel in Italy was drilled by a Tunnel Boring Machine (TBM). During the TBM drilling, four fault zones were encountered, each causing TBM stoppages. This paper describes the worst stoppage.

The method for repairing the fault zone was chemical grouting using resins. Prior to the injections, the hydraulic pressure on the TBM head was lowered by a drainage system, increasing the efficiency of the resins. The injection of the mono-component resins filled the cavities and protected the TBM from the subsequent injections of structural resins. The bi-component resins for structural purposes were used to consolidate

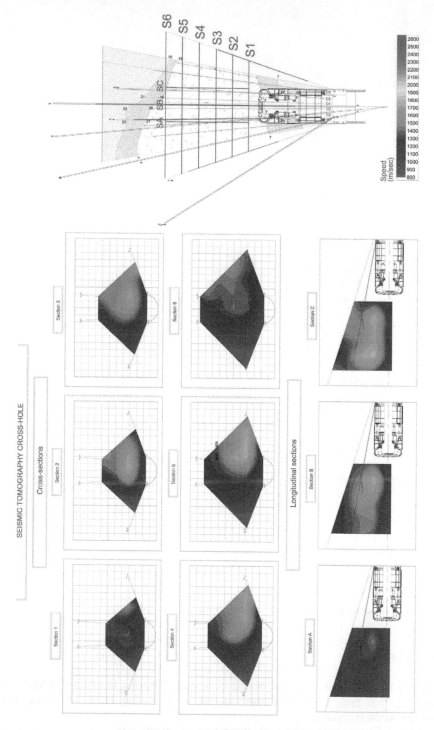

Figure 13.   Schematic results of the seismic tomography tests of the first step.

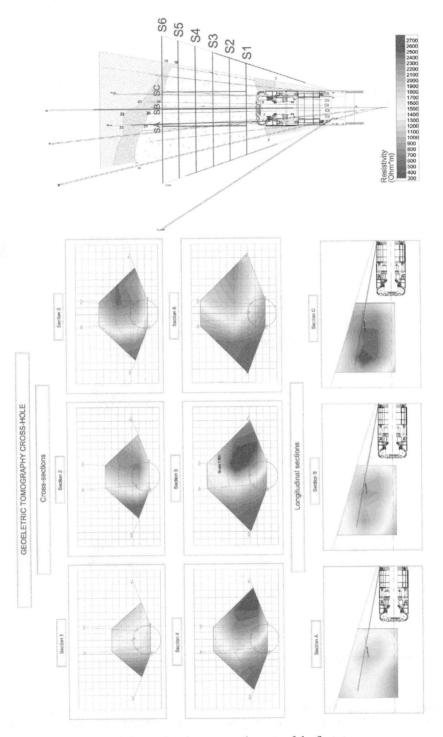

Figure 14. Schematic results of the geoelectric tomography tests of the first step.

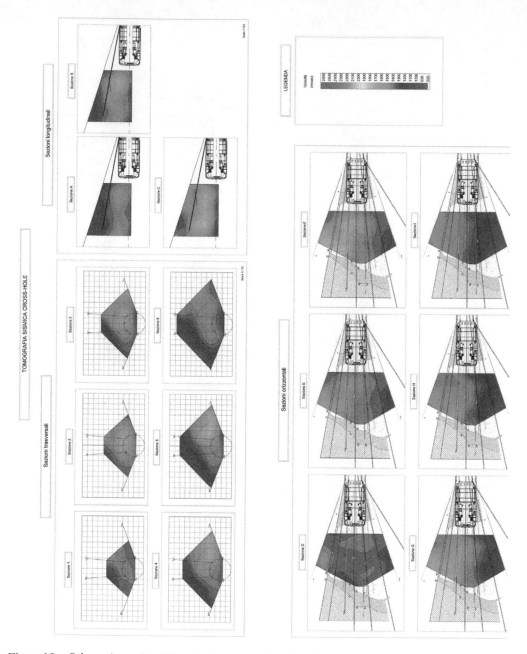

Figure 15. Schematic results of the seismic tomography after the second step.

the altered and collapsed soil that in the presence of water would have prevented the TBM passage. All the combined techniques allowed the TBM to overcome the fault.

REFERENCES

Bringiotti, M. 2003. *Guida al Tunneling 2nd ed.* PEI.
Bringiotti, M., and Bottero, D. 1999. *Consolidamenti e fondazioni 2nd ed.* PEI.

Tunnels and Underground Cities: Engineering and Innovation meet Archaeology,
Architecture and Art, Volume 4: Ground improvement in
underground constructions – Peila, Viggiani & Celestino (Eds)
© 2020 Taylor & Francis Group, London, ISBN 978-0-367-46868-2

The technical management of the permeation grouting works in the execution of the new Milan Metro Line 4

A. Pettinaroli & P. Caffaro
Studio Ing. Andrea Pettinaroli s.r.l., Milan, Italy

M. Lodico & A. Carrettucci
MetroBlu s.c.r.l., Milan, Italy

ABSTRACT: The new Milan Metro Line 4 requires the treatment of alluvial soil by permeation grouting, in order to allow stations and service shafts excavation. The Line stretch underpassing the city involves the careful scheduling of site activities for minimizing the impact on the city life. An efficient grouting work management has been planned from the beginning. For each working site, geotechnical investigations have been executed, detecting stratigraphy and granulometric composition of the soil layers to be treated, showing their variability along the Line. First grout mixtures were set up in laboratory, and work organization tested on site. During the work, actually in progress, tube-a-manchette (TAM) meshes and grouting parameters are optimized for each site. Regular checks on mixture, daily examination of injection parameters and of surrounding buildings structural monitoring allow to implement a real-time management of the grouting activities tailored on each site, respecting the general work development and scheduling.

1 INTRODUCTION

The new Metro Line 4 in Milan runs from east to south west. It connects Linate Airport to the city center, continuing under the so called "Cerchia dei Navigli", an ancient water channel designed by Leonardo da Vinci, which is currently buried though it will be probably restored in the next years. The line consists of two main tunnels, excavated with EPB-TBMs, connecting 21 stations and including 26 service shafts. It can be divided into three stretches: the central segment, underpassing the city center and the "Cerchia dei Navigli"; the east segment, up to the east line terminus at Linate Airport Station; the south segment, ending at the south line terminus, San Cristoforo Station.

The tunnels of the peripheral stretches are excavated by 4 TBMs, 6,70 m diameter, pass through the stations.

The central stretch runs under an urban context of narrow and busy streets. In order to reduce the interferences of the works with the city viability, the dimensions of the station shafts have been minimized, including just the stairs and the access area.

The platforms are obtained inside the two tunnels, which are excavated with a diameter of 9,15 m, and run beside the shaft. The two TBMs are going to excavate the tunnels starting from the last shaft of the eastern stretch, Tricolore Station, to Solari Station (Figure 1).

The connection between the platforms in the tunnel and the shafts, in the city center, must be consequently excavated with the traditional method, being in presence of a hydraulic head from 8 to 15 m above the tunnel crown.

These difficult conditions require a consolidation and waterproofing treatment of the sandy-gravelly soil, typical of Milan subsoil, to be carried out mainly from the street level.

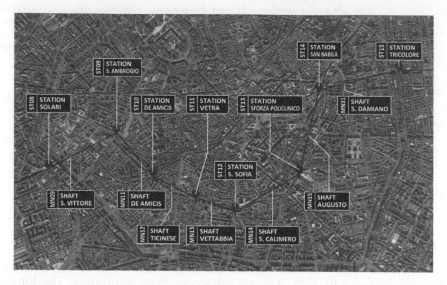

Figure 1.   Plan view of the Metro Line 4 City Center stretch.

Once the decision was taken to proceed by using the technology of the permeation grouting with TAMs, due mainly to the long-time experience gained in this context in Milan, all the activities were oriented and managed to minimize the impact of the works on the city-center life, according to the following steps: a detailed geotechnical investigation on each work site; the development of the grouting design, tailored on each site; an accurate method-of-statement defining the activities and the controls procedures for each stage of the work. The following chapters will describe those activities.

2   PRELIMINARY PHASE: INVESTIGATIONS AND GROUT SET-UP

The geotechnical characterization of the soil, focused on the use of the permeation grouting technology, has required the execution of additional borehole investigation on each site of the city center stretch (6 stations and 7 service shafts), including on site and laboratory tests.

Then an accurate stage of laboratory tests was started on the grout mixtures, in order to optimize the mix design as a function of those properties, which are necessary for the efficient treatment of the soils: penetrability, stability, workability, mechanical properties.

2.1   Geotechnical investigations

The Milan subsoil is composed by recent alluvium with widely variable alternations of gravel and sand, including "lenses" of silt, that may have an extension up to some dozens of meters; the silty components tend to increase with the depth, becoming prevalent around 40 m under the ground level. These surficial strata include the upper aquifer, having a water table level that lies about 13-15 m under the street level during the activities, with a possible seasonal fluctuation of about 0,70 ÷ 1,00 m.

A borehole has been carried out for each site, recovering several specimens at different depth, on which granulometric analysis were performed. Therefore, it has been possible to obtain a detailed stratigraphy of the ground. The general tendency has been approximately confirmed, with local exception. The silty layers broadly start from 36-38 m of depth. "Lenses" of silty sand, 2 ÷ 3 m thick, were detected between 20 m and 25 m, predominantly in the central stretch of the line.

The granulometric curve has been compared with the standard injectability curve of two different grouts: the fine-cement grout (curve "c" in the following Figure 3) and the silica grout (curve "s"). These are indicative references of the optimal granulometric composition of a soil that can be properly treated by permeation grouting. The efficient diameter d_{10} of the soil is usually assumed as the critical parameter for the injectability, equal to 0,2 mm for the curve "c" and 0,02 mm for the curve "s".

The soil layers were classified on the basis of the granulometric analysis, referring to their injectability. As shown in Figure 2, the normally injectable layers were marked in green ($d_{10} \geq 0,2$ mm) and yellow ($d_{10} \geq 0,02$ mm), while the not injectable in blue. In red are marked the layers in which the injection must be carried out adopting a particular care in the management of the operative parameters on site, as described in a further chapter.

The section in Figure 2 shows the case of a site in which the arch and the sides of the drift to be excavated lie in sandy gravelly strata, including few thin layers with finer soil, where the silica grout is definitely necessary for homogenously penetrating the ground. The layer just under the invert, composed by sandy gravel, is well treatable; beneath it lies a silty-clayey layer, classified as not groutable.

In chapter 5 some case histories describe how the permeation grouting has been executed in different stratigraphic conditions.

2.2 Grout mixtures set-up

Once the granulometric composition of the layers to be treated is known, the suitable grouts must be set-up. The injection works generally foresee 3 subsequent stages, using the appropriate operative parameters as well as mixtures having growing penetrability properties. The 1st and the 2nd grouting stages are carried out using a stable fine cement mixture, in order to permeate the soil, filling the voids of large and medium size (gravel and medium sand). The 3rd stage is carried out using a silica mixture suitable to permeate and consolidate the finer soil fraction (medium and fine sand).

The set-up of the grout properties (rheology and mechanical properties) has been carried out before the beginning of the works, by executing some laboratory test.

As widely experienced with the injection works in Milan subsoil, a grout with cement and bentonite, with a cement/water ratio c/w=0,4 and a stabilizing, superplasticizing admixture, allows to obtain optimal results in terms penetrability and stability. This can be evaluated by executing standard test of viscosity with the Marsh funnel, bleeding test and stability test with filter-press under a pressure of 0,7 MPa. The frequent presence of sandy layers (yellow marker – Figure 2) in the ground to be treated leads to maximize the penetrability performance. Several laboratory tests were carried out, using different types of bentonite and admixtures. The chart

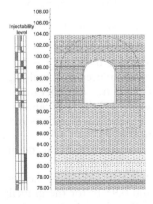

Figure 2. Soil stratigraphy and relative injectability level.

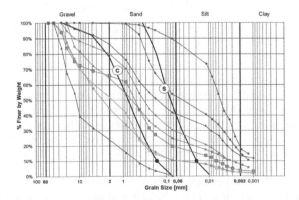

Figure 3. Granulometric curves of soil specimens and standard groutability curves: cement grout (c) – silica grout (s).

1461

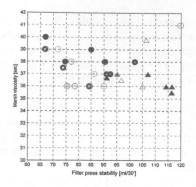

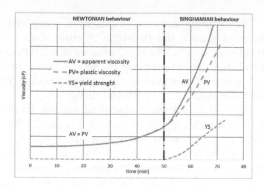

Figure 4. Tests on cement grout mixtures:
Marsh viscosity vs. Filter press stability.

Figure 5. Tests on silica grout mixtures: viscosity vs.
time - groutability time.

in Figure 4 shows the filter press stability versus the Marsh viscosity for several mixtures made varying the components and the dosage of bentonite and admixture. The best results were obtained with a highly fine bentonite (circle symbol), which works properly with all the admixtures: a good viscosity, $35,5 \div 36$ seconds with Marsh funnel, and 75-85 cm^3 of filtrate after 30 min at the filter-press test. The choice of the admixture has fallen upon the one that allowed to minimize the dosage of the bentonite and the admixture itself. The triangle-shaped data refer to a standard bentonite for grouting mixtures.

The silica grout is a highly penetrability mixture, based on silica solution and inorganic reagents that, mixed together, react producing a crystalline, stable structure, not affected by syneresis. The reaction starts at the beginning of the mixing. In the initial phases the grout maintains a low, stable viscosity (with a Newtonian behaviour), which then increases with time, gradually showing a Binghamian behaviour up to the final setting. The occurrence of the change of rheological behaviour determines the groutability time of the grout. This period must be long enough in order to allow the complete injection of the grout into the ground (Figure 5).

Preliminary test carried out in laboratory allowed to set up the rheological properties of the mixture, which evolves during the reaction. The chosen grout had an initial viscosity of $5 \div 7$mPa*s and a groutability time varying between 50 and 80 minutes. Laboratory grouting tests on standard monogranular sand column were carried out; UCS test on several samples of the grouted columns gave strength resistance results between 1,2 and 1,8 MPa.

3   DETAILED DESIGN OF THE TREATMENTS

The detailed design of the treatment for each site has taken into account two basic aspects:

- the very complex local context of the work sites in the historical city center of Milan;
- the specific geotechnical soil conditions.

The layout of the treatments (geometry, length and inclination of the drillings) has been defined as function of the effective work site area as well as of the interferences with buildings (underground floor of existing buildings, old masonry), urban roads and traffic, underground facilities (aqueduct, sewerage, gas pipes, electric, lighting and data ducts) as shown in Figure 6.

The interpretation of the previously described results of the geotechnical investigations gave information such as the soil stratigraphy, the granulometric composition of the various and significant soil layers and their relative injectability levels. This information allows to define the grouting operative parameters.

The effective permeation radius of the grouting has been evaluated on the basis of the percentage of the fine grain size (silt and clay) of the soil, varying from 0,95 m for the gravelly sandy soil (green marker) down to 0,80 m in the layers classified with red marker (with finer fraction around 25%). The drilling meshes were consequently drawn.

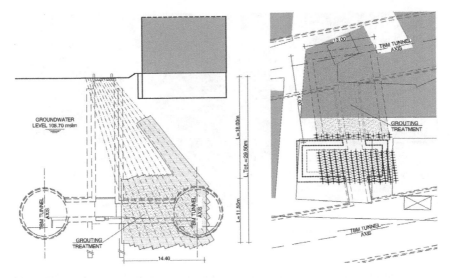

Figure 6. Grouting treatment of the ground for the junction tunnels excavation: section type and plan view.

The planning of the quantity of cement and silica grouts to be injected has been strictly correlated to the different geotechnical soil conditions.

In case of gravelly sandy soil, the volume of injected stable cement grout is preponderant compared to the silica one (ratio $V_s/V_c \sim 0{,}6 \div 0{,}8$). Otherwise, in case of sandy soil, or soil with fraction of silt and clay up to 25%, a higher volume of silica grout is necessary ($V_s/V_c \sim 1 \div 1{,}2$) in order to penetrate the finer fraction of the soil.

The maximum injection pressure has been fixed for each granulometric condition and for each grouting stage, in order to avoid the hydro-fracturing phenomena (claquage), which may lead to possible uplift of the ground during the injection, as well as to a lack of treatment homogeneity, with high risk of piping during the excavation.

It must be pointed out that all the execution parameters foreseen in the detailed design phase (volume, pressure, etc.) are verified and optimized in real time during the works.

4 THE MANAGEMENT OF WORKS ON THE SITES

The activities on site start with the drilling and the installation of the sleeved pipes (or TAMs: tube a manchettes), regularly checking the sheath grout used for the embedment and the sealing of the pipes in the boreholes (Figure 7). The behaviour of the buildings beside or above the working zone is monitored by regular topographic survey.

The rigs for the injection (the mixing units for the bentonite slurry and the grouts, the grouting pump skids, the data logger for the control and record of the grouting operative parameters) are installed in the yard. Due to the coincidence in time of the works for different shafts, it is very common that the teams, the rigs and the supplier of the grout components may vary from site to site. Therefore, it is planned to proceed in any case with a preliminary phase for setting up the mixtures (cement- and silica-based), using the yard's own equipment; a particular care is given to the calibration of the automatic control and record unit of the grouting operative parameters (pressure, quantity, flow rate).

The cement grout requires, in this phase, a careful final test, in order to verify the good activation of the bentonite and the correct quantity of admixture necessary to respect the design parameters of stability and viscosity. A correct initial set-up usually allows to forge ahead with the work during the injection phase. Quality controls are then carried out on the cement grouts during the injection stages, by daily measuring the density, the Marsh viscosity and the bleeding; furthermore, a weekly check of the grout stability is carried out by using the filter

Figure 7. Drilling and grouting works (respectively on the left and on the right).

press. Silica-based grout is very sensitive both to temperature of the air and of the liquid components. The mix-design is consequently calibrated preliminary on each site, on the basis of the average temperature of the current season and as a function of the expected productivity on site, optimizing the groutability time of the mixture. The setting time is evaluated during the works for each mixing using the quick method of the Cup test (a cup half-filled with grout can be tilted 90° without flushing the set mixture). Regular controls are then carried out on the density, and by measuring the groutability with the rheometer. Sometimes it is necessary to adjust the mix design, usually because of the variation in temperature of the air, more frequently in mid-seasons.

The injection operative parameters are controlled and recorded by data loggers (Figure 8.a). The model of the latter may vary from site to site, so that a preliminary calibration is necessary before the start of each work. For each grouting stage the grout quantity to be injected in a single sleeve, the limit pressure and the maximum instant flow-rate must be set (Figure 8.c). A specific setting, to be calibrated from time to time, allows to manage the reduction of the instant flow rate whereas the grouting pressure rises. The injection is automatically stopped when the limit pressure value is reached.

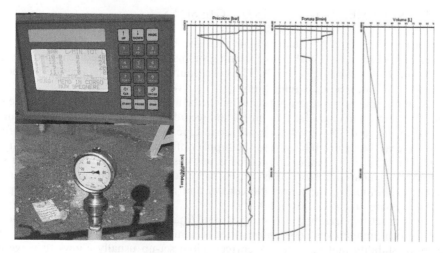

Figure 8. a-Data logger for the grouting parameters recording (in the upper left). b-Manometer at the head of the injecting borehole (in the lower left). c-Charts of the recorded grouting parameters (on the right).

Each injection line, connected to a single tube, is always equipped with a manometer at the head of the borehole (Figure 8.b), so that it is possible to control directly on site the injection development, and to verify the properly functioning of the data acquisition system. If abnormal values (particularly about the pressure) are pointed out after a first check of the operative parameters, additional investigations could be necessary (for instance, check of possible presence in the subsoil of facilities, obtacles, old wells or other structures not previously indicated). It may be also possible that during the works an alert is given by the structural monitoring system. Then it would be necessary to modify the injection procedures, by varying the grouting sequence of the holes or by reducing the limit pressure or the grout quantity per sleeve.

The data processing are daily updated, by plotting charts (Figure 9) showing the pressure values recorded at the end of each stage of injection for each sleeve of the TAM. The cromatic scale adopted for the plot allows to easily check the injection pressure reached in the soil, pointing out also the sleeves where the limit pressure has been reached and the grouting stopped.

In detail, Figure 9 illustrates the pressure of injection at the end of the 1st and 2nd grouting stages of a certain section. The charts show that the pressure is increased during the 2nd stage; in fact, the medium values of pressure rise from $4 \div 10$ bar in the 1st stage (predominance of cyan, grey and green colours) to $8 \div 14$ bar in the 2nd stage (predominance of orange and green colours).

After the completion of a grouting stage, the analysis of those diagrams allows to evaluate where the soil has been already satisfactorily treated and where to proceed with an intagrative injection of grout, in order to reach a good homogeneity of the treatment. If the grouting pressure doesn't reach an adequate value (approximately $8 \div 12$ bar, depending on the grouting stage) in some sleeves, additional grout is then injected. This evaluation is carried out taking into care the structural monitoring data.

As a result, the total quantities of cement and silica grouts injected in the soil may then differ, more or less, from the volumes predicted by the design. It has been observed that at the moment, for the Metro Line 4 in Milan, this difference can be up to $\pm 5 \div 7$ %, because of the good accuracy of the project method.

The following chapterwil describe the grouting works carried out in a couple of sites of the Line 4, putting in evidence the aspect here above expressed.

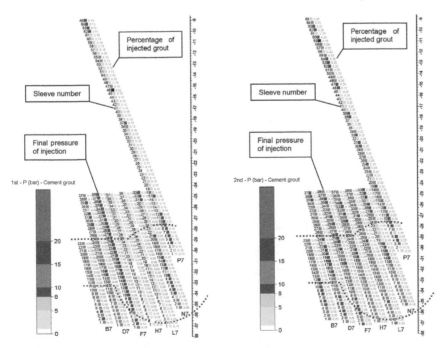

Figure 9.   Grouting pressure-volume diagrams: 1st grouting stage on the left and 2nd on the right.

5 GROUTING CASE HISTORIES

5.1 *Site A: grouting in gravelly sandy soil*

This first case history illustrates how the grouting works have been managed and carried out in a site which is mainly characterized by gravelly sandy soil.

As shown in Figure 10, the soil to be treated around the tunnel at the invert and the sides consists of sandy gravel with a fine fraction (silt and clay) below 10%.

Moving to the upper part of the treatment, the percentage of sand increases slightly, whereas the one of the gravel declines; however, the fine fraction in those layers is lower than 20% and the efficient diameter $d_{10} > 0.02$mm.

Therefore, the soil interested by the grouting treatment in this site can be globally classified as "well injectable" (green marker). The Figure 10 shows that not injectable layers (blue marker) are present right above the treatment volume (silty sand) and below it (sandy silt).

These assessments led to prescribe the injection in the ground of a slightly greater quantity of cement mixture compared to the silica one: a ratio $V_s/V_c \sim 0.85$ has been predicted, taking into account the sandy layers at the top of the treatment.

The analysis of the pressure and volume diagrams after each grouting stage has led during the works to prescribe additional cement and silica grouting, respectively at the end of the 2nd and 3rd stages.

As a result, a total quantity of cement and silica grout of about 30% of the theoretical volume of soil to be treated has been injected, with an effective ratio $V_s/V_c = 0.90$, close to the design hypothesis.

5.2 *Site B: grouting in sandy soil with presence of finer fraction*

The second case history illustrates the management of the grouting works in a site which is characterized by a slightly more complex geotechnical condition (referring to the injectability level) compared to the previous one. In fact, the investigations have detected the presence of a sandy soil layer with 25% of silt and clay.

In detail, as shown in Figure 11, the treatment at the level of the tunnel sides has interested a sandy layer (percentage of sand greater than 75%) with a quantity of fine fraction between 10% and 25%.

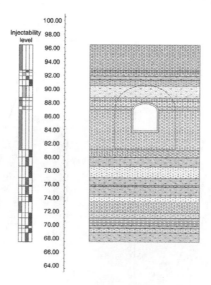

Figure 10. Site A: soil stratigraphy and relative injectability level.

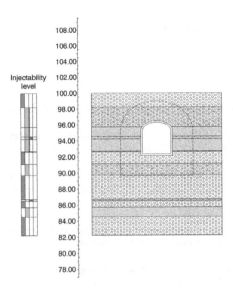

Figure 11.   Site B: soil stratigraphy and relative injectability level.

The soil layers above the crown tunnel and at the bottom of the treatment are characterized by sandy gravelly soil with a percentage of silt and clay of about 20%. Gravelly layer has been detected only at the tunnel invert.

In general, as regard to the injectability level, the soil interested by the treatment (Figure 11) can be classified as "injectable" (yellow marker), often needing a particular care in the grouting management where the fine fraction is higher (layer marked also with a thin red line).

Therefore, a higher quantity of silica grout has been predicted in order to permeate the sand and the finer fraction and to obtain a homogeneous soil consolidation. In detail, a ratio $V_s/V_c = 1$ has been prescribed.

During the 2nd stage, the injection through several sleeves reached the limit pressure; a lower volume of cement grout has been absorbed by the layers with a rather remarkable fine fraction (fine sand and silt). The silica grout instead permeated regularly the soil, and at the end of the 3rd stage additional quantities were still injected.

The volume of grouts injected on this site has been 28% of the volume of soil treated. The ratio between silica and cement grout volume has been $V_s/V_c = 1.2$. This value, as described in the previous paragraphs, fully accords to the granulometric composition detected by the investigation.

6   CONCLUSIONS

The permeation grouting treatments necessary for the construction of the new Metro Line 4 in the city center of Milan require to operate from several sites, in correspondence of the stations and the service shafts. The soil injectability conditions, which are function of the granulometric composition of the same, are rather variable from yard to yard.

The preliminary phases of investigations were followed by the design of the treatment, the set-up of the grouts as well as the working procedures that all together have defined in details the activities to be carried out. The works are now ongoing. The injections are managed on each site by mean of the daily analysis of the grouting data (pressures and quantities) and the simultaneously control of the monitoring system of the existing buildings. The processed diagrams allow to verify the evolution of the grouting in progress and to give an overall vision of the soil treatment outcome. Good correlations are generally obtained between the recorded grouting parameters and the granulometric soil characteristics.

Therefore, all the grouting activities are carried out in safety conditions, in compliance with the general timetable of the works.

REFERENCES

American Petroleum Institute 2003. *Recommended practice for field test water-based drilling fluids.* ANSI/ API 13B-1/ISO 10414-1.
Balossi Restelli, A., D' Alò, G. & Pettinaroli, A. 2002 Due differenti metodologie di avanzamento nell'ambito di un lotto della linea 3 della Metropolitana Milanese. *Atti del XXI Convegno di Geotecnica,* L'Aquila, 437–446
De Paoli, B., Bosco, B., Granata, R. & Bruce, D.A. 1992. Fundamental Observation on Cement Based Grouts: Traditional Materials. *Grouting, Soil Improvement and Geosynthetics Proceedings, GT Div/ ASCE, 25-28 February 1992.* New Orleans, 474–485
Fraccaroli, D., Grosina, S., Balossi Restelli, A., De Sanctis, F. & Galvanin, P. 2013. Linea Metropolitana M5, Milano. Esperienze di scavo e di monitoraggio strutturale in prossimità di edifici a torre. *Congresso Società Italiana Gallerie: Gallerie e spazio sotterraneo nello sviluppo dell'Europa, 17-19 ottobre 2013.* Bologna, 424–435.
Mongilardi, E. & Tornaghi R. 1986. Construction of large underground openings and use of grouts. *International Conference on Deep Foundations, 1-5 September 1986.* Beijing.
Tornaghi, R. 1978. Iniezioni. *Atti del seminario su consolidamento di terreni e rocce in posto nell'ingegneria civile, 26-27 May 1978.* Stresa.
Tornaghi, R., Bosco, B. & De Paoli, B. 1988. Application of recently developed grouting procedures for tunnelling in the Milan urban area. *Fifth International Symposium Tunnelling 88, 18-21 April 1988.* London.

Tunnels and Underground Cities: Engineering and Innovation meet Archaeology,
Architecture and Art, Volume 4: Ground improvement in
underground constructions – Peila, Viggiani & Celestino (Eds)
© 2020 Taylor & Francis Group, London, ISBN 978-0-367-46868-2

Design and construction of composite cut-off system at Grand Ethiopian Renaissance Dam, Ethiopia

G. Pietrangeli, G. Pittalis, M. Rinaldi & R. Cifra
Studio Ing. G. Pietrangeli s.r.l., Italy

ABSTRACT: The Grand Ethiopian Renaissance Dam (GERd) Project is located along the Blue Nile River in Ethiopia. This paper is focused on the main design and construction aspects of the composite cut-off system of the Saddle Dam (60 m high and 5 km long, 15 Mm^3 of embankment volume), executed all along the central portion of the dam founded on residual soils (about 4 km).

Composite cut-off, constituted by pressure grouting and plastic diaphragm panels, was conceived to address two different requirements: permeability and erosion control.

This paper covers the investigations carried out to assess the erodibility of foundation material and the design elements conceived to counteract the unlikely sequence of events that, if occurring in progression, would have the potential of evolving into a progressive erosion process. Moreover, the criteria adopted to define the extension of diaphragm panels and the continuous refinement process of diaphragm geometry following the construction progress and the acquisition of additional geotechnical information are discussed in details. Finally, the assessment of composite cut-off system is provided based on the data acquired during the implementation of the works.

1 GEOLOGICAL SETTING

The Saddle Dam of GERd Project is located in a contact zone where the meta-sediment (low metamorphic) collided with the basement rocks (high metamorphic). The compression stress is given by an orogenic cycle known as Mozambique Belt. The boundaries between the gneissic and volcano-sedimentary sequences are typically of tectonic origin. The geomorphology indicates that between the two dam shoulders, constituted of schist on the left side and metagranite on the right, the foundation material derives from the highly decomposed base rock.

At site scale the geological setting of the Saddle dam foundations is composed of five main geological units characterized by metamorphism at different grades of both volcanic, igneous and sedimentary rocks. The identified units are stacked by a pre-cambrian compression regime also witnessed by the presence of folds, faults and boudinage structures.

The Figure 1 outlines a schematic geological plan reporting the main geological formations. In particular, the Saddle dam is founded on rock (weathered, suffix "w" and fresh) in left and right abutments, and on residual soils (i.e. decomposed rock, suffix "d") of variable depth in the central part.

From left to right bank the following five main geological units are present:

- Schist (SCH)
- Femic (FEM)
  - Meta-gabbro (mG)
  - Meta-basalt (mBa)
- Phyllite (PH)
- Marble (M)
- Metagranite (GR)

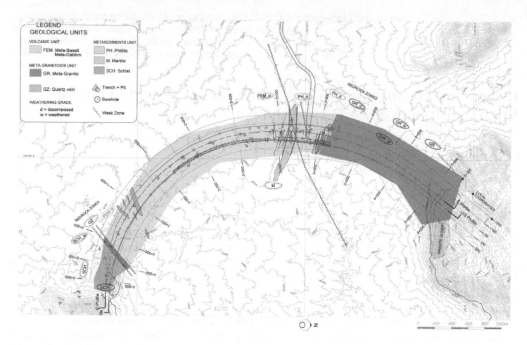

Figure 1.   Saddle Dam. Schematic Geological Plan.

Minor quarzitic veins are found in the left and central portion of the foundation. Hence, the dam is founded on residual soil for a length of 3.6 km and on rock for the remaining portion of 1.4 km.

The boundaries between the above-mentioned geological units are often characterized by tectonic contacts giving rise to the following two weak zones (wz):

– wz_0+400 (Qz);
– wz 3+200 (M/PH_d) to wz 3+450 (deep PH_d).

These zones are continuous in the upstream to downstream direction crossing the entire dam foundation footprint and are characterized by high permeability and by the presence of potentially erodible material. The wz 0+400 is characterized by adjacent slivers of very strong quartz, marble and decomposed metagranite with a metric band of tremolite. In the central-right portion of the Saddle dam foundation, the area between wz 3+200 (M/PH_d) and wz 3+450 (deep PH_d) is characterized by sharp lithological contacts between phyllite and marble units and a more than 30 meters deep weathering profile where decomposed to highly weathered phyllite is found.

The excavation of the inspection gallery trench along the upstream toe outlined a complex geological setting made of sub-vertical slivers of material at different grades of weathering. A detailed geological mapping has been carried out along the plinth and the inspection gallery footprints following the progress of the excavation and concreting activities. At some place the excavation of the diaphragm panels in the upstream area has been inspected and mapped.

2   SCHEMATIC LAYOUT OF THE SADDLE DAM

The Saddle dam of the Grand Ethiopian Renaissance Project is a rockfill dam provided with upstream concrete facing (CFrd). Main features are resumed hereinafter:

– crest length   [m]   4865
– maximum height (H)   [m]   ~ 65
– U/S slope   [h/v]   1.3:1

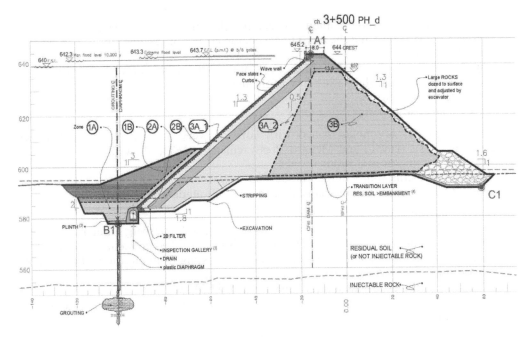

Figure 2. Saddle dam typical section on residual soil.

– D/S slope	[h/v]	1.3:1
– crest width	[m]	8

At Full Supply Level the overall reservoir volume of the Project, is about 80'000 Mm^3, 65'000 Mm^3 of which are stored above the minimum foundation level of the Saddle Dam.

As described in the previous paragraph, the dam is founded on rock in correspondence of the abutments and in residual soil in the central portion.

According to the local conditions of foundation, No. 2 main typical sections have been designed:

1. for the portion founded on rock, approximately 1'400 m long, the dam is provided with an upstream plinth and the seepage barrier is constituted by grout injections in correspondence of the U/S toe plinth.
2. for the portion founded on residual soil, approximately 3'600 m long, the dam is provided with an upstream inspection gallery and plinth, and the seepage barrier is constituted by composite cut-off (i.e. the combination of plastic diaphragm and grout injections)

No. 5 transversal galleries are provided to access from downstream the inspection gallery.

3 MAIN INVESTIGATIONS

3.1 *General*

An extensive investigation campaign has been carried out in order to characterized the foundation material of the Saddle Dam and ensure a safe design.

The different methods used during the first phase of the investigation campaign (carried out before the detailed design) are listed hereafter, grouped for different levels of detail:

– MACRO scale: approximately 12 km of geophysical lines processed with the tomographic approach allowed to identify the thickness of low velocity layers and the location of the main weak zones;

- MEDIUM scale: excavation of about 15 deep trenches (up to 17 m of depth) allowed to:
  - observe the structure and the mechanical characteristics of residual soils;
  - sample soil specimens;
  - estimate the minimum shear resistance of soil by back analysis of walls stability;
  - measure the residual shear resistance of the materials;
  - estimate the soil stiffness;
  - execute large scale permeability tests.

- SMALL scale: 45 boreholes drilled with core recovery and water tested; Marchetti Dilatometer tests (DMT) and Menard Pressuremeter (PMT), allowed to:
  - measure the thickness of residual soils;
  - evaluate the permeability of the foundation;
  - evaluate the stiffness and mechanical properties of foundation materials.

- MICRO scale: laboratory tests allowed to:
  - execute identification tests on the materials (grading and Atterberg Limits);
  - evaluate resistance and stiffness of residual soils in controlled boundary conditions;
  - evaluate variation in soil stiffness and resistance passing from unsaturated to saturated condition.
  - assess the swelling/shrinkage behaviour of the residual soils;
  - assess the behaviour of the residual soils with respect to still and flowing water (dispersivity and erodibility);

Moreover, before the excavation of the plastic diaphragm in the central part of the foundation where residual formations were encountered, borehole drilling for primary holes of curtain grouting were carried out with continuous core recovery every 12 m. This investigation provided detailed geotechnical information on the subsurface conditions, allowing to accurately define the characteristics of the foundation and the geometry of the diaphragm.

3.2 *HET*

In order to assess the risk connected to potential erodibility of the foundation material and design relevant countermeasures, Hole Erosion Tests (HET) were carried out on undisturbed samples collected at various depth from the excavated trenches.

Hole erosion test is a method for evaluating the erodibility of cohesive soils basing on the effects of water flowing through a hole predrilled in the specimen.

It is therefore conceived to investigate the erodibility of a cohesive soil subjected to concentrated leaks. Application to highly weathered rock formations, with large coarse grain fractions and residual structure is "stretched" and understandably problematic.

The HET is conducted in the laboratory using an undisturbed tube sample or a soil specimen compacted into a Standard Proctor mild. A 6-mm diameter hole is predrilled through the centreline axis, and the specimen is installed into the test apparatus in which water flows through the hole under a constant hydraulic head that is increased incrementally until progressive erosion is produced. Once erosion is observed, the test is continued at a constant hydraulic head for up to 45 minutes, or as long as flow can be maintained.

HET have been carried out on No. 21 undisturbed samples prepared by resizing blocks of residual soils sampled from the trenches excavated in the saddle dam foundation area (from 4.5 m to 11 m of depth).

Results of HET were quite scattered due to the anisotropy (orientation of the foliation of the parental rock with respect to the drilled hole) and inhomogeneity (weathering degree, local plane of weakness) of the residual soil at the scale of the sample. Therefore, the definition of critical gradient value for the different formations of the residual soil resulted problematic and scarcely significant.

Consequently, in order to protect the foundation from the potential risk of progressive erosion, it was adopted a "multiple lines of defence" approach, tailored for the different stretches of foundation depending on local geotechnical/geometrical/hydraulic characteristics and boundary conditions.

4 COMPOSITE CUT-OFF CONCEPT

Composite cut-off is constituted by pressure grouting and plastic concrete diaphragm panels. It has been executed all along the central portion of the dam founded on residual soil while on the left and right banks, characterized by the presence of schist and metagranite respectively, only pressure grouting has been carried out.

This system was conceived to address two different requirements:

– permeability correction;
– erosion control.

This last requirement was generally met by deepening the diaphragm down to a level of non-erodible rock or, in case of continuous at depth potential erodible material, down to a level where the corresponding seepage gradient at the U/S toe results lower than the critical gradient.

Construction of the diaphragm wall was preceded by the execution of the curtain grouting which allowed to acquire, together with the examination of U/S trench excavation slopes, detailed geotechnical information on the foundation materials. The bottom limit of the diaphragm (named as Objective Level or OL in the prosecution of this paper) was therefore defined in subsequent steps:

– tentative OL
  based on the analysis of primary grouting holes 12 m spaced and drilled with core recovery

– refined OL
  review of tentatively OL based on:

  – examination of U/S trench excavation slopes
  – analysis of grout takes and check holes results

– actual OL
  as-built of diaphragm wall.

In general, the excavation was carried out with a clamshell up to refusal conditions and then completed with the hydro-mill in order to socket the diaphragm into competent rock (the hydromel required a minimum pre-excavation depth of about 4 m). Positive or negative deviations between tentative and actual OL were observed along the foundation. Special cases, requiring specific evaluation, were constituted mainly by the following two categories:

– clamshell refusal above 4 m (with OL > 4 m). In such condition it was not possible to use the hydro-mill, being the pre-excavation trench too shallow;
– sharp lateral variation of panel depth (reflecting the geological structure of the foundation) which could result in a non-complete removal of portions of weak rock (overshadowed by sound rock).

These cases were duly recorded and analysed also by means of visual inspection of excavated panels, surface mapping of the trenches, and verification of hydro-mill digging parameters (recoded in continuous by the hydromill apparatus). In some cases, additional excavations were carried out by means of staggered panels or using heavy chisels. In other cases, special additional grouting injections were executed.

It is highlighted that, in addition to the composite cut-off system (diaphragm and pressure grouting), several active and passive design elements have been conceived to counteract the unlikely sequence of events that, if occurring in progression, would have the potential of evolving into a progressive erosion process:

– DEEP U/S TRENCH

A deep trench (from 10 to 20 m) has been excavated in the U/S toe of the dam, where higher hydraulic gradients are expected, in order to found the plinth and the gallery on a material with better geotechnical characteristics and therefore less susceptible to erosion.

– PLINTH and GALLERY (with drains)

Plinth and gallery, with an overall width of about 10 m, reduce themselves the gradient to values that are generally accepted for intensely weathered and fractured rocks. The gallery equipped with drains and piezometers will allow to monitor the efficiency of composite cut-off and, if necessary, execute additional corrective works.

– FILTER in the U/S trench

A filter has been placed above the portion of residual soil foundation between the inspection gallery and the top of the upstream trench which would counteract continuation of erosion process.

– TRANSITION layer

A transition layer has been placed in the remaining portion of the dam footprint on residual soil.

– U/S BACKFILL

The upstream fine grained backfill, extended to about one third of the local height of the dam, will tend to clog potential pervious water conduits (flow-limiting and crack-filling actions).

5   ASSESSMENT OF COMPOSITE CUT-OFF

5.1   *Foundation permeability before composite Cut-Off*

Permeability of Saddle dam foundation have been investigated at great detail by means of investigation holes generally spaced 36 m, drilled with core recovery along the cut-off axis and water tested (before the execution of grouting) with standard Lugeon methodology following the progress of drilling works (downward, with stage lengths of 5 m, 5 steps of ascending and descending pressure, each lasting 10 minutes, and a maximum pressure of 10 bars).

More than No. 120 investigation holes have been executed for an overall length of about 5'500 m. The graph in Figure 3 illustrates the distribution of Lugeon values below the invert level of gallery/plinth for the No. 4 main geological units (a fictitious value of 100 UL was assigned to the stages where it has been observed bypass, or water-absorption exceeding the pump capacity).

Main observations are the following:

– general distribution of Lugeon values with depth is similar in the different foundation stretches;
– foundation is impervious at large: 68% of the values is less than 5 UL;
– higher permeability is generally limited to the upper portion of foundation (i.e. from 0 to 20 m below plinth level). An exception is represented by the foundation stretch between ch. 3+050 and 3+500 characterized by the presence of deep-decomposed phyllite and marble.

No. 58 wash-out stages over an overall number of No. 1070 tested stages have been recorded, which corresponds to about 5%. About 70% of these stages are concentrated in the first 15 m below the gallery/plinth invert level. These stages are distributed almost evenly among the different foundation stretches. Anomalous behaviours associated to high Lugeon Units and low grout take are found in about 10% of the tested stages.

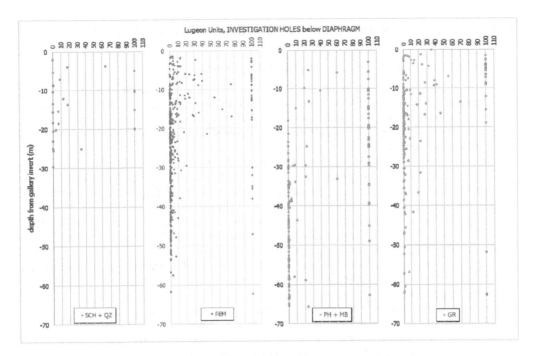

Figure 3.   Investigation Holes, Lugeon Units VS depth below gallery/plinth invert.

5.2   Grouting

About 3'300 grouting holes, for an overall length of about 100 km have been executed along the Saddle dam foundation profile. The overall net quantity of cement injected in the sums to about 1'500 tons. The graph in Figure 4 reports the spatial distribution of net take of each grout hole series from primary (red area on the background) to quaternary (grey area on the foreground).

The main observations are the following:

– globally, the net average grout take reduces from 23 kg/m, recorded in the primary holes to less than 6 kg/m in quaternary holes;
– higher grout takes are observed in correspondence of most pervious areas (Quartz lineament at ch. 340; marble area and contact between PH/FEM and PH/GR between ch. 3+000 and 3+300);
– along the above areas the average grout take is considerably higher, reaching more than 100 kg/m;
– reduction of take from primary to quaternary hole series is generally observed along the entire foundation profile.

5.3   Diaphragm

The as-built profile of diaphragm panels was included in detailed analytic profiles reporting all the main information related to the composite cut-off system:

– Position of investigation holes (the ones water tested are highlighted with a blue hatch);
– Position of check holes executed after grouting;
– Tentative Objective Level;
– As-built diaphragm panels;
– Location of additional treatment areas and relevant check holes;
– Schematic geological profile, as surveyed on the foundation level of plinth and gallery;
– Excavation methodology of diaphragm panels (hydromill or clamshell);

- Difference between tentative Objective Level (OL) and Diaphragm bottom level (DL);
- Section type.

As far as the diaphragm panels is concerned it is observed that:

- Diaphragm panels extend from ch. 0+200 to 3+650, with a max. depth of 30 m.
- The excavation of the panels has been completed always with the hydromill where the pre-excavation executed with the clamshell exceeded 3–4 m of depth (which is the minimum required to operate the hydromill). Few spot-points where this criterion was not respected, because of local conditions, were recorded.
- Hydromill excavation has been generally continued until a digging load higher than 200–250 kN (maintained for 10–15 minutes) or a drilling speed lower than 2.5 cm/min (for 10–15 minutes) have been reached. Excavation parameters have been carefully recorded continuously and in real time by the digital control apparatus of the hydromill.
- As built diaphragm resulted shallower than tentative OL along about 550 m over 3'450 m (i.e. 15%). The average deviation along this 550 m long stretches is just 3 m with a maximum of 30 m in correspondence of Quartz sliver at ch. 0+345.
- Specific inspections have been always carried out where diaphragm panels resulted shallower than the tentative Objective Level or if the excavation has been executed entirely with the clamshell.
- Sharp lateral variation of the weathering grade of the rock foundation are often reflected by the difference in depth of the adjacent diaphragm panels.
- Comparing the diaphragm profile with the results of water tests executed on along No. 90 check holes carried out after grouting it is observed that about 90% of the water-tested stages below the bottom of the diaphragm have less than 5 UL

5.4 Foundation permeability after composite Cut-Off

Permeability after grouting works have been investigated by means of check holes generally spaced 36 m, drilled and tested with the same methodology of the Investigation Holes.

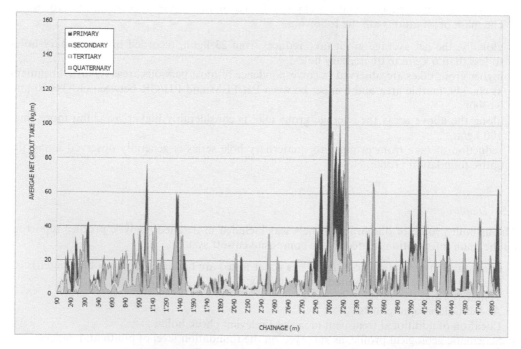

Figure 4.   Distribution of average NET grout take (Primary to Quaternary holes).

More than No. 120 check holes have been executed for an overall length of about 4'200 m.

Figure 5 illustrate the distribution of Lugeon values (below the bottom of the diaphragm) relevant to main geological units. Comparing the water test results before and after the execution of cut-off works it is observed that:

- the general distribution of Lugeon units shows reduced and less dispersed permeability values for the four different foundation stretches;
- the percentage of stages with permeability less than 5 UL results increased from 68% to 87%;
- higher permeability is generally limited to the upper portion of foundation (i.e. from 0 to 20 m below plinth level). Anomalous deep high-absorption stages are almost absent after cut-off works;
- the percentage of stages with wash-out behaviour is almost the same before and after cut-off works (i.e. about 5%) and are distributed almost evenly among the different foundation stretches;
- after cut-off works it is observed that the Lugeon values associated to wash-out behaviour are generally lower and that the shallower and most critical wash-out stages are almost disappeared.
- the anomalous stages with high Lugeon Units and low grout take are reduced from about 10 %, in the Investigation holes, to 5% in the Check holes.

5.5 *Additional Treatments*

Additional treatments have been prescribed in the portions of foundation below the plinth/diaphragm where it was observed:

- High Lugeon Units in the check holes;
- High Lugeon and low grout-take in the Investigation holes and in the adjacent grout holes;
- Diaphragm level shallower than tentative Objective Level;

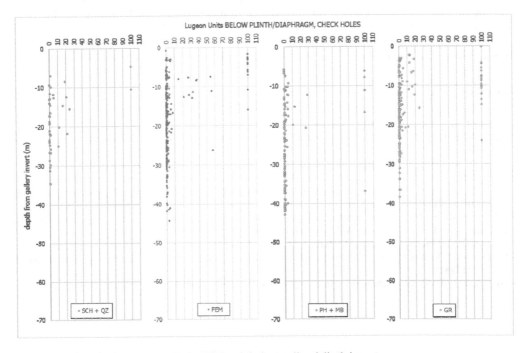

Figure 5.   Check Holes, Lugeon Units VS depth below gallery/plinth invert.

- Extremely shallow diaphragm panels (excavated with clamshell only);
- Sharp lateral variation of diaphragm panels depth.

The additional grouting treatments have been generally carried out using the local available cement milled with the equipment available at site with the aim to improve the efficacy of grouting treatment by increasing of fineness of cement and reducing the size of larger particles diameter (that mainly control the penetration of the grout).

The overall length of the foundation stretches interested by additional grouting treatments summed to more than 1'500 m.

Moreover, in correspondence of the two main weak zones:

- wz_0+400 (Qz);
- wz 3+200 (M/PH_d) ... wz 3+450 (deep PH_d).

continuous in U/S - D/S direction and characterized by high permeability and by the presence of potentially erodible material, the composite cut-off system has been integrated with a PVC blanket extending upstream for about 150 m in order to reduce the resulting D/S gradients to acceptable values.

6 CONCLUSIONS

Main results of composite cut-off works in terms of permeability correction are summarized in Figure 3 and 5. After composite cut-off works it is observed that:

- the Lugeon Values are generally reduced and less dispersed in all the foundation stretches;
- the percentage of stages with permeability less than 5 UL increased from 68 % to 87 %;
- the percentage of stages with wash-out behaviour is limited to about 5 % of the entire dataset. Shallower and most critical wash-out stages are almost disappeared.

Additional grouting treatments have been prescribed following the criteria reported in chapter 5.5 and included more than 1'500 m of foundation stretches.

In correspondence of the two most critical weak zones a PVC blanket extending about 150 m in the upstream direction have been introduced in order to limit the hydraulic gradients.

The analysis of composite cut-off system, together with the other defensive design elements (i.e. deep U/S trench, plinth and gallery, filter and layer, U/S backfill, horizontal blanket) allows to meet the Design's requirements of permeability correction and erosion control.

REFERENCES

Blight, G.E. & Leong, E.C. (2012). Mechanics of residual soils. Rotterdam: Balkema.
Bonelli, S. (2013). Erosion in geomechanics applied to dams and levees. Hoboken: John Wiley & Sons, Inc.
Cruz, P.T., Materón, B. & Freitas, M. (2009). Concrete face rockfill dams. Rotterdam: Balkema.
Fell, R. & Fry, J.J. (2007). Internal Erosion of Dams and their foundations. Rotterdam: Balkema.
Fell, R., MacGregor, P., Stapledon, D., Bell, G. & Foster, M. (2015). Geotechnical Engineering of Dams 2nd Edition. Rotterdam: Balkema.
Houlsby, A.C. (1990). Construction and design of cement grouting: a guide to grouting in rock foundations. Hoboken: John Wiley & Sons, Inc.
ICOLD (2018). Cut-offs for dams. Bulletin 150. Paris: International Commission on Large Dams.
ICOLD (1985). Filling materials for watertight cut-off walls. Bulletin 51. Paris: International Commission on Large Dams.
Wesley, L.D. (2010). Geotechnical Engineering in residual soils. Hoboken: John Wiley & Sons, Inc.

Tunnels and Underground Cities: Engineering and Innovation meet Archaeology, Architecture and Art, Volume 4: Ground improvement in underground constructions – Peila, Viggiani & Celestino (Eds)
© 2020 Taylor & Francis Group, London, ISBN 978-0-367-46868-2

Design of artificial ground freezing for an access tunnel of a railway station in Switzerland

E. Pimentel & G. Anagnostou
Institute for Geotechnical Engineering (IGT), ETH Zurich, Switzerland

ABSTRACT: The paper presents the static and thermal design of an artificial ground freezing (AGF) measure with brine for an access tunnel of a railway station in central Switzerland. The geology of the first 45 m of the access tunnel consist of sediments (silty sand to sandy silt) followed by sandstone. The tunnel crosses under the railway lines and train platforms of one of the train stations with the highest traffic. At any time during construction of the tunnel traffic of trains cannot stopped. Due to the sediments aforementioned and since the water table lies slightly above the crown of the tunnel, AGF is the only suitable auxiliary measure for the tunnel. For achieving an adequate frozen body above the crown water table must be raise up by watering the ground locally with drains. The paper presents also the dimensioning and optimization of these drains. The strength parameters were estimated from literature and verified with uniaxial tests with frozen soil samples under constant temperature. Due to project requirements the core of the tunnel is also frozen. Implementation of the AGF measures starts from a shaft. Due to spatial conditions given by the location of reinforcement of the diaphragm wall of the shaft and of a reinforcement beam the distribution of the freeze pipes was demanding. The goals of the thermal design were the generation of a frozen body with uniform thickness around the tunnel and to avoid generation of excess pore pressure in the core of the tunnel during freezing.

1 INTRODUCTION AND SITE CONDITIONS

In order to increase the capacity of a train station with the highest traffic in Switzerland two caverns will be constructed beneath the existing station. The access tunnel to the west side of these caverns is located under the railway lines and train platforms and will be excavated conventionally from a shaft located outside the area of rail traffic (Figure 1).

The ground in the area of the shaft consists of sedimentary deposits. In the excavation direction (towards the caverns) a moraine layer followed by sandstones appears with an inclination of about 15°–20°. The cross section of the access tunnel is located completely in the sedimentary layers in the first 26 m (starting from the shaft) and completely in sandstone after approximately 40 m (Figure 2).

The sedimentary layers are heterogeneously distributed. According to the borehole profiles of exploratory drillings close to the shaft, silty sand and sandy silt layers are expected. In the excavation direction, the layers become coarser, changing to coarse sand and gravely sand. The ground is water bearing.

As the available geological-geotechnical data were incomplete in the design stage of the project, design was based upon working hypotheses, which were afterwards verified by field and laboratory investigations.

The design presented in this paper is the basis for the tender. Construction of this part of the project will start in spring 2019.

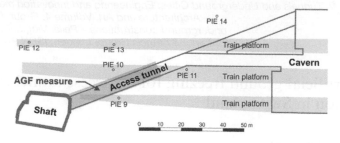

Figure 1. Top view of the access tunnel.

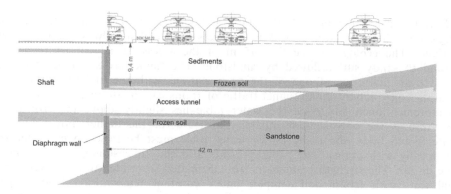

Figure 2. Longitudinal profile of the access tunnel.

2 SPECIAL PROJECT REQUIREMENTS

2.1 General aspects

Train traffic and station operation must not be interrupted at any time during construction of the access tunnel; therefore, excavation methods like cut and cover or top-down construction were discarded. Due to the presence of water up to the crown level of the access tunnel jet grouting can also not be applied. Grouting or drainage were also excluded due to environmental aspects. The quality of the groundwater must be preserved and a draw-down of the water table was also not an option. Therefore, artificial ground freezing (AGF) as auxiliary measure was selected in order to impermeabilize, stiffen and strengthen the ground around the tunnel and thus mitigate the potential hazards of water inrush or inadmissible settlements of the railway lines.

2.2 Water table

Due to the uncertainty concerning the elevation of the water table and a potentially significant seepage flow, piezometers were installed close to the project area (PIE9 to PIE14 in Figure 1). According to the measurements, the water table varies seasonally and is in average about 0.6 m above the crown of the tunnel. For the application of AGF, it is mandatory to have a degree of saturation of minimum 50–70% (Andersland & Ladanyi, 2004). Thus, the water table should be raised up artificially during the freezing stage (Section 4). According to the measurements, there is no evidence of seepage flow in the area of the planned access tunnel.

2.3 Tunnel face stability

Ground reinforcement by bolts was not considered adequate for stabilizing the face under the expected conditions (sandy silt to silty sand). Thus, great part of the face should also be frozen for ensuring its stability.

2.4 Location of the freeze pipes

In the AGF method, pipes are installed in the ground, and a liquid refrigerant circulates through the pipes and extracts heat from the surrounding soil (Figure 3). After a certain time period, a frozen soil body develops. The freeze tubes used in the AGF method consist of two concentric pipes (Harris, 1995). The outer pipe is closed at the freeze end; the inner one is open.

Implementation of the AGF measures starts from a shaft. The shaft is supported by a reinforced diaphragm wall and two horizontal beams. The lower beam is located beneath the crown of the access tunnel and the upper one approximately at the highest point of the planned frozen body (Figure 4). The horizontal beams cannot be perforated for installing freeze pipes. The reinforcement of the diaphragm wall was distributed in such a way that its perforation is only possible along specified vertical lines (turquoise dot-dashed lines in Figure 4). Thus the boreholes for the freeze pipes can be located only along these vertical lines and, since the boreholes must be drilled using a preventer in order to avoid drainage of the ground at the crown area of the tunnel, not closer than 30 cm to the horizontal beams.

3 DESIGN OF THE AGF MEASURE

3.1 Static design

The following hazard scenarios were considered for the static design of the frozen body:

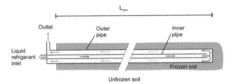

Figure 3.   Freeze pipe.

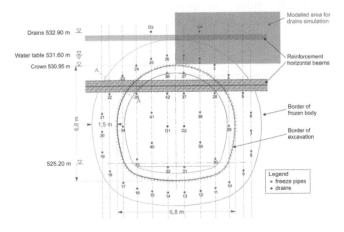

Figure 4.   Location of the freeze pipes.

1481

- large deformations due to creep of the frozen body;
- overstressing of the frozen body due to high shear stresses; and
- overstressing of the frozen body due to high normal stresses and bending moments.

The first hazard scenario can be excluded if the stresses in the frozen body are low and the loading period (*i.e.* the time between excavation and development of the temporary support resistance) short. In the present case the advance per round is one meter and temporary support consists of steel arches and 40 cm reinforced shotcrete that is applied immediately after each excavation round.

The other two hazard scenarios were evaluated by means of numerical stress analyses of a plain strain model, considering the frozen body as an elastic material and checking the elastically determined stresses. The strength of frozen soil depends strongly on temperature and is very low for temperatures higher than -2 °C. Therefore, only the region of the frozen body with temperature lower than -2 °C was considered as bearing in the numerical analyses (the thickness of the frozen body was taken equal to 1.25 m instead of the actual 1.50 m). For the sake of simplicity, a homogeneous frozen body was considered, with parameters corresponding to the mean temperature of -10 °C. Based on experience of projects in similar ground, conservative parameters for the frozen body were selected (uniaxial compressive strength of 2.3 MPa and shear strength of 1 MPa). The results of the stress analysis indicate that:

- the normal stresses are lower than 50% of the uniaxial compressive strength (global safety factor ≥ 2);
- tensile stresses do not develop; and
- shear stresses are lower than 16% of the shear strength (global safety factor ≥ 6).

The assumed strength parameters were verified later by means of laboratory tests on remoulded samples obtained in the area close to the shaft. The samples were compacted to the same density as in-situ, saturated and frozen to temperatures of approximately -10.5 °C. The results of the first two samples show uniaxial strength higher than 5.3 MPa (Figure 5). Although these values are twice as high as the values that were adopted in the stress analysis, the thickness of the frozen body was not reduced.

3.2 Thermal design

3.2.1 AGF method and goals
Taking the costs into consideration and the absence of relevant seepage flow, AGF with brine instead of liquid nitrogen was chosen. Thermal design was based upon typical parameters for this method (freeze pipe temperature of -30 °C and freeze pipe diameter of 4 inches). The thermal design aims to determine: the number and distribution of the freeze pipes; the required refrigeration time for the growing stage; and the necessary cooling power. In AGF

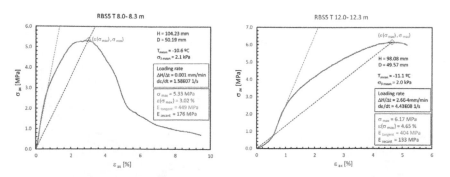

Figure 5. Results of uniaxial tests on frozen samples.

the growing stage denotes the refrigeration stage where the frozen body increases its dimensions continuously. After having achieved the planned frozen body size, excavation of the tunnel starts and refrigeration of the pipes outside of the tunnel continues but with reduced cooling power just in order to maintain the thickness of the frozen body (holding phase). The goals of the AGF measure are, as mentioned above, achieving a thickness of the frozen body around the tunnel of at least 1.5 m and freezing a great part of the face within a reasonable refrigeration time. An additional goal of the thermal design is to find a freeze pipe distribution that allows dissipation of excess pore pressures especially during freezing of the face area. Specifically, due to the volume expansion of frozen pore water, excess pore pressures develop, which reduce the resistance of the ground to shearing. Therefore, the design must allow dissipation of the excess pore pressures. For this purpose two drains are installed in the centre of the access tunnel. In order to avoid drainage of the tunnel area to be frozen these two drains are connected to vertical tubes with a spillover at the level of the drains for watering (Section 4). Prior excavation the vertical tubes are removed so that drainage is allowed. Thermal design was carried-out using the 3D numerical code FREEZE (Sres, 2009), which was developed at the ETH Zurich for simulating coupled thermohydraulic processes and verified against the results of large scale model tests (Pimentel *et al.* 2007). As mentioned above there is no evidence of seepage flow in the area of the project and thus only thermal simulations were performed.

3.2.2 *Thermal parameters*
In order to estimate more realistically the thermal parameters, the mineralogical composition of the remoulded samples (Section 3.1) was determined with X-ray diffraction. The average values of the mineral contents from the three samples are listed in Table 1. The samples consist mostly of quartz and calcite, which are minerals with high thermal conductivity and thus advantageous for the application of AGF.

The unfrozen water content for temperatures below the freezing point ($T_0 = 0$ °C) was determined after Anderson & Tice (1972) and the corresponding parameters were taken from literature for a silty sand to sandy silt ($\alpha = 0.016$ und $\beta = -0.608$; Frivik, 1981) (Figure 6 left). The soil to be frozen is considered as saturated. The heat capacity of the soil is determined as the arithmetic mean of the heat capacity of its constituents, weighted by their fraction. Analogously the thermal conductivity of the soil is calculated as the geometric mean of the thermal conductivity of its constituents weighted by their fraction. The parameters for the mineral fraction are evaluated in the same way, *i.e.* as arithmetic or geometric mean of the different mineral constituents weighted by their fraction. The temperature dependency of the thermal parameters of the mineral fraction is neglected, what is not the case for water and ice. In Figure 6 the heat capacity and thermal conductivity of the soil over the temperature are plotted.

3.2.3 *Required refrigeration time for the growing stage*
The optimum distribution of the freeze pipes was determined iteratively by means of successive simulations. The optimization goal was to find the minimum number of freeze pipes which allow: (a) achieving an approximately uniform frozen body with a thickness of 1.5 m; (b) freezing of the ground inside the tunnel cross-section; and, (c) rapid dissipation of the excess pore pressures that develop due to volume expansion of frozen pore water during the freezing stage. These simulations were performed taking additionally into consideration the required cooling power (Section 3.2.5).

Figure 7 shows the temperature field after 9.33, 11, 13, 15, 17 and 19 days of refrigeration for the optimized freeze pipe distribution. The isotherms for $T = 0$ °C (border of the frozen

Table 1. Mineralogical composition.

Mineral	quartz	calcite	plagioclase	K-feldspar	muscovite	dolomite	chlorite
Content [%]	42.0	33.27	9.83	4.40	3.57	3.20	2.73

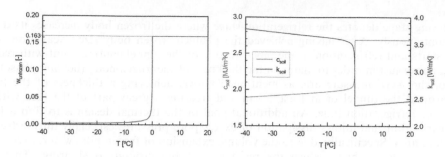

Figure 6.   Thermal parameters of the soil (l.h.s.: unfrozen water content; r.h.s.: specific heat capacity and thermal conductivity).

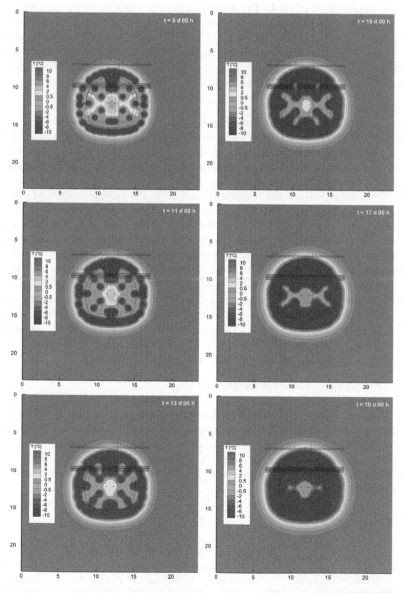

Figure 7.   Simulation results after 9.33, 11, 13, 15, 17 and 19 days refrigeration.

body), -2 °C and -10 °C are marked. The green and red lines represent in all diagrams of Figure 7 the border of the planned frozen body and the border of the tunnel excavation respectively. The freeze pipes and the two drains in the area to be excavated are plotted as filled green and blue circles.

Closure of the frozen body is achieved after 9.33 days refrigeration. After further 4 days, *i.e.* 13 days of refrigeration in total, the frozen body has reached the planned thickness of 1.5 m except at the area around the reinforcement beam; radially draining paths are formed and remain open during two further refrigeration days. After 17 days refrigeration, the frozen body has reached the planned thickness everywhere. Finally, after 19 days refrigeration, only a small area (about 2 x 2 m) of the tunnel cross section remains unfrozen and its temperature is not higher than 0.5 °C. Since this small area of the face is considered to remain stable due to the apparent cohesion of the ground, the growing stage can now be finished, taking 19 days in total. In order to evaluate the effective thickness of the frozen body (region with $T \leq -2$ °C; *cf.* Section 3.1) and its mean temperature, the temperature distribution along the most critical line crossing the frozen body (line A-A in Figure 4) is evaluated for different refrigeration times. The results are plotted in Figure 8, whereby the origin of the x-axis is located at the intersection of line A-A with the line connecting the adjacent freeze pipes nr. 22 and nr. 23 (Figure 4). The area marked blue represents the extension of the planned and effective frozen body. The temperature at the outer and inner border of the frozen body after 19 days refrigeration amounts to -3.2 °C and -19.4 °C, respectively, and is thus lower than -2 °C. The average temperature is equal to -14.3 °C, which is lower than the -10 °C assumed in the static design.

3.2.4 *Influence of borehole deviations on the refrigeration time*

All boreholes must be drilled by means of horizontal directional drilling. This technique limits, but does not eliminate the deviations completely. The prescribed tolerance is 20 cm at any point of the boreholes. In order to quantify the influence of possible borehole deviations on the necessary refrigeration time, the most critical freeze pipes (nr. 22 and nr. 23 in Figure 4) were moved vertically by 20 cm in the opposite direction. These pipes (as well as pipes nr. 4 and 5 due to symmetry) are critical because they are located close to the horizontal beams, where drilling restrictions (Section 2.4) lead to a relatively big spacing of 1.35 m (nominal position). It is assumed that this spacing increases to 1.76 m due to drilling inaccuracy. The numerical simulation was performed for the same time-development of the freeze pipe temperature as in Section 3.2.5). The frozen body reaches the planned size (green line in Figure 9) after 19.5 days refrigeration (Figure 9 shows the temperature field). Thus the refrigeration time needed for the growing stage increases due to borehole deviations only slightly. Analogously to the case without borehole deviations (nominal position), the temperature distribution along the line A-A (Figure 4) was examined (Figure 10). After 19.5 days refrigeration the temperature at the outer

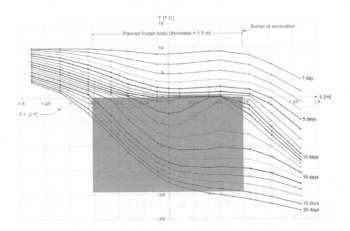

Figure 8.   Temperature distribution along line A-A (*cf.* Figure 4).

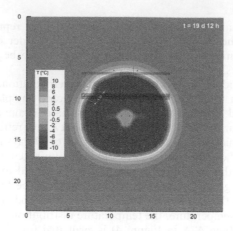

Figure 9. Simulation results after 19.5 days refrigeration time considering a possible deviation of the boreholes.

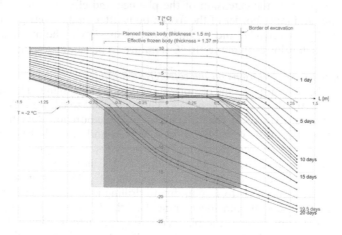

Figure 10. Temperature distribution along line A-A (*cf.* Figure 4) considering a possible deviation of the boreholes.

border of the frozen body amounts obviously 0 °C, while the temperature at the inner border reaches -18.2 °C. The effective thickness of the frozen body for this refrigeration time and its average temperature amounts to 1.37 m and -11.9 °C. The assumptions of the static design (1.25 m and -10 °C) cover thus also the case of drilling inaccuracies.

3.2.5 *Required cooling power*

Due to the big temperature gradient between freeze pipe and surrounding soil the maximum cooling power is required immediately after starting refrigeration with a constant very low temperature of the freeze pipes (*e.g.* T_{pipe} = -30 °C). With increasing refrigeration time the required cooling power for maintaining this low temperature at the freeze pipes decreases continuously. Thus, one possibility for reducing the cooling power that must be installed is to start at a moderate low freeze pipe temperature (*e.g.* T_{pipe} = -5 °C) followed by a continuous decrease until reaching the final lowest freeze pipe temperature (-30 °C in the present case). Three cases were simulated (reaching the lowest temperature after 5, 7 or 8 days refrigeration time). For each of these cases the cooling power was read out at each time step and for each freeze pipe. The total net cooling power is determined by multiplying the sum of cooling power of all freeze pipes by their average length of 40 m. Figure 11 shows the total net cooling power (l.h.s. axis) and the freeze pipe temperature (r.h.

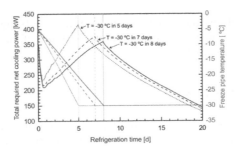

Figure 11.   Total required net cooling power for the AGF.

s. axis) over time for the three analysed cases. Cooling down in 8 days results in the lowest power requirements; the required net cooling power amounts to 350 kW in this case.

4   RAISING UP OF THE WATER TABLE ABOVE THE TUNNEL CROWN

The performed piezometric measurements show that the water table is located 0.9 m lower than the highest point of the planned frozen body. The water content in the crown area of the planned frozen body should be increased artificially by watering the ground locally with two horizontal drains. The horizontal drains should be placed about 25 cm above the highest point of the planned frozen body (Figure 4). The effectiveness of this measure was checked numerically using the commercial FE-code Plaxis. Specifically, the 2D seepage flow through the partially saturated soil in the tunnel crown region was analysed considering an isotropic and homogeneous ground obeying the van Genuchten model. The soil parameters were taken from the literature for loamy sand (van Genuchten, 1980), whereby the computations were performed for two permeability values ($k_f = 10^{-5}$ m/s and $k_f = 10^{-6}$ m/s). Due to symmetry conditions only half of the crown region of the planned frozen body was modelled (red area in Figure 4). The drains were considered as line elements with constant water input quantity q. The length of the drain element corresponds to the perimeter of the real drain with diameter $D = 102$ mm. In the numerical computations, the location of the drains and the water input quantity were varied and the watering was simulated until achieving steady state, *i.e.* until the saturation degree reaches its final value. Figure 12 shows the contour lines of saturation degree at steady state for the relatively high permeability of 10^{-5} m/s and an input quantity per meter drain of $q = 0.2$ m³/day. Steady state is reached in this case after less than 5 days watering. Consequently watering can start together with refrigeration. The results for the lower

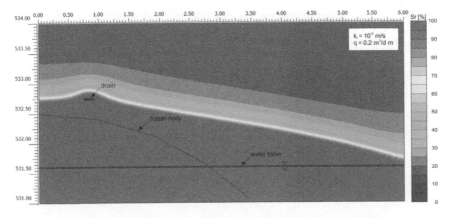

Figure 12.   Saturation degree in the crown area of the frozen body at steady state.

permeability ground are about the same, the only difference being that the input quantity should be one order of magnitude smaller (the required amount of input water increases linearly with the permeability of the ground).

5 SUMMARY AND CONCLUSIONS

Artificial ground freezing with brine as an auxiliary measure fulfils the project requirement of constructing the access tunnel under the railway lines and train platforms without causing interruptions of the train traffic and station operation. The ground is water bearing but the water table in the crown area of the planned frozen body must be raised up with two drains. The effectiveness of these drains could be proven by unsaturated seepage flow analyses. The strength and deformability parameters were estimated for the static design and successfully verified with laboratory strength tests on frozen samples. Due to face stability requirements, the major part of the ground inside the tunnel cross section must also frozen. For the chosen thermal parameters (freeze pipe temperature and diameter of T_{pipe} = -30 °C and D_{pipe} = 102 mm) the required thickness of the frozen body amounts to 1.5 m. The project conditions impose limitations to the the location of the freeze pipes. An optimized distribution of the freeze pipes could be achieved after several iterations. With this setup, consisting of 42 freeze pipes and 2 drains in the tunnel, excess pore pressure due to volume expansion of frozen pore water can be dissipated and a uniformly frozen body can be created. The refrigeration time needed for creating a frozen body according to the project requirements (thickness of 1.5 m and freezing of the tunnel cross-section) amounts to approximately 20 days even if considering deviations of the boreholes within the allowed tolerances. For an optimized operation of the cooling units the required net cooling power amounts to 350 kW.

REFERENCES

Andersland, O. & Ladanyi, B. 2004. Frozen Ground Engineering – SecondEdition. American Society of Civil Engineering, John Wiley & Sons Inc., Hoboken, New Jersey.
Anderson, D.M., & Tice, A.R. 1972. The Unfrozen Interfacial Phase in Frozen Soil Water Systems. Physical Aspects of Soil Water and Salts in Ecosystems, pp. 107–124.
Sres A. 2009. Theoretische und experimentelle Untersuchungen zur künstlichen Bodenvereisung im strömenden Grundwasser. Veröffentlichungen des Instituts für Geotechnik (IGT) an der ETH Zürich, Volume 234.
Pimentel, E., Sres, A. & Anagnostou, A. 2007. 3D-Modellierung der Frostkörperbildung beim Gefrierverfahren unter Berücksichtigung einer Grundwasserströmung. Proc. 22. Christian Veder Kolloquium, Graz, No. 30, 2007, pp.161–176.
Frivik, P. 1981. State-of-the-Art Report. Ground Freezing: Thermal Properties, Modelling of Processes and Thermal Design.
Harris, J.S. 1995. Ground freezing in practice. Thomas Telford, London.
van Genuchten, M.TH. 1980. A closed-form equation for predicting the hydraulic conductivity of unsaturated soils. Soil Sci. Soc. Am. J. doi:10.2136/sssaj1980.03615995004400050002x.

Tunnels and Underground Cities: Engineering and Innovation meet Archaeology,
Architecture and Art, Volume 4: Ground improvement in
underground constructions – Peila, Viggiani & Celestino (Eds)
© 2020 Taylor & Francis Group, London, ISBN 978-0-367-46868-2

Discussion of selection of ground improvement parameters from back analysis of monitored excavation in Singapore

G. Pittaro
Mott MacDonald Singapore & National University of Singapore, Singapore

ABSTRACT: Deep excavations in soft ground often require additional stabilization through ground improvement (GI). Some of the common methods to improve the ground are Jet Grouting Piles (JGP), Deep Soil Mixing (DSM) or Wet Speed Mixing (WSM). JGP, DSM and WSM are achieved by mixing the soil with cement and water, generating a structure that performs well under compression forces but not under tension forces. These ground improvement blocks provide larger passive resistance and thereby reduce wall displacements. For Ground Improvement properties different design codes along different regions and countries define different approaches and nowadays there is not a unique strategy when it comes to define the strength and stiffness properties of GI.

This paper focuses on the selection of appropriate ground improvement parameters during design phase. The discussion is supported by a monitored excavation of a TBM shaft and a back analysis using FEM 2D models.

1 INTRODUCTION

1.1 *Deep Excavation in Singapore Marine Clay*

Deep excavations with retaining structures in soil formations with low geotechnical parameters (strength and stiffness) often require ground improvement techniques. There are several reasons for this; decreasing forces in the retaining structure, preventing large wall displacements, reducing ground heave, etc. This is applied mostly in urban areas because it is important to avoid damage to existing structures such as surrounded building, underground utilities, adjacent roads, etc. In Singapore, one of the methods that GI techniques are used is by form of temporary props below and above the formation level of excavations.

During design stage, GI strength and stiffness parameters need to be chosen to model the behavior of the ground improvement.

1.2 *Deep Excavation in Singapore Marine Clay*

Eurocode 7 defines the characteristic (or design) value of a geotechnical parameter as a "cautious estimate of the value affecting the occurrence of the limit state". It also adds that the characteristic value can be lower values, which are lower than the most probable value or upper values, which are higher than the most probable parameter. The previous definitions are clear by itself, however it remains to the engineer the task to define the design value of a geotechnical parameter.

In order to define what parameter needs to be selected for design it is important to understand how much ground is involved to the occurrence of the limit state (Bond and Harris, 2008). This means that if the limit state is governed by the failure of the mass of the ground, the engineer should choose a parameter close to the average because the failure of full mass of the ground will lead to the limit state, on the other hand, if the failure depends on the strength of a small volume of soil, the value of the parameter to be selected should be a significantly lower than the average.

If statistical methods are used, EC7 section 2.4.5.2.11 defines the previous two scenarios described above as:

-A cautions estimate of the mean value is a selection of the mean value of the limited set of geotechnical parameter values with a confidence of 95%.

-Where local failure is concerned, a cautious estimate of the low value is a 5% fractile.

Bond (2011) has proposed a method to define the geotechnical design parameters based on statistics as:

$$Xk, \text{inf} = mx - kn.sx \tag{1}$$

Where
mx = parameter mean.
kn = statistical coefficient.
sx = standard deviation.

This definition of characteristic value also matches with the definition provided by Liu et al. (2015), where

$$Qd = Q - \beta\sigma_Q \tag{2}$$

Where
Q = parameter mean.
σ_Q = standard deviation.
β = Reliability index related to the percentile p.

The previous definition was defined for soils, but when it comes to ground improvement there is not clear definition whether the above equations can be used. On the other hand, BS EN 1990 2002 "Basis of structural design" in section 4.2.3 defines concrete properties to be 5% fractile value. The below equation from Trevor (2017) is applied for concrete materials:

$$Xk = Xmean - 1.645_{Sx} \tag{3}$$

Difference between equations 1 or 2 and 3 is that in equation 1 and 2 there is a reduction in the variance of the parameter. This reduction in the variance accounts for the failure mechanisms to be considers in design. Still there is no clear definition about design parameters in GI using Eurocode.

1.3 *Definition of GI parameters from Federal Highway Administration (FHA) Design Manual: Deep Mixing for Embankment and Foundation Support values according to (2013)*

This chapter discusses the FHA approach to define ground improvement properties for design. These are used as guidelines in the USA for embankments and foundation support. The design strength for deep mixing method is calculated by:

$$sdm = 1/2 \, fr \, fc \, qdm, spec \tag{4}$$

Where
sdm = Design value of shear strength.
qdm,spec= Trial value of the 28-day strength (based om background information)
fc = Factor depending on time
fr = Factor depending on post peak strength

Once sdm is defined, a factor defined as fv is applied. Fv accounts for the variability that typically exists in the strength of the GI. Table 12 in section 6.1.3 of this design code defines the factor fv considering 10, 20 or 30% fractile and considering the failure mechanisms with the variation of the ground improvement. Subsequently, it recommends that midranges of inputs with 20% fractile can be considered for a well-qualified deep mixing contractor.

1.4 *Definition of GI parameters from The Deep Mixing Method (2013)*

This chapter discusses the approach discussed in the deep mixing method by Masaki Kitazume & Masaaki Terashi (2013). These are used as guidelines in Japan for embankments, foundation support, liquefaction mitigation, etc. The design strength for deep mixing method is calculated by:

$$quck \leq quf - K \cdot \sigma \qquad (5)$$

Where
quck = Design standard strength.
quf = average unconfined compressive strength of in-situ stabilized column
σ = Standard deviation of field strength
K = Coefficient

In this definition it is not specified the parameter K to be used. As an interpretation from the author, the strength is specified as 15% to 20% fractile of the field values. This interpretation comes from the correction factor for strength variability of 0,5 to 0,6 that is applied in the internal failure check (allowable strength of stabilized soil) in chapter 5.3.5.3 of this book.

1.5 *Discussion of design codes*

As stated in the abstract of this paper, there is not a unique way of calculation the design parameters of ground improvement. From the above bibliography selected along Europe, USA and Japan, it is on the author's opinion that in Europe the values potentially selected for the properties of ground improvement are closer to the mean value (if global failure is concerned) than USA and Japan. In other words, in USA and Japan more conservative values are selected. It is important to highlight that for this discussion, the EC7 approach has been used as BS EN 14679-2005 (2005) refers to EN 1997-1 when it comes to design

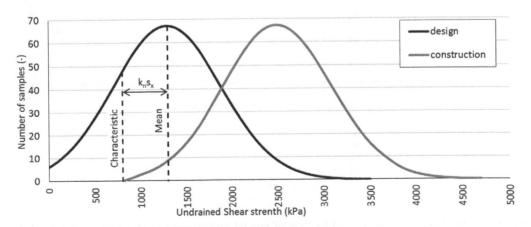

Figure 1. Typical statistical distribution of Undrained Shear strength.

principles. This means that GI has been treated as soil when it comes to selecting design parameters.

1.6 Common Practice in Singapore

Usually, in Singapore, for ground improvement, the design values defined during design stage are taken as "minimum" during construction phase, hence when it comes to the sampling and testing the supervision team makes sure that all the samples are higher than the minimum or design value by the designer. This leads to a different approach between design and construction. The above stated can be depicted in Figure 1 in the case on undrained shear strength.

1.7 Objective of this paper

The objective of this paper is to show which approach could be used when selecting GI properties, this is showed with a back analysis of a monitored excavation.

2 MONITORED EXCAVATION IN SINGAPORE

2.1 Ground Improvement in Marine Clay

In Singapore, Kallang Formation is of marine, alluvial, littoral and estuarine origin. The most important unit of the Kallang Formation is the Marine Clay, which is an under to normally consolidated soft, silty, kaolin-rich clay. In general the clay content is high at around 60 to 70%. Ground improvement is typically required in the Kallang formation, Figure 2 shows the geological soil profile and the location where the ground improvement has been installed and the samples have been retrieved.

2.2 Unconfined Compressive Strength Results

Unconfined Compressive Strength (UCS) tests were carried out to define the strength and stiffness properties of the ground improvement samples. During the sampling of the ground improvement, a total of 150 coring points were chosen including construction closer to the

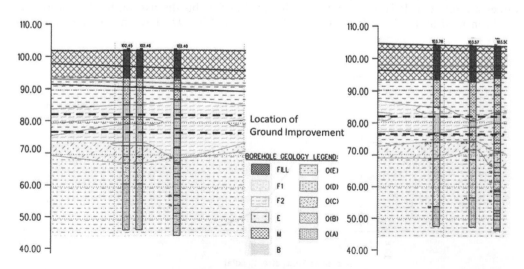

Figure 2. Geological soil profile and location of ground improvement block where the samples have been retrieved.

monitored excavation. 3 samples were tested for each coring points. This resulted in a total of 450 tests.

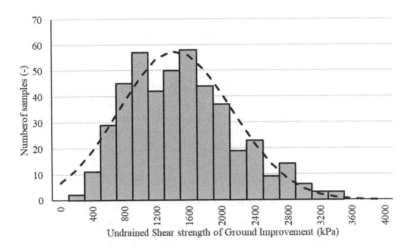

Figure 3.   Undrained Shear Strength results from UCS tests.

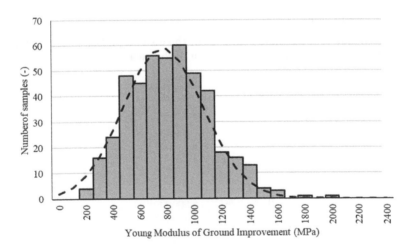

Figure 4.   Young Modulus from UCS tests.

Table 1.   Statistics parameters from the UCS tests.

Parameter	Undrained shear strength	Young Modulus
Mean	1471 kPa	772 MPa
Standard deviation	692 kPa	296 MPa
Coefficient of Variation COV (-)	0.48	0.38

2.2.1 Unconfined Compressive Strength Results

The results obtained are presented in the form of histogram showing the strength of the ground improvement.

The next statistic parameters are obtained from the previous tests.

2.3 Definition of Characteristic values

Using Bond (2011), the characteristic values are calculated as follows:

$$Xk, \inf = mx - kn.sx$$

The kn parameter is a statistical coefficient that depend on number of data available. For a more detailed discussion about this parameter please refer to Bond (2011). For this case study with 450 samples the kn is calculated using a 95% confident and 50% fractile, the 50% fractile is chosen because deep excavation involves high volume of ground improvement. In other words, we need to be 95% confident on the mean value obtained.

$$kn = tn - 1^{95\%}(1/n)^{1/2} = 1.645. \, (1/450)^{1/2} = 0.08$$

$$Cu, \inf = 1471 \, kPa - 692 \, kPa.0.08 = 1377 \, kPa$$

$$Eu, \inf = 772 \, MPa - 296 \, Mpa.0.08 = 749 \, MPa$$

It is important to highlight that in this paper are calculated just the properties using EC7. This is discussed in section 3.2.

2.4 Excavation under study

This sections shows the geometry and sequence of the excavation that is used for discussion in this paper, the below figure depicts a cross section of the excavation:

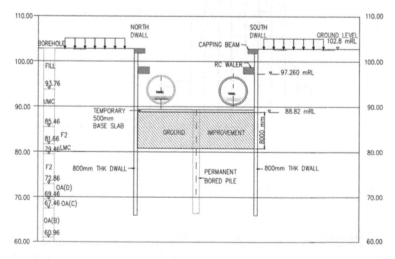

Figure 5.   Cross section of shaft exaction in Kallang Formation used.

In the figure above it is presented with hatch the location of ground improvement below the formation level. Figure 6 shows the excavation when the formation level has been reached and the temporary slab casted.

The construction sequence of the excavation under study is a typical bottom up construction with installation of diaphragm walls, and subsequent casting of supports (capping and

Figure 6.   Excavation of the shaft. Picture above showing excavation to Formation level and picture below showing casting of base slab.

Table 2.   Geotechnical Parameters used in FEM analysis.

Soil Layer		γ_s (kN/m³)	φ' (°)	c' (kPa)	C_u (kPa)	E_u (kPa)	E' (kPa)	K_0 (kPa)	k (m/s)
Fill		20	30	0	-	-	12,000	0.5	1×10^{-05}
Kallang Formation	Fluvial Sand (F1)	20	30	0	-	-	560 (113.5-Z)	0.5	1×10^{-05}
	Fluvial Clay (F2)	19	24	0	1.25. (120-Z) *	375 (120-Z)	E_u/1.2	0.7	1×10
	Estuarine Deposits €	19	22	0	0.75. (126.67-Z) *	187.5 (126.67-Z)	E_u/1.2	0.63	1×10^{-08}
	Upper Marine Clay (UMC)	16	22	0	0.33. (161.67-Z) *	99 (161.67-Z)	E_u/1.2	0.63	1×10^{-09}
Old Alluvium	O(D) [10<N<30]	20	32	1	100	30,000	E_u/1.2	0.8	1×10^{-08}
	O(C) [30<N<50]	20	35	5	200	80,000	E_u/1.2	0.8	1×10^{-08}
	O(B) [50<N<100]	21	35	5	30C	1,50,000	E_u/1.2	0.8	1×10^{-08}
	O(A) [N>100]		35	10	40C	2,00,000	E_u/1.2	0.8	1×10^{-08}

* Note = Z denotes the depth in reduced level.

1495

waler beam) while excavating. The soil parameters used in the analysis are presented in Table 2.

Fluvial Clay Layer (F2), Marine Clay layer (UMC) and Estaurine layer (E) are modelled using Mohr Coloumb Undrained type B behavior and the rest of the soil layers are modelled using Mohr Coloumb drained behavior. The ground Improvement is modelled using Mohr Coloumb with non porous behavior. Pore water pressure is modelled using hydrostatic type.

The contact between soil and plate was modelled using interface elements with a Rint value of 2/3.

3 BACK ANALYSIS

The displacements obtained from inclinometer and original design are shown in the below picture. Both graphs presented in the below figure are representative of the moment when reaching formation level.

It is important to mention that the original ground improvement properties for design where defined as:

$$Cu = 800 \text{ kPa}$$

$$Eu = 280 \text{ MPa}$$

When comparing the values obtained from UCS tests and the original properties, it can be concluded that, on the one hand the strength properties used are conservative and on the other hand, the stiffness values are very conservative.

3.1 Back Analysis using EC7 characteristic properties from UCS tests

The values of ground improvement properties obtained in section 2.3 are used. Below the values used:

$$Cu, \inf = 1471 \text{ kPa} - 692 \text{ kPa}.0.08 = 1377 \text{ kPa}$$

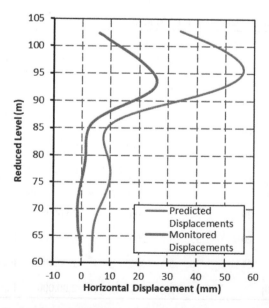

Figure 7.   Inclinometer displacements and predicted displacements.

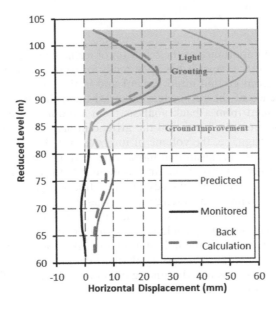

Figure 8.   Inclinometer displacements, predicted displacements and back analysis.

$$Eu, inf = 772 \, MPa - 296 \, Mpa.0.08 = 749 \, MPa$$

On top of the updated values of ground improvement, it has been included light grouting properties as Eu = 40MPa, φ = 35deg. These properties come from empirical correlations from SPT tests carried out in this portion of improved layer. The layer of light grouting is presented in the Figure 8.

3.2   *Discussion*

The above figure shows a straightforward comparison between the original predicted wall displacements, the monitored displacements and the back calculation using the updated ground improvement properties using EC7. What the author wants to highlight with this comparison are three main points:

-Performance of excavation using ground improvement parameter calculated using EC7 show satisfactory match with monitored excavation.

-The original values of ground improvement are very conservative, tough these values are similar to design values used along other Singapore projects and abroad. Ground Improvement has already been used for several decades, the author believes that there is room to make the future designs more efficient with the use of past experiences and the experience of this project.

-As the comparison with EC7 parameters is considered to be satisfactory, no comparison has been made with the values defined by FHA (2013) or Deep Mixing Method (2013), however both of these codes define GI properties more conservative as discussed previously. No merit on the use of these parameters has been made and it has been avoided for the sake of the extension of this paper.

As mentioned in EC7, the characteristic value depends on the failure mode, more attention needs to be paid and more study needs to be done on the failure mode for the excavation types shows in this example. The wall displacements below the GI slab (80 mrl to 65 mrl) also need a further study, the author believes that the wall movement using FEM analysis could potentially be reduced by using other model type in the soft clay and Old Alluvium (i.e. Cam Clay model and Hardening soil model respectively).

4 CONCLUSION

This paper discusses the importance of the selection of realistic properties of ground improvement. On the first part of the paper it has been analyzed some results from unconfined compressive strength tests from ground improvement and in the second part of the paper it has been studied the influence of the parameters obtained through a back analysis. Satisfactory results have been obtained from a back analysis from a monitored excavation and a FEM analysis. This is important since, in general the ground improvement quantities depend from the properties used in design, i.e. the volume of improved GI slabs is a function of the input from design; in other words, this mean that more realistic values during design can potentially reduce the volume of ground improvement and subsequently the cost of the construction.

The author believes that a more detail study on the stiffness of ground improvement is necessary, a lot of attention is paid to the strength properties, however the Young Modulus could potentially gain importance as more realistic values are used. This is also necessary, as higher values of GI used in design potentially could reflect in higher stresses attracted by the GI slabs.

Also, further studies considering the failure modes of GI slabs used in excavations must be done. If EC7 approach is used, it is necessary that the designer understand which the failure mechanism for the excavation using GI (either global or local failure). This would have a significant impact on the properties selected.

REFERENCES

Bond and Harris (2008). Decoding Eurocode. Taylor & Francis.
Eurocode 7 (1997). Geotechnical Design – Part 1.
Bond (2011). A Procedure for determining the characteristic value of a geotechnical parameter. ISGSR 2011 – Vogt, Schuppener, Straub & Brau (eds).
BS EN 14679-2005 (2005). Execution of special geotechnical works – deep mixing. British Standars (English Version)
Hashimoto et al., (2016). Long term strength characteristics of the cement treated soil after 30 years. 19 Southeast Asian Geotechnical Conference & 2nd AGSSEA Conference, Kuala Lumpur.
Kitazume and Terashi (2013). The Deep Mixing Method. Taylor & Francis Group, London, UK
Liu et al., (2015). Effect of spatial variation of strength and modulus of the lateral compression response of cement-admixed clay slabs. Geotechnique 65. No 10.
Liu et al., (2016). Towards a design framework for spatial variability in cement treatment for underground construction. Geotechnical Engineering Journal of the SEAGS & AGSSEA Vol. 47 No 3.
LTA, 2010. Civil Design Criteria.
Trevor (2017). Defining and selecting characteristic values of geotechnical parameters for designs to Eurocode 7. Georisk: Assessment and Management of Risk for Engineered Systems and Geohazards.

Tunnels and Underground Cities: Engineering and Innovation meet Archaeology,
Architecture and Art, Volume 4: Ground improvement in
underground constructions – Peila, Viggiani & Celestino (Eds)
© 2020 Taylor & Francis Group, London, ISBN 978-0-367-46868-2

Water loss solutions in construction of pressure shafts for Uma Oya Project, Sri Lanka

A. Rahbar Farshbar & F. Foroutan
Farab Co. Tehran, Iran

D. Eccleston
Poyry Switzerland AG, Zurich, Switzerland

D. Dodangeh
Farab Co. Tehran, Iran

ABSTRACT: Uma Oya Multipurpose Development Project is under construction in Sri Lanka. This project involves 2 Roller-Compacted Concrete (RCC) Dams connected through a link tunnel, Headrace Tunnel, 3 shafts as escape shaft (3.7m diameter in depth of 90m) in MAT, Surge Shaft (3.7m diameter in depth of 90m 135m) & Pressure Shaft (3.2m diameter in depth of 90m 618m) on Headrace Tunnel. The water will be transferred through a 15 km headrace tunnel to the top of a 650 m deep drop shaft that feeds the high pressure water to an underground powerhouse and turbine chamber for generation of 120 MW of electricity. A RBM 600 machine from Herrenknecht has been used in Shafts constructions by raise-boring method. Methodology was based on drilling the pilot hole from the top to the bottom and reaming from the bottom to the top. For having the best accuracy, RVDS system used in this project to minimise the deviation in Surge & Pressure Shafts. During the pilot drilling of pressure shaft (618m), two sudden water lost have been encountered, which made good experiences. In this paper the method statement, problems & solutions will be reviewed for all 3 shafts, with more emphasize on pressure shaft in this project.

1 INTRODUCTION

Uma Oya Multipurpose Development Project is under construction in Sri Lanka. This project involves two Roller-Compacted Concrete (RCC) Dams connected through a link tunnel, a Headrace Tunnel, 2 shafts as Escape Shafts (3.7m diameter in depth of 90m, called ES) in the MAT (Main Access Tunnel), Surge Shaft (3.7m diameter to a depth of 135m, called SS) and a Pressure Shaft (3.2m diameter in depth of 618m, called PS). The water will be transferred through the 15 km headrace tunnel to the top of a 618 m deep pressure shaft that feeds the high pressure water to an underground powerhouse and turbine chamber for generation of 120 MW of electricity (Figure 1).

A RBM 600 machine from Herrenknecht has been used for the shafts construction adopting the raise-boring method. Methodology was based on drilling the pilot hole from the top to bottom and reaming from the bottom to top. For having the highest accuracy, an RVDS system was used in this project to minimise the deviation in the Surge & Pressure Shafts. During the pilot drilling of pressure shaft (618m), two sudden water losses were encountered [Rahbar, A. 2017].

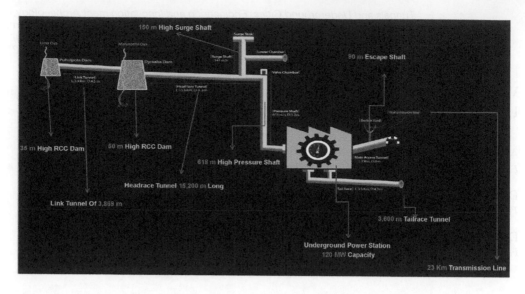

Figure 1. Uma Oya Project Plan and cross section.

2 GEOLOGICAL CONDITIONS

In the design study phase of Uma Oya, there were totally three deep investigation boreholes drilled in the pressure shaft area of which B.H. ST01 borehole (194.5m from PS Axis) is located in the pressure shaft area and B.H. PH01 (83.7m from PS Axis) and B.H. PH02 (134.7m from PS Axis). The boreholes are in the power cavern area, but currently the power cavern boreholes are practically closer to the pressure shaft.

The geological conditions of pressure shaft and surge shaft area are characterized as complex, due to varying lithology of metamorphic rocks even at a small scale. The expected rock types for the vertical PS mainly comprise of massive marbles, calc-silicate gneisses, banded to massive gneisses with various mineralogical to pure quartzite. The rock mass conditions are described with general massive appearance, with a minor dilution potential of blocks for the shaft and local fractured zones with a non-defined dilution potential.

Hydrogeological testing of rock mass permeability during the investigation phase reveals some distinct zones with a minor permeability based on indicative Lugeon tests. The geotechnical investigations consists of the in-situ investigations and laboratory testing which were carried out in the area of the surge shaft and the pressure shaft. These investigations have been summarized in Table 1.

The bedrock is formed by poly-metamorphic gneisses and marbles. The petrographic units may change quickly (within m-scale) or may remain unchanged for many tens of meters. The very local occurrence along the shaft axis cannot be predicted in detail. Three main groups of petrographic units are to be distinguished, whereas the groups 1 and 2 are predominant over group 3:

• Marble, coarse grained and massive; often present as thick bodies
• "Granitic" gneisses with varying content of quartz, feldspar, garnet and hornblende/pyroxene
• Quartz-rich gneisses and pure quartzite.

Table 1. Main geotechnical investigations in the surge shaft and pressure shaft.

	Borehole name	Depth (m)/Incl.	Position of bore hole	Investigations done
1	ST1	851.5(90)	Near to surge shaft, about 30 m out of section	• Permeability(LU)
2	PH1	750.2(90)	Power Cavern Western End (near to the pressure shaft)	• Groundwater monitoring (piezometer) • RQD and core recovery determination • Mechanical parameters of intact rock • Physical parameters of intact rock
3	PH2	750.55(90)	Power Cavern Eastern End (near to the pressure shaft)	• Hydraulic fracturing (HF) in ST1 and PH1 • Hydraulic test on pre-existing fractures (HTPF) in ST1, PH1

3 CONSTRUCTION METHOD

Taking due regard to the depths, sizes, time schedule and access to bottom of the shafts, the raise boring method has been considered as the favored construction method. For a 618m deep shaft of relatively small diameter, only raise boring could realistically be considered as the excavation method. Marti Shaft Co. was nominated as the contractor for excavation of the shafts with a raise boring rig model RBR600VF from Herrenknecht which is designed for long raises and large reaming diameters with a mechanized wrench and rod handling system for improved work safety [Rahbar, A. 2017]. The diameter of the pilot was 15 inch. Final diameter of the shafts after reaming is 3.2m for PS and 3.7m for SS & ES.

4 PRESSURE SHAFT (PS)

4.1 Pilot Drilling First Stage

Drilling of the pilot hole started on 25 October 2014. During the pilot hole drilling, the first distinctive zone with strong water losses was encountered at a depth of 63m (50% water loss) and at 64.5m (100% water loss). The drilling activity was suspended at the depth of 65m on 30 of October due to ongoing water losses.

All the rods and the bits were removed from the pilot hole and attempts were made to inspect the inside the pilot hole with a down-the-hole camera.

According to the contractor's proposal, a geotextile packer (G Packer) with a length of 2m and diameter of 38cm was assembled at the job site, installed and packed at 58m in the pilot hole. The water level was approx. at 60m depth and the packer was set above the underground water table. After that, to seal the geotextile packer, the pilot hole was filled with grout (mix water/cement ratio 0.5) up to 7m depth. This was to provide some sealing of the water loss area. Grouting was performed through the packer and after waiting for the final setting, re-drilling started. The grouting was not successful, water losses were again experienced as the water loss position was in fact discovered under the underground water table. Drilling of the pilot was re-started and reached 90m. Water losses however continued. At this point it was decided to change the method of setting the packer and it was decided to install a double inflatable packer (two inflatable sleeves connected with a perforated steel rod approx. 3m apart) and try to inject the grout and seal the water loss zone section by section. The intention with using an inflatable double packer is to isolate the particular zone of interest and to perform targeted grouting. The water loss zones however could not be sealed.

4.2 Additional Exploratory Borehole

Due to the encountered geological situation, to understand potential difficulties during excavation, as well as review the necessity of later rock support and the final lining and to minimize the encountered risks, the entire shaft length needed to be investigated by cored drill holes.

Exploratory boreholes for shafts should not be drilled within the footprint or excavation area of the future shaft, as there is always the possibility that the drilling bit and rods are stuck within the borehole. In such cases the shaft excavation by raise drill method would greatly hindered as the steel parts could damage the raise drilling cutter bits and the shaft axis must be relocated. Boreholes always deviate from the vertical axis to varying degrees, depending on geological and drilling parameters. Therefore a precise and safe vertical drilling is required or the drilling location has to be placed at a safe distance from the shaft area.

To investigate further the encountered water loss zones and for the continuation of encountered weak zone along the pressure shaft and additionally to have the possibility to improve the bedrock mass with pressure grouting around the pilot hole, exploratory borehole(s) were planned. These borehole(s) were to recover drill cores over the entire length. The boreholes were drilled with 76mm diameter.

The drilling rig chosen to be used for the new and future exploratory boreholes near the shaft areas must be able to drill without problems several 100 m's deep, with the minimum borehole diameter core recovery and to drill with wireline system to speed up core recovery time.

Due to the special geological conditions in the project area, the axis of the exploratory boreholes should be located as close as possible to the planned shaft excavation area and deflect only negligibly from verticality; core recovery must be high as possible (90%) and fracturing of cores by drilling must be kept as low as possible to retrieve as much information as possible.

4.3 Considerations for Shaft Construction

Considering the cores obtained from exploratory boreholes around the pressure shaft and images taken by the down-the-hole camera in the pilot hole, one of the issues regarding water loss at depth of 50m to 57m was the possibility of the contact of marble layers and their adjacent layers. Images taken at these depths are indicative of caves with considerable dimensions and the volume of grout and grouted concrete introduced is very noticeable. A number of options were considered for dealing with the issue of water loss.

Three solutions were considered for dealing with the water loss issue.

4.3.1 Solution 1 - Reaming

The risk of water ingress below the aquifer during reaming exceeding volumes of more than 1 m3 was likely to occur, taking the encountered voids in a diameter up to 3 m within the profile of pilot drill into account. Considering an inflow rate of 2.5 m^3 from the assumed catchment area over an overall period of 183 days it can be concluded, with constant water inflow, the duration of the inflow can vary from several days and weeks of inflow to even a longer time span than anticipated.

This inflow would have severely hindered the other construction activities. Within this period practically no works are possible, also within the lower bend of the shaft and the shaft. Nevertheless, the water could be diverted through the Tailrace Tunnel (TRT), restricting further works within the adjacent connection galleries of the lower shaft bend to the TRT. Extensive grouting works within the shaft will anyway be necessary to seal the shaft walls for commencing further works for lining installation. It shall be mentioned that grouting works within the shaft will be challenging, since all grouting works must be conducted under flowing water conditions. Improper installation of primary lining, due to inflow effects of high water is very likely.

4.3.2 *Solution 2 - Grouting operation prior to shaft reaming operation*

The execution of a grouted zone in the periphery of the shaft would help to decrease inflow during shaft reaming operation as well as to stabilize the shaft prior to lining. The requirements for the drilling and grouting equipment are very high, since the desired depth of the drilling and grouting is significant. The grout curtain will seal, to a large extent, the water inflow towards the shaft. The number of grouting holes needed will depend on the actual conditions. Grouting and drilling works in these geological conditions demand specific experience and special equipment to successfully perform the works. Drilling to depths up to 100 to 120 m requires specialized drilling equipment, especially for works below the groundwater table and to deal with voids up to 3m. Considering the design, primary and secondary holes and partially tertiary holes are required to seal off the water loss or ingress zones. Localized water loss or ingress towards the shaft after reaming operation cannot however be excluded due to imperfections in the grout curtain (Figure 2).

The reduced water inflow rate, will allow a direct treatment of the ground from a suspended platform. Minor water ingress can be channelled and diverted to the lower shaft bend. Drainage holes have to be filled in the course of back filling and final grouting operations for the lining.

4.3.3 *Solution 3 - Dewatering of area*

The installation of an active drainage system close to shaft area allows for improved post grouting operation upon reaming of the shaft. Due to the limited space within the shaft head cavern a maximum of 3 – 4 active drainage systems can be installed. Considering the limited pumping capacity of an active drainage system (12.5m³/h) the time span to significantly lower the groundwater head. A time period of 3 months to lower the groundwater head would lead to an insignificant lowcring of the groundwater head of only 0.2 % of the assumed reservoir.

4.4 *Summary of risk and recommendations*

Solution 1 and Solution 3 would highly depend on the size of the catchment area. The actual delay could not be predicted, since a detailed geological assessment to predict the inflow rates cannot be investigated in a timely way so as not to create additionally delay to the construction works.

Further risks for these methods apply in terms of welding operation, where dripping or flowing water cannot be tolerated. Flowing water exceeding the potential volume for drainage will negatively affect the concrete quality either for back filling or the reinforced concrete lining.

Solution 2 holds the least risk for the supplementary works within the shaft or adjacent structures. The required time for the completion of works can be estimated at 2–3 months

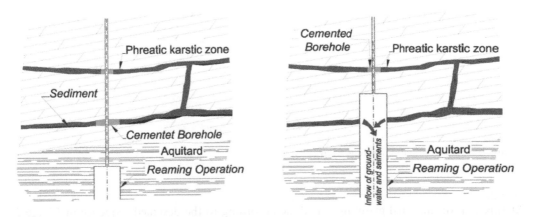

Figure 2.   Reaming through the water loss zones.

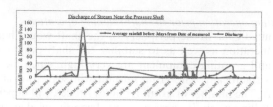

Figure 3. Shaft Stream Monitoring.

including only a minor risk for small delays. The conditions within the shaft will allow for more adequate working conditions within the shaft to prepare the conditions to commence with supplementary works. Also with this solution ground improvement will be done and shaft reaming will be done with minor problem.

After a Panel Review of the options by all the parties, it was finally decided to move the RBM to start the pilot drilling of the Escape Shaft and in the meantime to perform the exploratory investigations around the pressure shaft. This also allowed progress to be made in the overall shafts execution time schedule. Next action for the treatment and improvement of the area was removing the RBM to make a room for the grouting activity.

4.5 *Surface Water Monitoring*

Monitoring of a nearby stream was also started. There is just one stream near the PS. The results of the stream monitoring vs the rainfall are included in the Figure 3 (CECB, 2016–2017).

5 GROUTING PROGRAM

After moving the RBM to start the Escape Shaft, the grouting of the pressure shaft area could start.

Step 1:
For this first step of grouting borehole, exploratory boreholes were drilled to investigate the ground condition. The deviation of the drill rod was also checked by an inclinometer. The first hole was treated as a test panel and all required parameters like grout-ability, W/C ratio and grouting pressure were optimized in this hole (Figure 4).

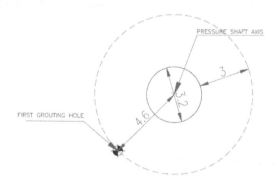

Figure 4. Pressure Shaft Grouting, Step 1.

Step 2:
At this step drilling and grouting was done according to the designed sequence in Figure 5. Drilling of a hole was not permitted before the completion of grouting in the adjacent hole.

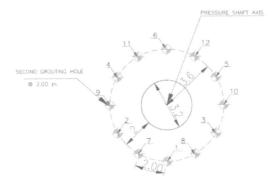

Figure 5. Pressure Shaft Grouting, Step 2.

Step 3 & 4:
At this step, third level holes were drilled to check the efficiency of the previous grouting step. If the executed grouting is not satisfactory, further grout injection was done from these holes (Figures 6 & 7).

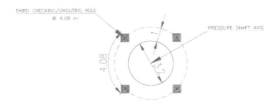

Figure 6. Pressure Shaft Grouting, Step 3.

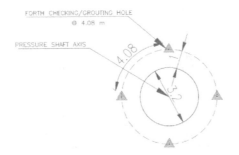

Figure 7. Pressure Shaft Grouting, Step 4.

6   PS GROUTING – STAGE 1

The grouting campaign detailed plan is included in Table 2 (Figure 8). The grouting started on 20 March 2015 and 23 grout holes were completed on 6 September 2015. 2'420 meters were drilled in rock, 1'790m re-drilling cement, 1'270 m^3 including more than 1'240 tons of OPC cement injected at this stage. An anti-wash out polymers additive was added to the grout mix in an attempt to reduce wash-out of grouting. The task was complicated by the depth of the stage to be grouted but some success was seen.

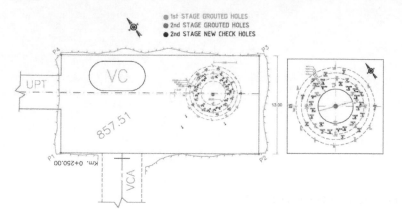

Figure 8.   Valve Chamber Plan & the final grout hole positions.

Table 2.   PS Grouting in Stage 1 (Farab).

No	Borehole	Drilling in rock (m)	Volume (m³)	Drilling in Cement
1	PS.P-01	93.4	68.5	181.9
2	PS.P-02	90.4	9.05	83.5
3	PS.P-03	94.4	70.4	234.3
4	PS.P-04	95	46.8	106.7
5	PS.P-04B	78.7	1.5	–
6	PS.P-05	93.7	52.6	104.7
7	PS.P-07	84.7	270	74.2
8	PS.P.09	84.7	201	213.1
9	PS.P.11	89.2	124.2	114.4
10	PS.P.12	95.2	134.7	205.3
11	PS.P-C01	93.4	45.6	72.6
12	PS.P-C02	90.4	1.2	–
13	PS.P-C03	90.7	78.3	100.5
14	PS.P-C04	90.7	43.2	129.6
15	PS.P-D01	90.7	2.2	–
16	PS.P-D02	88.9	4.8	35.3
17	PS.P-D03	90.7	49.4	56.6
18	PS.P-D04	90.7	31	12.7
19	PS.P-CH01	90.7	0.5	–
20	PS.P-CH02	93.5	2.4	20
21	PS.P-CH03	74.7	3.1	23.4
22	PS.P-CH04	87.6	21	5.6
23	PS.P-CH05	87.1	8.5	13.6
	Total	2'059.20	1'269.95	1'788

The personnel involved as a site manager for 183 man days, Site Engineer for 62 days, Senior Driller for 261 days, 344 Man/Day of Driller and 191 days of grouting technicians.

7   ESCAPE SHAFT (ES) AND SURGE SHAFT (SS)

As the problems with water loss continued in the Pressure Shaft, and while grout treatment was ongoing, all focus was concentrated on the other shafts. During the drilling and grouting around the PS and according to the experience in the PS, exploration bore holes were drilled in the vicinity of the Escape Shaft and Surge Shaft, which showed a similar water loss concern up to 7m in ES and 15m in SS. It was planned to excavate to these depths and to backfill these

Table 3. SS Grouting Plan (Farab).

No	Borehole	Drilling in rock (m)	Volume (m^3)	Drilling in Cement
1	SS.P-01	40	25.2	38.8
2	SS.P-02	39	1.7	6.5
3	SS.P-03	40	17.4	39.2
4	SS.P-04	39.8	1.5	9.8
5	SS.P-05	4.	16.4	24.4
6	SS.P-06	39.3	3	16.5
7	SS.P-07	40	15.4	14.5
8	SS.P-08	39.6	10.1	22.4
9	SS.P-09	40	0.4	
10	SS.P-10	38	0.8	
	Total	395.7	91.9	172.1

zones with concrete for both shafts. Check holes were subsequently drilled through these zone to control the water loss. Results showed water loss in the contact between the concrete and rock. These check holes were later used as grouting holes. Later results show the very good workability and plan for this two shafts and there was no delays thereafter regarding the geological conditions. In the SS, regarding the depth of the shaft some exploratory borehole were also planned. The detailed grouting plan is shown in Table 3. The grouting started on 2 May 2015 and was completed with 10 grout holes on 31 May 2015. 395 meters were drilled in rock, 172m re-drilling in cement, 91 m^3 grout of OPC cement injected in this stage. Site Engineer was needed for 31 days, 62 man/day of driller and 31 days of grouting technicians were involved.

8 RESTART OF PRESSURE SHAFT PILOT

After grouting works were completed for the Pressure Shaft, and completion of the Escape and Surge Shafts, re-drilling of the PS pilot started. These results were good and drilling of the pilot was ongoing until the water loss again happened at a depth of 379m. All the rods were removed over some days and the pilot checked with a down-hole camera. As there were sediments of the pilot drilling material in the water column, the bottom of the pilot hole could not checked. However the RBM parameters show that the rock was very dense and strong. It was possibly a new joint for water loss. After controlling the entire pilot hole it showed that there were no new geological conditions in the pilot hole. Monitoring of water table of pilot hole show that it was the same as the first water loss.

Video Control, RBM parameters and underground water table reinforced the hypothesis that the water loss again happened at the same depth as before. So using the G-Packer, again the previous zone was sealed and the zone checked. The results did confirm that the water source was from the previous zone. Finally after significant amounts of grouting to stabilise the area around the PS and to control water loss, the pilot hole breakthrough occurs on 10 October 2016.

The grouting results of this stage are summarized in Table 4. The grouting plan started on 26 December 2015 and was completed with 44 grout holes on 14 June 2016. 3'930 meters were drilled in rock, 1'340m re-drilling in cement, 536 m^3 grout including more than 543 tons of

Table 4. PS Grouting in Stage 2 (Farab).

No of Boreholes	Drilling in rock (m)	Volume (m^3)	Drilling in Cement
44	3'923	536.4	1'333

OPC cement were injected. Resources were site manager for 167 man days, site Engineer for 152 days, senior driller for 183 days, 542 Man/Day of driller and 319 days of grouting technicians.

9 PRESSURE SHAFT REAMING

After the pilot hole breakthrough, the reaming was performed with the 3.2m reamer. The concern that more grouting may be needed to control water loss or water ingress did not materialise and the reaming was successfully completed on 24 February 2017.

10 LESSONS LEARNT

The lessons learnt for grouting for stabilisation and water control for such a deep shaft can be summarized that certain thorough geological information is needed beforehand to understand the fracturing of the ground.

Measurement of the groundwater table before starting the pilot hole provides a datum for future aquifer behaviour, understand and control.

The presence of a permanent groundwater head generally refers to a phreatic zone (see Figure 2 for definitions). The complete distribution of dissolution cavity prone areas within the project perimeter is not documented. An estimation of a potential groundwater catchment area is not possible under this circumstances, obeying detailed prediction of water inflow, especially during reaming.

Measurement of water flow rate in the borehole would indicatively help to get a better under-standing of the general conditions within the system.

Additives such as anti-wash out polymers are more stable within such adverse conditions and can be used. The addition of the polymer must be adopted to the depth of application to avoid clogging of lines and requires the presence of experienced site personal to control the setting.

REFERENCES

Rahbar Farshbar, A. & Ahadi Manafi, B., Stakne, P. & Dodangeh, D. 2017. Experiences and challenge in shaft construction at the Uma Oya project, Sri Lanka, Hydro 2017; Proc. intern. symp. Sevilla, Spain, 9–11 October 2017.
CECB 2016–2017, Central Engineering Consultancy Bureau (CECB), Streams Monitoring and re-monitoring of streams, Volume 1 to 5 & Monitoring of Domestic Wells Monthly reports. Farab Co, Uma Oya Project Documentary.

Tunnels and Underground Cities: Engineering and Innovation meet Archaeology,
Architecture and Art, Volume 4: Ground improvement in
underground constructions – Peila, Viggiani & Celestino (Eds)
© 2020 Taylor & Francis Group, London, ISBN 978-0-367-46868-2

Constructing ground struts by means of ground improvement underneath existing box elements in large-scale renovation of a subway station under operation constructed by the Caisson method

S. Sakata, K. Okanoya & Y. Arai
Tokyo Metro Co., Ltd., Tokyo, Japan

M. Nishiaoki & T. Kondou
Taisei Corporation, Tokyo, Japan

ABSTRACT: This construction project involved renovating Minami-sunamachi station on the Tokyo Metro Tozai Line by converting it from a single island platform station with one set of tracks on each side into a two-platform station with three sets of tracks. Minami-sunamachi Station was constructed using the caisson method in what was a canal at the time of construction; even now, the ground surrounding the box elements is very soft. Thus, measures to minimize displacement of the surrounding ground and existing box elements were required for each process of renovating the station. As for the construction work, excavation was performed to a depth of 14 m using the cut-and-cover method. For earth retention, diaphragm walls were used in an effort to make use of existing structures. For reinforcement, ground struts were installed below the base of the excavation area and performed ground improvement to improve the base.

1 INTRODUCTION

Since Minami-sunamachi Station on the Tozai Line opened in 1969 (Figure 1), the construction of major commercial facilities, condominiums and other buildings adjacent to the station have driven increases in the population along the line to the point where passenger congestion on the station platform during commuter rush hours has become problematic. The congestion lengthens the amount of time required for passengers to board and disembark trains, and as a result, the station and line suffer from chronic train delays.

Tokyo Metro planned major renovation work at Minami-sunamachi Station to prevent delays for trains heading toward central Tokyo (toward Nakano from Nishi-funabashi). The plans called for the construction of a new platform to the south of the existing platform and the addition of a second set of tracks for Nakano-bound trains to convert from single island platform station with one set of tracks on each side into a two-platform station with three sets of tracks. The plans also called for the widening of the existing platform, the relocation of station facilities, and the addition of elevators, escalators and entrances.

2 WORK STEPS

Figure 2 shows a plan view and a cross-section of the present and planned state of Minami-sunamachi Station. The total length of the construction area was roughly 430 m, and the cut-and-cover method was used over the entire area. As shown in Figure 4 of the state of Minami-sunamachi Station during construction, the station was constructed using the

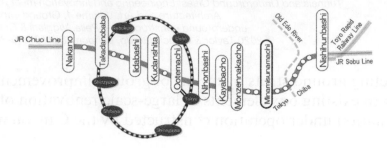

Figure 1.   Tozai Line Overview.

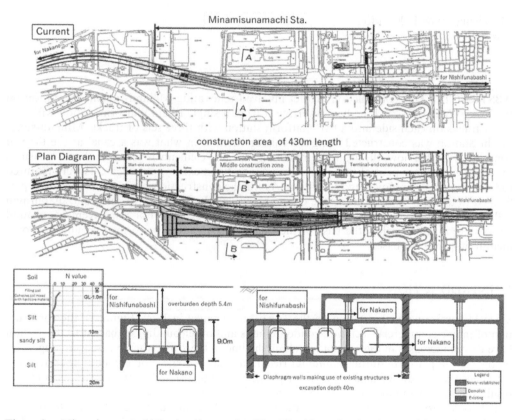

Figure 2.   Minami-sunamachi Station Renovation Plan View/Cross-Section (Present/Planned State).

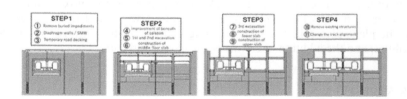

Figure 3.   Work Steps.

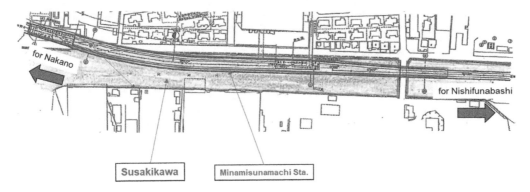

Figure 4.   Minami-sunamachi Station and Environs During Construction.

caisson method in what was then a canal called Susakikawa; thus, even now, the ground surrounding the station is very soft (N = 0–1).

Figure 3 shows the work steps. The first steps were to Obstacle underground utilities from the street level and to perform earth retaining piling work including the construction of diaphragm walls. Next, ground struts were installed and ground improvement performed as a measure against heaving, this was followed by sequentially excavating earth above and to the sides of existing box elements, and constructing new structures completely surrounding the existing structures. The final steps were to remove any existing structures intruding on the structure gauge and change the track alignment to convert to a two-platform station with three sets of tracks.

3   DIAPHRAGM WALL WORK

3.1   *Work Plan*

Existing diaphragm walls that serve as structural sidewalls doubled as retaining walls. The diaphragm walls are 1.0 m and 1.2 m thick, and can be separated into two types of element: Primary element (roughly 2 m wide, roughly 44 m deep) and secondary element (roughly 6 m wide, roughly 17 m deep). A bucket excavator (MHL) was used to perform the excavation. Figure 5 shows the element layout.

For the Primary element, the MHL excavator was used to excavate three trenches, with the third and central trench being the deepest. Then, reinforcement cages were inserted into the trenches using a crane.

When concrete was placed into the Primary element, it was only placed on the inside of an area surrounded by partitions and canvas sheets (1.9 m wide) (Figure 6).

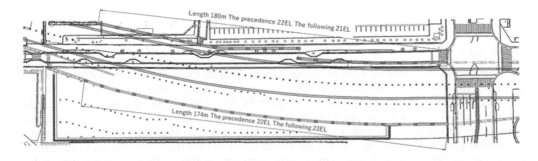

Figure 5.   Excavation Cross-Section.

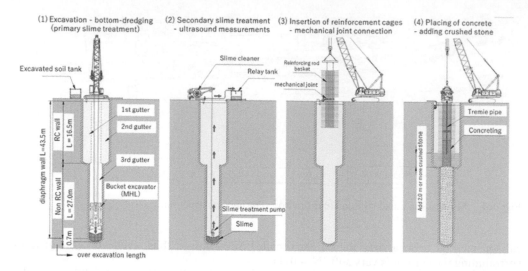

Figure 6.   Forward Element Work Procedure.

For the secondary element, a single trench was excavated in the earth remaining between the Primary element, and the reinforcement joints of the Primary element were cleaned before reinforcement cages were inserted. The reinforcement cages for the secondary element were wrapped with distribution bar, and the partitions that protruded from the Primary element were inserted into the distribution bar before the remaining concrete was placed.

3.2   Challenges and Countermeasures

In this section, the challenges faced during the course of diaphragm wall work that affected existing box elements through displacement of the surrounding ground are described, as well as the measures taken to resolve them.

The ground around the construction site contains a thick layer of alluvial cohesive soil that is very soft (N = 0–1) and only gently supports the existing box elements. These conditions and the fact that element excavation was to be performed to a depth of 44 m prompted concern over slurry wall collapse in the ground behind the excavation areas. To address that concern, the stability of the slurry walls was calculated before beginning element excavation, and it was decided to construct soil walls (using the SMW method with no core material) to protect the slurry walls. There was also concern that the excavation of the ground around the station would cause horizontal displacement of that ground, which would in turn displace existing box elements (Figure 7). To address that concern, clinometers were installed and settlement gauges 2 m away from diaphragm walls in the space between existing box elements and the diaphragm wall work locations in order to confirm ground displacement, and measured four track elements (gauge, cross level, alignment and longitudinal level) overnight to confirm a lack of track distortion as the work progressed.

3.3   Results

Clinometer measurements of ground displacement revealed that excavation caused horizontal displacement of 4 mm toward the excavation, and concrete placement caused 1 mm of displacement back toward the original position (Figure 8). In addition, settlement gauges recorded settlement of up to 0.8 mm during concrete placement. Thus, the work caused minimal displacement to the surrounding ground. In addition, nightly track measurements during the construction period produced results within Tokyo Metro standards for track maintenance (less than 17 mm elevation difference and less than 20 mm alignment within a 10-m section); thus, it was confirmed that the work had no effect on the tracks.

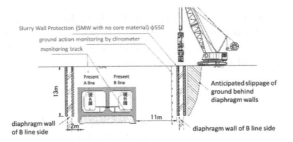

Figure 7. Slurry Wall Protection/Clinometer Installation Cross-Section.

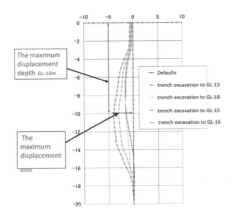

Figure 8. Holizontal Displacement of monitoring by Clinometer.

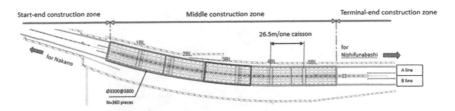

Figure 9. Ground Improvement Layout.

4 GROUND IMPROVEMENT WORK UNDER OPERATING LINE STRUCTURES

4.1 Work Plan

The work of excavating to a depth of 14 m below the surface of very soft (N = 0–1) ground required reinforcement in the form of installing ground struts below the base of the excavation area and performing ground improvement to improve the base. It is worth noting that the ground to be improved is a silt layer with cohesion exceeding 50 kN/m2.

Installing ground struts directly beneath the existing box elements required us to bring machines into the station and down to the track level to insert rods through the platform slabs and lower floor slabs, as shown in Figure 10, in order to create an improved structure down to the bottom surfaces of the existing box elements.

The plan called for two machines to perform ground improvement work directly beneath structures from the track level and atop the platform in a total of 306 locations (Figure 9, Figure 10, Figure 11). It is worth noting that temporary enclosures were built to store one

1513

machine on the platform and the other on the tracks. At night, the machines were transported from the enclosures to the worksites. Accounting for machine transportation and set up as well as teardown and cleanup, actual ground improvement work could only be conducted within an extremely tight window of 80 minutes each night.

4.2 *Challenges and Countermeasures*

Due to the high viscosity of the target ground, blockage caused by sludge was anticipated to occur during structure improvement, which prompted concern over the displacement of the

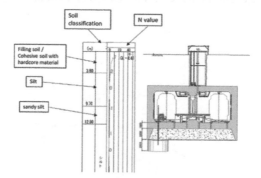

Figure 10. Ground Improvement Cross-Section.

Figure 11. Ground Improvement Under Structures.

existing box elements and tracks directly above it. To fully understand deformations of the box elements, measuring instruments (settlement gauges, strain gauges, clinometers and joint meters) were installed in various places on the elements before work commenced, and a system was prepared that made it possible to confirm measured values of heaving or settlement of the elements online in real time.

After actual construction began, and as overnight work progressed using a single machine at a single location, a tendency for existing box elements to settle 0.1–0.2 mm per day was noted. With two machines performing ground improvement in two locations, the settlement speed increased to 0.4–0.5 mm/day, which had a more pronounced effect on ground improvement.

Among the many possible causes of the settlement are (1) bleeding causing a soft layer to form at the top of improved ground; (2) prejetting and ground improvement dispersing excess pore water pressure into the natural ground surrounding that area, causing sediment consolidation; (3)

prejetting to cut into natural ground before improved ground has solidified, causing a temporary drop in bearing capacity; and (4) injection air gathering at the top of improved ground (Figure 12). Of these possible causes, the third in particular results in gradual settlement between the end of structure improvement during overnight work and the beginning of daytime train operation, but because the settlement is resolved from that point in time, this is likely the most significant factor.

Although it is impossible to completely prevent ground improvement under structures from causing settlement under existing box elements, this could be combated by taking care to avoid causing localized displacement at the elements and by analyzing and assessing measurement results to control settlement during work. Specifically, control values for measurement values were managed through determinations based on relative displacement rather than absolute displacement, and the locations and sequence of the work was varied. For example, on a given night when performing structure improvement with two machines, at least two caissons' worth of separation between the two would be ensured, or structure improvement would be performed at a location far removed from the previous day's work.

As for the method of calculating the relative displacement values used for measurement control, settlement gauges were installed at each end of the caissons as shown in Figure 13, and the relative displacement of each caisson span (28 m) was then calculated to derive the control values. For example, the relative displacement between the settlement gauge at A9 and A13 is the difference between a straight line between those two points and a straight line between A9 and A11.

Example: Relative displacement of A11 = $(\Delta A9 + \Delta A13)/2 - \Delta A11$, where Δ are absolute values.

Note that for settlement control values, relative displacement of -10.7 mm was used, which represents the strain at the yield point of the distribution bar in the lower floor slab concrete of the caisson joints, as the secondary control value, and -8.5 mm, which is 80% of the secondary control value, as the primary control value.

As for other settlement measures, improved crown heights were confirmed the day after improvement and additional injections were promptly performed as necessary to avoid settlement, and caisson joints were checked for cracks to fully understand localized displacement of box elements in more detail (Figure 14).

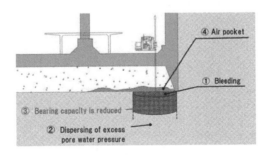

Figure 12.　Causes of Settlement.

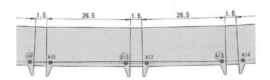

Figure 13.　Image of Relative Displacement Calculations (Longitudinal Diagram).

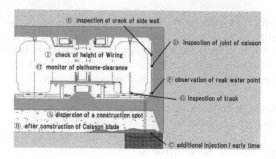

Figure 14.   Plan Diagram of Settlement Countermeasures.

4.3 *Results*

Figure 15 shows measurements of box element displacement before and after ground improvement. 1BL (the leftmost section outlined in red, see Figure 9) on the plan view is the location of the intensive construction work that was performed in the initial stage of this project. Absolute displacement values reveal settlement of more than 15 mm. As a countermeasure, work was dispersed to limit localized settlement in spite of overall settlement. These countermeasures were successful, and it was possible to keep relative displacement below the primary control value through the end of the work. Nightly track measurements were also performed during the ground improvement period just as they were for the diaphragm wall excavation, and these measurements produced results within Tokyo Metro maintenance standards; thus, it was confirmed that the work had no effect on the tracks.

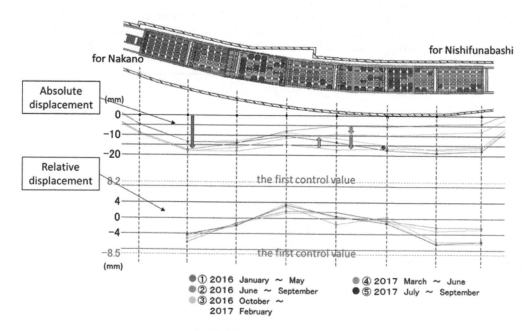

Figure 15.   Box Element Displacement Results.

5 CONCLUSION

The following knowledge was gained from this work.

It was possible to complete the bucket excavation portion of the diaphragm wall work while limiting horizontal displacement of the surrounding ground and station box elements to 4 mm in the direction of the excavation, and settlement to 0.8 mm, which satisfies the standard.

In addition, the settlement speed of existing box elements was limited to 0.1–0.2 mm when improving one location at a time with a single machine each day, and to 0.4–0.5 mm when improving two locations with two machines each day. thus, it was possible to progress through and complete the structure improvement while considering work locations, and without causing major deformations to box elements.

REFERENCES

K. Okanoya, H. Yanagisako, K. Yamanaka. 2012. Improving Minami-sunamachi Station on the Tozai Line by Creating Two Platforms and Three Tracks, *Collection of Summaries of the 67th Annual Academic Lecture Meeting of the Japan Society of Civil Engineers*, VI-580: pp. 1159–1160.

R. Okada, R. Fukuda, K. Yamanaka. 2013. Improving Minami-sunamachi Station on the Tozai Line by Creating Two Platforms and Three Tracks, *Proceedings of the Symposium on Underground Space vol. 18*: pp. 119–122.

A. Numata, T. Hirano. 2013. Large-scale Project to Improve Tokyo Subway Network Efficiency—Minami-sunamachi Station Improvement Plan—, *WTC2013*: pp.427–434.

M. Nishiaoki. 2016 Improvement Work on the Existing Caisson Structure of a Station Constructed on Very Soft Ground (Diaphragm Walls Making Use of Existing Structures/Ground Improvement Inside Station)—Renovation of Minami-sunamachi Station on the Tokyo Metro Tozai Line—, *DOBOKU Tokyo Civil Engineering*, Vol.65: pp.6–14.

N. Shiroishi, Y. Kawagishi, T. Moriya, M. Nishiaoki, T. Kondo, Y. Hirakawa. 2016. Changes of Construction Method and Test Construction of Ground Improvement Under Structures in Subway Stations, *Collection of Summaries of the 71th Annual Academic Lecture Meeting of the Japan Society of Civil Engineers*, VI-204: pp. 407–408.

Y. Hirakawa, M. Nishiaoki, T. Kondo, Y. Kawagishi, T. Moriya, T. Suzuki. 2016. Foundation Improvement Work Carried Out From Inside a Station Under an Operating Subway Line, *Collection of Summaries of the 71th Annual Academic Lecture Meeting of the Japan Society of Civil Engineers*, VI-373: pp. 745–746.

K. Okanoya, Y. Kawagishi, R. Fukuda. 2017. The Removal of Buried Impediments During Improvement Work on the Minami-sunamachi Station on the Tozai Line, *Foundation Engineering & Equipment Monthly*, Vol.45, No.2: pp.32–36.

H. Kuwamoto, T. Kondo, S. Komori, Y. Kawagishi, R. Fukuda, Y. Tsuda. 2017. Ground Improvement Under Subway Structures to Control Structural Deterioration of an Operating Subway Line, *Collection of Summaries of the 72nd Annual Academic Lecture Meeting of the Japan Society of Civil Engineers*, VI-887: pp. 1773–1774.

K. Okanoya, S. Konishi. 2018. Tokyo Metro Project: Efforts to Reduce Congestion on the Tozai Line—Renovation of Minami-sunamachi Station and Kiba Station—, *Geotechnical Engineering Magazine*, 66–5 (724): pp. 12–15.

M. Iizuka, R. Okada, R. Fukuda, W. Kanza. 2018. The Construction Method for Ground Improvement Under Structures in Subway Stations, *Collection of Summaries of the 73rd Annual Academic Lecture Meeting of the Japan Society of Civil Engineers*, VI-986: pp. 1971–1972.

Tunnels and Underground Cities: Engineering and Innovation meet Archaeology, Architecture and Art, Volume 4: Ground improvement in underground constructions – Peila, Viggiani & Celestino (Eds)
© 2020 Taylor & Francis Group, London, ISBN 978-0-367-46868-2

Rational design of jet grouted bottom plugs: The example of Tribunale station of Naples Metro Line 1

M. Saviano & A. Flora
DICEA, University of Naples Federico II, Naples, Italy

F. Cavuoto
Project Manager of Metro Line 1, Naples, Italy

ABSTRACT: When planning an open excavation in granular soils below ground water level, one of the main design concerns refers to water seepage from the bottom. Such a seepage may cause local piping, subsidence of the surrounding area with undesired settlements of nearby buildings, or a water inflow which may interfere with constructional operations. The adoption of continuous jet grouted bottom plugs is by far the most effective way to solve this critical design issue, and its use is constantly increasing in underground works. In this paper, the example of Tribunale station is reported and discussed. It is shown that rational design can be best carried out using a deterministic approach to cope with code indications on equilibrium (i.e. uplift and structural cracking), while the choice of the spacing among columns cannot be but probabilistic. By using Jetplug, a numerical code developed at the University of Naples Federico II to simulate probabilistically the behavior of a plug made of any number of columns, it was possible to estimate the water inflow through the plug as a function of the spacing among columns. By assigning the maximum tolerable inflow, the spacing could be selected on a cost effectiveness basis, assuming a risk compatible with Eurocode's indications.

1 INTRODUCTION

1.1 *Jet-Grouting*

The jet grouting technology is based on the high-velocity injection from one or more fluids (Grout, air, water) into the subsoil. The fluids are injected trough small-diameter nozzles placed on a pipe that is first drilled into the soil and is then raised towards the ground sur-face during jetting. The jets propagate orthogonally to the drilling axis inducing a complex mechanical phenomenon of soil remolding and permeation, with partial soil removal. The injected water-cement (W-C) grout cures underground, eventually producing a body made of cemented soil. Most of the time, the treated volume has a quasi-cylindrical shape and is thus named 'Jet-grouted column'.

1.2 *Bottom plug*

When planning an open excavation in granular soils below the ground water level, some of the main design concerns refer to the excavation procedure to adopt with reference to water flow from the bottom. Such a flow may possibly produce local piping and subsidence of the surrounding area, with undesired settlements of nearby buildings. As a consequence, it is often the case to first seal the excavation bottom and then excavate. Such a sealing is best obtained by using jet grouting Plugs: partially overlapped jet grouted columns are realized to produce a massive treatment at the excavation bottom, able to completely seal it. Figure 1

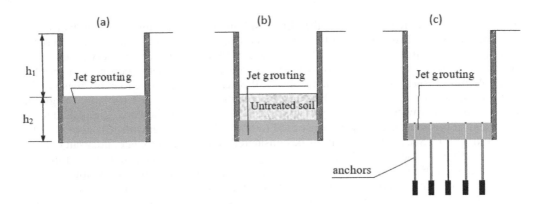

Figure 1.   Possible schemes of jet grouted impermeable bottom barriers.

shows some possible solutions for the jet grouted barrier. During excavation, this barrier has the further positive effect of bracing the retaining structure, playing therefore a positive effect also from a static point of view. The barrier must also be heavy enough to sustain uplift due to water pressure.

With reference to the simple scheme of Figure 2.a (with water level at ground level), the problem is that, because of defects, the true thickness $h2^*$ may be smaller than the nominal one (Flora et al., 2012).

In principle, the design of these impermeable barriers is an easy task (choose diameter and length of the columns, assign spacing to get a continuous barrier). However, it often happens that the bottom sealing fails, and unexpected and undesired water inflows take place. This malfunctioning of impermeable barriers is mostly due to columns defects, and in practice most times an overconservative design approach is adopted to avoid it (very small spacing between the column).

To improve the design procedure, columns defects can be easily taken into account with a probabilistic approach. Even though the use of jet-grouting for soil consolidation has greatly increased in recent years, at the design stage there is still a relevant degree of uncertainty due to the lack of reliable methods for predicting geometrical (diameter and position of each single column) and mechanical properties of the columns. Jet grouted bodies have geometrical defects and their properties differ from the ideal ones indeed. The diameter D of jet grouted columns is usually not constant: as known the diameter D of jet grouted columns is rarely constant, being the results of the interaction between the erosive action of the jet and the resistance of the natural soil to cutting (Croce and Flora, 2000, Modoni et al., 2016). Experimentally it has been observed that, at the same depth, the diameter of the columns of Jet varies according to a law of normal probabilistic function. Based on these experimental data, it can be assumed that the frequency distribution is symmetric around the mean value (which

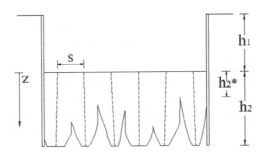

Figure 2.   Nominal (h2) and effective (h2*) length of the impermeable barrier; s: spacing of the columns.

may be considered to reduce with depth), thus resulting into a gaussian statistical distribution of column diameter. Instead, it can be shown a constant decrease of diameter with depth as result of soil shear strength increases. The relevance of these defects depends very much on soil properties, but it cannot be overlooked (Croce et al 2004). Another defect concerns the position of the columns, as their axis may deviate from the theoretical position due to many reasons. The situation becomes even more critical because the columns have to be injected from the ground surface prior to excavation, even in perfectly positioned columns, however, the axis deviates from the ideal position, as clearly demonstrated by some field measurements (Croce et al., 2004, Flora et al., 2011). Such a deviation must be considered, as the overlapping of adjacent columns and the continuity of jet grouted structures strongly depends on it. The azimuth α is randomly variable between 0° and 180°, and may be considered as having a uniform distribution. As a consequence, its statistical distribution needs no parameters to be defined. On the contrary, the angle β (deviation from ideal axis position in the vertical plane) seems to have a normal distribution (Croce et al., 2004), with a null average value (theoretical axis position) and a standard deviation depending on column length, soil properties and drilling accuracy. To implement the probabilistic approach to the Jet grouted bottom plug, and generally to the design of a jet grouted structure, the Montecarlo procedure can be used. Montecarlo methods are a class of computational methods based on random sampling of data to obtain numerical results and are used to derive estimates through simulations. It is based on an algorithm that generates a series of unrelated numbers, which follow the probability distribution that is supposed to have the phenomenon (or the variable) to be investigated.

The Monte Carlo simulation then calculates a series of possible realizations (or values) of the variable, with its weight of the probability. Once this random sample has been calculated, the simulation performs a 'measurement' of the quantities of interest on that generated sample (Croce et al., 2004, Saurer et al., 2011).

2   CASE STUDY: PROBABILISTIC MODELLING OF TRIBUNALE STATION PLUG

2.1   *Jetplug code*

The used Matlab code using Montecarlo approach was developed to probabilistically model a case study's plug. It allows to evaluate the void's area into the plug (depending on depth), due to the irregular shape of column, and to determine the water flow coming up through the head of the plug.

The code calculation steps are described in the following. Jetplug acquires data input by a user-friendly Excel file as shown in Figure 3.

The input data to be inserted in the Excel file are the following:

– plug thickness and start depth of the plug;
– K: coefficient of vertical permeability of the ground;
– Dm: average diameter that is obtained through the chosen injection parameters Flora et. al 2013;
– grid unit (x, y): discretization interval in the horizontal plane;
– grid unit (z): represents the evaluation interval along the vertical, of the sections of interest;
– center coordinates of the design jet grouted column at ground level, easily self-extracted by a .dwg file; in this way, the spacing between the columns are implicitly defined;
– coordinates of the plug's vertex;
– CV (D): coefficient of variation of the diameter;
– DS (α): standard deviation of inclination of the column axis respect to the vertical line;
– Pf: failure probability: that in civil application is assumed, typically, equal to 5%;
– n: number of iterations to be performed within the Montecarlo procedure, in order to obtain a certain level of confidence in the results, choosing the probability of failure.

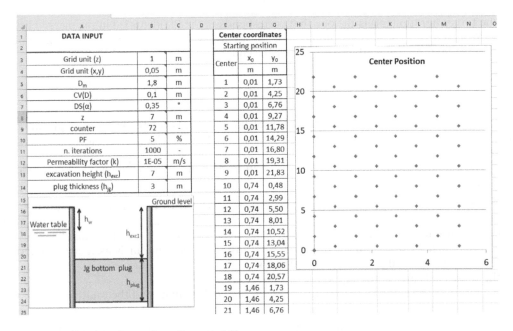

Figure 3. Jetplug interface, .xls preformatted file.

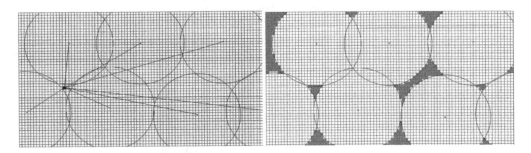

Figure 4. Horizontal control grid and voids area identification.

Furthermore, for each column, the code generates several values, equal to the number of iterations, of D_m, α and β. At the end of the iterations, Jetplug produces, for each depth, a horizontal control grid trought which identify, by a specific control procedure, the voids area (Figure 4).

Assuming that the column diameter is kept constant with depth (the approximation is appli-cable because the diameter variability is already taken into account through the use of the probability distribution for the diameter), Jetplug evaluates the position of the centers at various depths, assuming a linear variation law, because it is assumed that the center of the column follows the vertical trend of the injection rod.

Considering a variation range of α between (-90°, +90°) and β between (0°, 180°) the application of a linear variation formula, produces the positions of the centers at depth z.

Since all the voids position have been identified for each iteration Jetplug calculates the total area of the voids multiplying the area of the square unit of the grid by the number of voids found. Since Jetplug assigns to all the column and all iterations the same probability distribution, all iterations have the same happening chance, so the voids area value, referred to the chosen failure probability, is easily identifying.

Jetplug orders all the iterations depending on the decreasing value of the voids area, the first iteration will be the larger voids area.

1521

The target iteration (nt) is the one that corresponds to the Pf, identified in the merged list as $nt = Pf \cdot n$, where n is the total number of the iteration. Fixed the target iteration, Jetplug determines a volume of voids in the plug by the equation 1:

$$V_{tot} = A_{voids\,tot,1} \cdot \frac{g_z}{2} + \sum_{i=2}^{m-1} A_{voids\,tot,i} \cdot g_z + A_{voids\,tot,m} \cdot \frac{g_z}{2} \qquad (1)$$

where $A_{voids\ tot,1}$ is the voids area at the top of the plug, g_z is the thickness of depth discretization in m equal size, $A_{voids\ tot,i}$ is voids area at the head of the generic step; $A_{voids\ tot,m}$ is voids area at the plug's bottom.

At the end of calculation, Jetplug, produces an Excel results file containing the main following data:

- a graphic representation of the plug head section related to the selected Pf, showing the void area by specific contour (Figure 6). A green color is assigned untreated soil area (voids), founded inside a boundary zone of the plug, identified by a width equal to the average diameter of the columns (border voids). Instead a red contour is assigned to internal remaining untreated soil area (internal voids);
- the total voids area at the top of the plug;
- A variation, with depth, of the ratio between the voids volume and the total volume of the plug (Figure 7).

The water flow growing up through the untreated soil evaluated by the equation 2:

$$Q_{tot} = A_{un.tot} \cdot K \cdot j \qquad (2)$$

$$j = \frac{h_1}{h_2} = \frac{h_w}{h_{jg}}$$

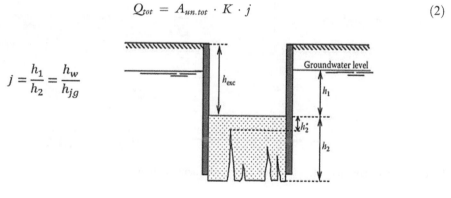

Figure 5.   Hydraulic gradient through the plug voids.

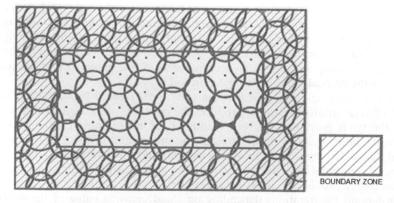

Figure 6.   Columns probabilistic arrangement at the plug head.

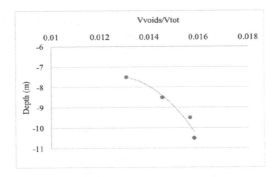

Figure 7. Variation, with depth, of the ratio between the voids volume and the total volume of the plug.

where K is the permeability of the soil, j is the hydraulic gradient in the voids of the plug, calculated, in the simplified hypotheses shown in the Figure 5.

2.2 *The case of Tribunale Station*

The Tribunale station is located among the Line 1 metro of Naples subway, near "Poggioreale" neighborhood. In order to define the stratigraphic and mechanical soil behavior, some geotechnical investigation was carried out between 2012 and 2015. The investigation results allow to define that the in-situ subsoils are originated by the pyroclastic Vesuvio activities, with the alternation of sandy and silty ashes and layers of yellow tuff's pozzolane. The water level was found at a depth of 2 m below the ground level. The station building has required a soil excavation characterized by narrow shape, reaching a depth of 7m from the free field. The northeast border of the station is defined by the Circumvesuviana building (Figure 8). The excavation retaining structures were designed to avoid damages to the Circumvesuviana building and to the other adjacent structures, limit the water inflow to the excavation bottom, avoid settlements of the surrounding areas and ensure the stability of the excavations.

For these reasons, the design needs the construction of a bulkheads along the northwest perimeter of the station and a bottom plug. The designed jet grouted plug has a variable thickness from 3 to 4m, starting at a depth of 7m below the ground level. It will consist of Jet-grouted columns with a design diameter of 1.8m, arranged according to a triangular mesh of 1.45m spacing. In order to validate the project plug parameters (Diameter, thickness, spacing etc.) a specific in-situ test field was carried out. In this field, 27 test columns were injected with 2 sets of different injection parameters (Injection pressure, water/cement ratio and mix flow).

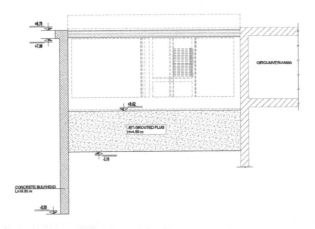

Figure 8. Tribunale station cross section.

The columns were injected from -1 to -11 m. To verify the effective diameters of the columns they were discovered up to 2m and the design and injection parameters of the columns were confirmed. Figure 8 shows a cross section of the station.

2.3 Modelling results

The probabilistic analysis, carried out with Jetplug code applied on the case study, was aimed to:

– evaluate the effect of the statistical variability of designs parameters of the column (such as average diameter, inclination of the columns) that compose the plug;
– estimate the water flow arising from the excavation due to the presence of untreated soil at the head of the plug. The maximum value of water flow is useful to identify the suitable pumping system (typically reasonable water flow is less than 1 lt/s);
– optimize the design's mesh of the columns, with the same average diameter of the plug, to reduce costs.

To optimize the computational times the entire plug was divided into 6 areas that were analyzed individually. For each area, 3 analyses have been performed, varying the diameters of columns. In particular:

– design diameter (1.8m);
– measured diameter at one meter from the ground level in the test field (2.2m);

Jetplug input parameters		
D_m	1.8	m
CV(D)	0.1	m
DS(α)	0.35	°
PF	5	%
n. iterations	1000	-
Excavation height	7	m
plug thickness	4	m
B	8	m
L	41	m

a)   b)

Figure 9.   Excel data input for Jetplug analysis (9.a) and design disposition of the column at free field (9.b).

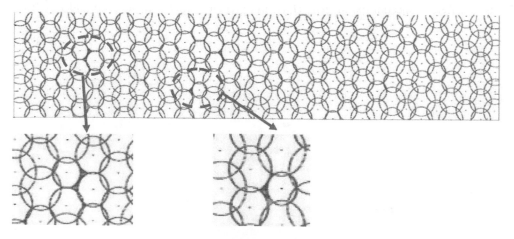

Figure 10.   Probabilistic (Pf = 5%) columns arrangement at the head of the plug. In red and green contour are represented untreated soil area (voids).

Grid Unit(x,y)	Area of voids	Untreated soil permeability factor (k_un)	i	Q
m	m²	m/s	–	l/s
0.05	0.53	0.000012512	1.25	0.0083

Figure 11. Evaluated water flow through the voids by Jetplug code in the modeled area.

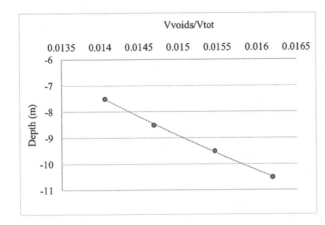

n	z
Vvoids/Vtot	m dal pc
0.002790	-7
0.003026	-8
0.003239	-9
0.003582	-10
0.003941	-11

Figure 12. Variation, with depth, of the ratio between the voids volume and the total volume of the plug (Pf = 5%).

– medium diameter Dm calculated with the formula proposed by Flora et al. 2013, (the formula was specifically calibrated on the soil mechanical characteristics and the columns injection parameters).

For example, in the following are presented the results of one analysis. In Figure 9 the input data are summarized, meanwhile Jetplug analysis results are represented in Figures 10–12.

As can be seen, comparing Figure 9.b and Figure 10, the differences between the design arrangement at the ground level columns and the one obtained by the probabilistic analysis at the head of the plug are considerable: the presence of the voids is graphically represented in the Figure 10 and the voids area is reported in the Figure 11. The estimated water flow rate is shown in the Figure 11 while in Figure 12 is reported (numerically and graphically) the trend of the index n with the depth; as previously illustrated also in that case the voids volume

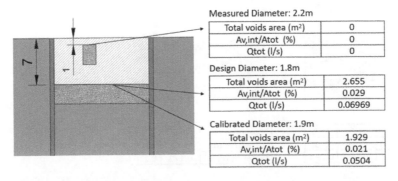

Measured Diameter: 2.2m

Total voids area (m²)	0
Av,int/Atot (%)	0
Qtot (l/s)	0

Design Diameter: 1.8m

Total voids area (m²)	2.655
Av,int/Atot (%)	0.029
Qtot (l/s)	0.06969

Calibrated Diameter: 1.9m

Total voids area (m²)	1.929
Av,int/Atot (%)	0.021
Qtot (l/s)	0.0504

Figure 13. Total results of the analysis Tribunale Station plug.

increases with depth due to the statistical variability of the design parameters of the columns. Finally, in Figure 13, are reported the total results of the analysis for the entire plug, obtained summarizing the single area's results.

As shown in Figure 13, the entire plug analysis carried out considering the diameter measured on site returns a total area of the voids at the head of the zero voids which then leads to a zero water flow. These results are not to be considered reliable because, due to the dispersion of the design parameters of the columns, this diameter, measured at 1 m from the ground level, is not the one expected at the head of the plug (at 7 m from the ground level). A more reliable prediction is represented by the modeling carried out with the diameter provided by the calibrated formula of the average diameter which shows how the expected flow rate is of the order of a few deciliter per second and therefore widely manageable on site.

As above illustrated, Jetplug allows to evaluate the effect of the statistical variability of the bottom plug design parameters and the seepage water flow from the excavation, according to the design parameters. Moreover, by using Jetplug code, it is possible to evaluate the effects of the variation of the design parameters (plug thickness, spacing and diameter of the columns) to identify the optimal one, statically and hydraulic compatible with the safety conditions and characterized by reasonable water flow manageable by usually pumping systems. That Jetplug application allows a consistent reduce of time and costs.

An example of that application is shown below. Jetplug analysis was carried out considering a variation of the design parameters set of the Tribunale station bottom plug. It was considered an equilateral triangular mesh of the columns with a center distance of 1.5m and 1.55m. In these analyses was implemented an untreated soil height on the plug and a reduction

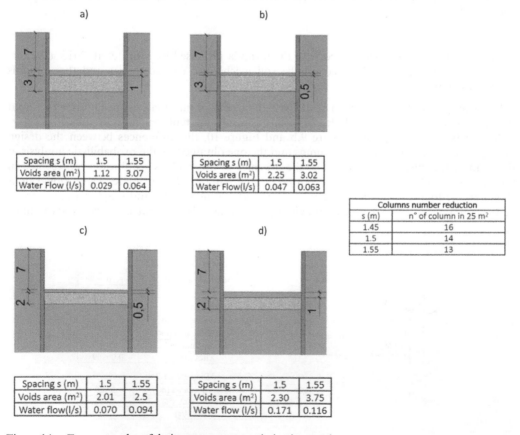

a)

Spacing s (m)	1.5	1.55
Voids area (m²)	1.12	3.07
Water Flow (l/s)	0.029	0.064

b)

Spacing s (m)	1.5	1.55
Voids area (m²)	2.25	3.02
Water Flow(l/s)	0.047	0.063

Columns number reduction	
s (m)	n° of column in 25 m²
1.45	16
1.5	14
1.55	13

c)

Spacing s (m)	1.5	1.55
Voids area (m²)	2.01	2.5
Water flow(l/s)	0.070	0.094

d)

Spacing s (m)	1.5	1.55
Voids area (m²)	2.30	3.75
Water flow (l/s)	0.171	0.116

Figure 14.   Four examples of design parameters optimization results.

of the treated soil height. The whole optimized design mesh respected the uplift equilibrium and the structural one. The application of this design strategy could had reduce the bottom plug realization costs, since the respect of the equilibrium to the lifting can be guaranteed by reducing the height of treated soil and calling to collaborate a thickness of untreated soil, which for the purposes of balance to lifting plays the same role, since the weight of the volume unit of treated and untreated soil is the same.

3 CONCLUSIONS

The paper shows a simple and rational way to design jet grouted impermeable bottom barriers, explicitly considering a risk level and considering defects in columns diameter and position of axis via a probabilistic approach. To evaluate the behavior of jet grouted bottom plugs, the Tribunale station of Metro Line 1 bottom plug was modeled. The analysis was developed by using the Monte Carlo probabilistic approach, implemented in a Matlab code named JetPlug; this code can be easily implemented and used in design practice. To evaluate the statistical variability effect of the jet grouted column design parameters (average diameter, inclination of the columns, etc..), synthetic results have been shown.

When defects within the barrier leave an untreated way through it, water flows into the excavation. Jetplug allows to evaluate the seepage water flow from the excavation varying the design parameters (plug thickness, spacing and diameter of the columns), verifying that it is compatible with the safety conditions and if it can be managed by pumping systems. Finally, by using Jetplug code, it was possible not to evaluate the effects of the variation of the design parameters (plug thickness, spacing and diameter of the columns) in order to identify the optimal one, statically and hydraulic compatible with the safety conditions, allowing a consistent reduce of time and costs.

REFERENCES

Croce, P. & Flora, A. 2000. Analysis of single fluid jet-grouting. *Geotechnique* Vol. 50, N. 6, 739–748, Thomas Telford Editore, Londra, Inghilterra, ISSN 0016-8505.

Croce, P., Flora, A. and Modoni, G. 2004. Jet grouting: progetto, esecuzione e controllo. *Hevelius editore, Benevento*, ISBN 88-86977-57-3.

Flora, A., Lirer, S., Lignola, G.P. and Modoni, G. 2011. Mechanical analysis of jet grouted supporting structures. *VII International Symposium on Geotechnical Aspects of Underground Construction in soft ground*, Viggiani G. Ed., *Rome*. ISBN: 978-0-415-68367-8

Flora, A., Lirer, S. and Monda, M. 2012. Probabilistic design of massive jet grouted water sealing barriers. *IV International Conference on Grouting and Deep Mixing (New Orleans)*, Vol. 2, 2034-2043, Johnsen L.F., Bruce D.A., Byle M.J. Ed, ASCE,. ISBN 978-0-7844-1235-0.

Flora, A., Modoni, G., Lirer, S. and Croce, P. 2013. The diameter of single, double and triple fluid jet grouting columns: prediction method and field trial results. *Géotechnique*, 63, No. 11: 934–945. DOI 10.1680/geot.12.P.062, ISSN 0016-8505.

Modoni, G., Flora, A., Lirer, S., Ochmański, M., and Croce, P. 2016. Design of Jet Grouted Excavation Bottom Plugs. *Journal of Geotechnical and Geoenvironmental Engineering*, ASCE, 142(7). DOI: 10.1061/(ASCE)GT.1943-5606.0001436

Tunnels and Underground Cities: Engineering and Innovation meet Archaeology,
Architecture and Art, Volume 4: Ground improvement in
underground constructions – Peila, Viggiani & Celestino (Eds)
© 2020 Taylor & Francis Group, London, ISBN 978-0-367-46868-2

Centrifuge model tests on shallow overburden tunnels with pre-ground improvement to clarify seismic behavior

Y. Sawamura, K. Konishi, K. Kishida & M. Kimura
Kyoto University, Kyoto, Japan

ABSTRACT: Due to the development of auxiliary construction methods, such as pre-ground improvement, the NATM is often used to excavate tunnels with a shallow overburden in a soft ground. The improvement stabilizes the tunnel face and suppresses the subsidence of the ground surface. In previous researches, although the optimum ground improvement area was examined through experiments and numerical simulations, the seismic behavior of the tunnels was not clearly discussed. In this study, dynamic centrifugal model experiments are conducted to clarify the dynamic behavior of shallow tunnels with pre-ground improvement. The experimental results indicate that when the entire ground around a tunnel is improved, the shear deformation of the tunnel can be suppressed. On the other hand, when the ground around the tunnel crown and top section is improved, the response of the tunnel is amplified by the concentration of weight at the upper part of the tunnel.

1 INTRODUCTION

Several problems arise when excavating a tunnel with a shallow overburden in a soft ground, such as the instability of the tunnel face and surface subsidence. In particular, because the loosening of the ground induced by the tunnel excavation expands to the surrounding ground, the influence will directly reach the ground surface. Traditionally, the cut and cover tunneling method has been widely used for the excavation of shallow overburden tunnels in soft grounds. Due to the technical development of auxiliary construction methods, however, the New Austrian Tunneling Method (NATM) is now often adopted for these shallow over-burden tunnel excavations. Although the NATM is more economical than the cut and cover tunneling method or shield tunneling method, the NATM presupposes the formation of a ground arch in the surrounding ground and the stability of the tunnel face. For this reason, when a shallow overburden tunnel is to be constructed by the NATM, safe and appropriate tunnel face stabilization methods are required for the excavation.

In the constructions of Tohoku Shinkansen (Bullet Train) and Hokuriku Shinkansen, several tunnels with a shallow overburden were constructed in sandy soil mountains due to the linear constraints of the topography. Therefore, in order to secure the stability of the tunnel face and to suppress the subsidence of the ground, pre-ground improvement was applied before the tunnel excavations to sections with few restrictions on the ground segments. Figure 1 and Table 1 show the physical properties of the ground around the tunnels to which the pre-ground improvement method was applied. The area and the strength of the improved ground varied depending on the conditions, such as the overburden, the geological conditions, and the allowable settlement. The ground around the tunnel crown and top section was improved in Ushikagi Tunnel, while the entire ground around the tunnel was improved in Akahira Tunnel (Nonomura et al., 2013). Figure 2 shows the construction process of the pre-ground improvement method. Firstly, the ground is excavated from the ground surface to the crown of the tunnel, and cement is mixed with natural ground around the side wall of the tunnel using the shallow or deep mixing stabilization method. Then, the excavated soil and

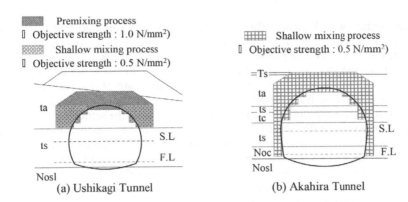

(a) Ushikagi Tunnel (b) Akahira Tunnel

Figure 1. Area and method of ground improvement around shallow overburden tunnel (After Nonomura et al., 2013).

Table 1. Physical properties of ground around tunnel with pre-ground improvement (After Nonomura et al., 2013).

	Symbol	N value	γ [kN/m³]	E [kN/m²]	ν	c [kN/m²]	ϕ [°]
Ushikagi Tunnel							
Volcanic ash	ta	2 – 4	14	3.5×10^3	0.35	36	0
Sandy soil	ts	10 – 15	18	2.0×10^4	0.35	20	30
Shallow mixing soil	-	-	19	5.0×10^4	0.35	144	30
Premixing soil	-	-	19	1.0×10^5	0.35	288	30
Akahira Tunnel							
Surface soil	Ts	-	15	5.0×10^3	0.4	13	-
Volcanic ash	ta	3 – 8	15	1.3×10^4	0.4	30	-
Sandy soil	ts	7	17	1.8×10^4	0.35	-	31
Cohesive soil	tc	3 – 4	15	1.0×10^4	0.4	57	-
Sandy soil	ts	7 – 22	18	3.0×10^4	0.35	-	31
Cohesive soil	Noc	5	17	1.3×10^4	0.4	31	-
Sandy soil	Nosl	16 – 50	20	1.3×10^5	0.3	-	38
Shallow mixing soil	-	-	19	1.0×10^5	0.35	288	30

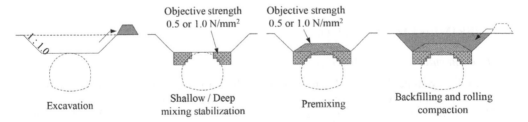

Figure 2. Construction process of pre-ground improvement method (After Kishida et al., 2016).

cement are premixed and backfilled. Finally, the excavated soil is compacted by rolling it on the ground surface. After these steps, the tunnel is excavated using the NATM. For the sections of these two tunnels for which the pre-ground improvement method was adopted, the ground surface subsidence was reported to have been suppressed and the tunnels were constructed securely (Nonomura et al., 2013). However, it should be kept in mind that the area and the strength of the improved ground were determined based on empirical judgment.

In previous researches, the optimum ground improvement area was examined through experiments and numerical simulations performed during the excavation process. Kishida et al. (2016) conducted a series of three-dimensional trapdoor experiments and corresponding FE analyses to evaluate the effect of the pre-ground improvement method during a tunnel excavation. In addition, the enhancement of tunnel stability resulting from the application of the ground improvement method was discussed. Based on the results, they reported that the advantages of the pre-ground improvement method can be presented as three issues, namely, the effect of shear reinforcement, the effect of earth pressure redistribution, and the effect of ground reinforcement. These three issues were seen as becoming even more effective as the width and the height of the improved ground increased. Cui et al. (2018) conducted 2D elasto-plastic FE analyses that simulated the excavation process for a tunnel with pre-ground improvement. They confirmed that the pre-ground improvement method was able to effectively prevent the settlement of the ground and the tunnel when the ground was improved down to the tunnel feet. Furthermore, the effect of the pre-ground improvement method was seen to slightly increase with an increase in the strength of the improved ground. However, the seismic behavior of the tunnel with pre-ground improvement was not clearly discussed in their study. It is thought that the difference in stiffness between a soft ground and an improved ground strongly affects the stability of a tunnel during an earthquake.

In general, it is thought that mountain tunnels are seismic-resistant structures because they are surrounded by strong ground such as rock. However, within the current century, several mountain tunnels have suffered damage, for example, in the 2004 Mid-Niigata Prefecture Earthquake (Konagai et al., 2009; Jiang et al., 2010), the 2008 Wenchuan Earthquake (Wang et al., 2009; Shen et al., 2014), and the 2016 Kumamoto Earthquake (Zhang et al., 2018). Yashiro et al. (2007) analyzed the damage done to mountain tunnels in Japan due to earthquakes and classified this seismic damage into the following three categories: (1) Damage to shallow tunnels, (2) Damage to tunnels in poor geological conditions, and (3) Damage to tunnels due to fault slides. From this classification, it is confirmed that the seismic behavior of shallow overburden tunnels must be considered. The ground around these tunnels is often loose due to the small confining pressure; and thus, large shear deformation occurs during earthquakes. In such cases, it is highly possible that bending cracks will occur at the shoulder parts of the lining due to the bending moments generated in the lining (Yashiro et al., 2007).

Many researches have been conducted on the earthquake behavior of underground structures, such as tunnels, under the influence of ground shear deformation. Earthquake-resistant design methods for underground structures, including the free-field method, have been proposed (e.g., Wang, 1993; Hashash et al., 2001) On the other hand, as mentioned above, in terms of tunnels with pre-ground improvement, the effect of the improved ground on the seismic behavior of the tunnels has not been clarified. In this study, therefore, dynamic centrifugal model tests were conducted to clarify the dynamic behavior of shallow tunnels with pre-ground improvement.

2 EXPERIMENTAL CONDITIONS

2.1 Experimental devices and objects

In this study, dynamic shaking table tests were conducted under a gravitational acceleration of 50G using a geotechnical centrifugal device at the Disaster Prevention Research Institute (DPRI), Kyoto University. Table 2 shows the specifications of the geotechnical centrifugal device. The experimental subjects were two Shinkansen tunnels constructed by the NATM. The thickness of the overburden above the tunnels was set to $0.5D$ (D: outer diameter of the tunnels) based on the construction records of NATM tunnels with pre-ground improvement. The improved ground patterns were the same as those adopted at actual construction sites, as shown

Table 2. Specifications of geotechnical centrifugal device at DPRI, Kyoto University.

Specific	Geotechnical centrifuge
Effective rotation radius [m]	0.20
Effective space for model installation [m]	0.80 (L) × 0.36 (W) × 0.80 (H)
Experiment capacity [G × ton]	24
Maximum centrifuge acceleration [G]	200 for static test, 50 for dynamic test
Maximum number of rotations [rpm]	260

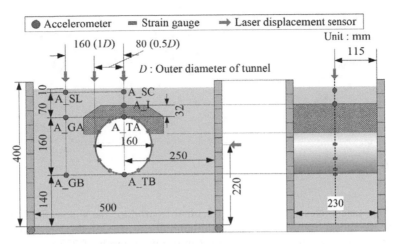

Figure 3.   Schematic illustration of experimental setup (Case-2).

Table 3.   Material properties of Toyoura sand.

Specific gravity G_s	2.64
Average diameter D_{50} [mm]	0.20
Internal friction angle ϕ [deg]	38.9
Cohesion c [kPa]	0
Maximum void ratio e_{max}	0.975
Minimum void ratio e_{min}	0.585
Unit weight (D_r =85 %) γ [kN/m³]	15.76

in Figure 1 and Table 1. Figure 3 presents a schematic illustration of the experimental setup. A flexible shear beam soil container, 500 mm wide, 230 mm deep, and 400 mm high, was used.

The container consisted of 10 shear beams with a height of 40 mm. Each shear beam could move independently during the shaking experiments. However, from the preliminary experiments using only soil samples, it was revealed that the reproducibility of the displacement of the side walls was poor. Therefore, the side walls were integrated with aluminum plates so that the soil container would permit the simple shear deformation of the model ground. Dry Toyoura sand was used as the ground material. Table 3 shows the physical properties of Toyoura sand. The model ground was constructed by compaction so that a relative density of 85% would be achieved.

2.2   Experimental cases

Figure 4 shows the three experimental cases. In Case-1, a tunnel without ground improvement was modeled despite a shallow overburden condition. Although this case would not be practically implemented, it is set for comparison with the two other cases with ground improvement. In Case-2, the ground around all the cross-sections of the tunnel was improved, while in Case-3, the ground around the crown of the tunnel and top section was improved. The improved ground patterns in Cases-2 and -3 were determined by referring to actual construction records (Nonomura et al., 2013).

As a prerequisite for the experiments, rock bolts and shotcrete were not modeled because their effects on the seismic behavior of a tunnel are small and they are difficult to model. Moreover, the stress release of the ground around the tunnel during its excavation was not considered.

2.3   Modeling of tunnels

The model tunnels were manufactured so that their size and flexural rigidity would be as consistent as possible with the typical NATM tunnels of the Shinkansen sections, taking into consideration the boundary effect from the soil container. In addition, the model tunnels were

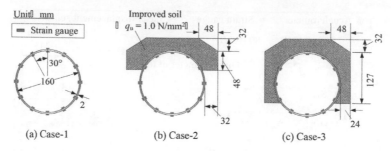

Figure 4. Experimental cases.

manufactured as true circles because it is difficult to reproduce the shape of an actual tunnel. Consequently, the model tunnels were made of aluminum, as true circles, having an outer diameter of 160 mm, a lining thickness of 2 mm, and a depth of 225 mm. Table 4 shows the relationship between an actual tunnel and a model tunnel.

2.4 Modeling of ground improvement

The pre-ground improvement was modeled with reference to the Japanese design manual for mountain tunnels. The improved ground was made by mixing Kasaoka clay ($w_P = 29.6\%$ and $w_L = 62.1\%$), high-early-strength cement, and water with the target of a uniaxial compressive strength q_u of 1.0 N/mm². The combined ratio of Kasaoka clay, high-early-strength cement, and water was 5: 1: 3.75 by weight.

Originally, when the pre-ground improvement method was adapted for the in-situ construction of a tunnel excavation, as shown in Figure 2, the ground around the tunnel was improved prior to the tunnel excavation. In this experiment, however, in order to simplify the experimental conditions and to increase the quality of the improved ground, the improved ground was cast to the model tunnels beforehand and then each model tunnel with an improved ground was put into the soil container. Figure 5 shows the procedure for making the improved ground. The improved ground was removed from the mold form 48 hours after casting. The ground was cured in water for 5 days after casting and then dried in air for 48 hours.

In Cases-2 and -3, wherein the improved ground was attached to the tunnel model, in order to strengthen the adherence of the tunnel and the improved ground, several screw holes were made at the end of the tunnel (avoiding the area where the instruments were equipped), and screws were attached from the inside of the model tunnel (Figures 5 (a) and (d)). A 5.0-mm-thick sponge and a Teflon sheet were attached to both sides of the improved ground in the depth direction to reduce the friction between the improved ground and the soil container. Moreover, for the wall of the soil container, Teflon sheets were also attached parallel to the direction of excitation.

Table 5 shows the material constants of the improved ground. Compared with the in-situ construction, as shown in Figure 1 and Table 1, the density of the improved ground in this experiment was about the same as that of the surrounding ground, although the improved ground at an actual site is heavier than the surrounding ground.

Table 4. Relationship between actual and model tunnels.

	Actual tunnel p	Model tunnel (prototype) m_p	Ratio p/m_p
Young's modulus E [N/mm²]	2.20×10^7	7.06×10^7	0.31
Outer diameter D [m]	10.1	8.0	1.26
Thickness t [m]	0.30	0.10	3.0
Axial stiffness EA [kN]	2.03×10^8	1.75×10^8	1.16
Bending stiffness EI [kN · m²]	2.44×10^9	1.37×10^9	1.78

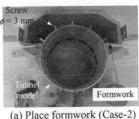

(a) Place formwork (Case-2) (b) Cast improved ground (Case-3)

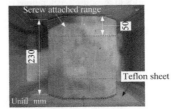

(c) Attach sponge cushion and Teflon sheet (d) Install into soil container

Figure 5. Casting-improved ground and installation into soil container.

Table 5. Material constants of improved ground.

Young's modulus E [kN/m^2]	2.92×10^5
Unit weight γ [kN/m^3]	15.67
Compressive strength f_c [N/mm^2]	1.01
Tensile strength f_t [N/mm^2]	0.29
Poisson's ratio ν	0.20

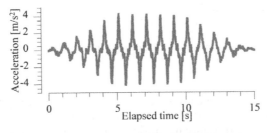

Figure 6. Time history of acceleration measured at shaking table (Case-1).

2.5 *Input waves and measurement items*

15 tapered sinusoidal waves, with a frequency of 1 Hz of the prototype, were input by controlling the displacement of the vibration table. The maximum acceleration of the input waves was about 4.0 m/s^2 in prototype scale. Figure 6 shows the time history of the input waves which were measured at the shaking table in Case-1. The response accelerations at the tunnel and the surrounding ground (right-hand side: positive), the strain occurring in the tunnel, and the horizontal displacement of the wall of the soil container (right-hand side: positive) were measured.

3 EXPERIMENTAL RESULTS

3.1 *Seismic responses of tunnel and surrounding ground*

The experimental results are expressed in the prototype scale unless otherwise noted. Figures 7 (a)-(f) show the time histories of the response accelerations of the tunnels and the improved ground, and the Fourier spectrums. In the figures, the response magnification (S_r) of

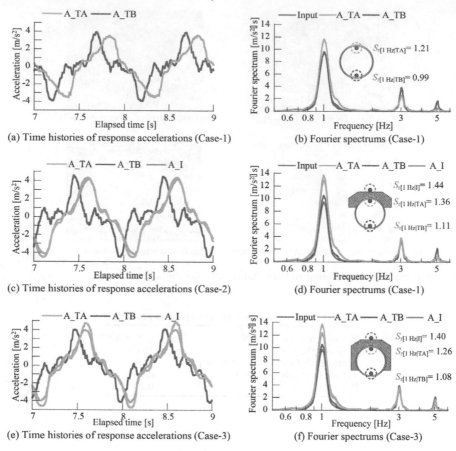

Figure 7. Time histories of response accelerations at tunnel and improved ground and Fourier spectrums.

the Fourier spectrums at a frequency of 1 Hz, defined by dividing the Fourier spectrum of the response acceleration by that of the input acceleration, is also shown. In terms of the Fourier spectrums and their response magnification, the response accelerations at A_TA are larger than those at A_TB in all cases, and the response magnification at A_TA in Case-2 was the maximum among all cases. It is considered that the improved ground behaves integrally with the tunnel because of the strong adhesion to the tunnel rather than to the surrounding ground. Therefore, in Case-2, where the ground around the crown of the tunnel and top section was improved, the tunnel became an unstable structure, with the weight concentrated at the upper part of the tunnel, and the inertial force acting on the tunnel increased.

Looking at the time histories of the response accelerations, A_TA responds with a delay behind A_TB in all cases. Figure 8 shows the relative response accelerations at A_TA with respect to A_TB. The maximum relative response acceleration is large in the order of Case-2 > Case-1

> Case-3. It is thought that when large relative acceleration occurs, large shearing deformation occurs between the crown and the bottom of the tunnel, so the generation of a large sectional force can be expected in Case-2. Regarding the response accelerations of A_TA and A_I in Figure 7, in Case-2, both the phase and the intensity of the response accelerations are in agreement. On the other hand, in Case-3, the intensity of the response accelerations of A_I is larger than that of A_TA, although these phases are almost the same. This is because, in Case-2, the response of the tunnel at the upper part is large, and the shear deformation of the

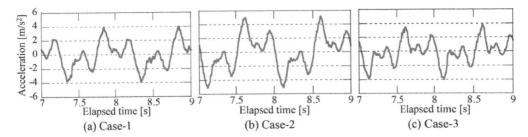

Figure 8. Time history of relative response accelerations (TB vs TA).

tunnel excels. On the other hand, in Case-3, the shear deformation of the tunnel is suppressed by highly rigid refiners; and thus, the rotational motion is superior.

From the above considerations, the seismic behavior of a tunnel with/without ground improvement is suggested as shown in Figure 9. When the ground around the tunnel crown and top section was improved (Case-2), the response of the tunnel was amplified by the concentration of weight at the upper part of the tunnel. On the other hand, when the entire ground around the tunnel was improved (Case-3), the shear deformation of the tunnel was suppressed by increments in the whole rigidity of the surrounding ground. Thus, the rocking motion will be dominant during earthquakes.

3.2 Section forces generated in the tunnel

Figures 10 and 11 show the distributions of bending moments and axial forces generated in the tunnels when the right-hand side shear strain of the ground reached the maximum. The shear strain of the ground was obtained by dividing the horizontal displacement of the soil container by the wall height. A positive bending moment was defined for the case where tension was generated outside the tunnel, and the compressive axial force was defined as a positive value. The initial and residual values are also plotted in these figures.

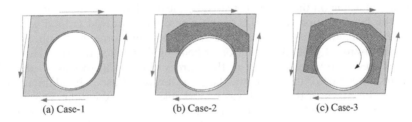

(a) Case-1   (b) Case-2   (c) Case-3

Figure 9. Expected earthquake response mode of tunnel in each case.

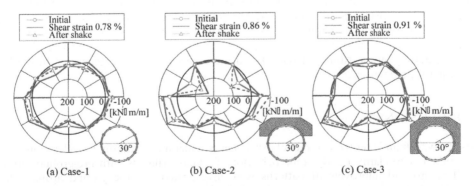

(a) Case-1   (b) Case-2   (c) Case-3

Figure 10. Distribution of bending moments generated in tunnel when right-hand side shear strain of ground reached maximum.

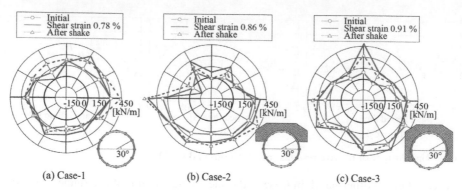

Figure 11.   Distribution of axial forces generated in tunnel when right-hand side shear strain of ground reached maximum.

In Case-1, the tunnel underwent oval deformation in the oblique direction by receiving shear deformation of the surrounding ground during excitation. At that time, large bending deformations occurred at the shoulder parts. At the left shoulder, positive bending moment occurred, whereas at the right shoulder, negative bending moment occurred. In addition, the axial force decreased at the left shoulder, and it increased at the right shoulder. Since this behavior is qualitatively in agreement with the behavior of a circular tunnel during an earthquake, discussed in a past research (e.g. Wang, 1993), it is confirmed that this experimental study can reproduce the seismic behavior of circular tunnels.

In Case-2, the bending moment in the region where the improved ground was placed was suppressed during vibrations. On the other hand, at the boundary between the improved ground and the surrounding ground, the sectional forces were concentrated due to the difference in rigidity, and large internal bending occurred. Moreover, the axial force decreased in these areas and tension (axial force was less than zero) occurred. In terms of the axial force generated in the lower part of the tunnel, it was smaller than in Case-1. This is considered to be due to the fact that the self-weight acting on the tunnel increased by the installation of the improved ground.

In Case-3, the bending moment in the region where the improved ground was placed was suppressed during vibrations, as it was in Case-2. The sectional forces were concentrated at the boundary of the improved ground, so that outer bending dominated the area. However, unlike Case-2, the changes in bending moment in these areas were small. In Case-3, the axial force generated at the lower part of the tunnel was large, as it was in Case-2. In addition, the increase in axial force at the crown of the tunnel was remarkable during excitation. This is because the self-weight acting on the tunnel had increased due to the installation of the improved ground. Moreover, the shearing deformation of the tunnel was suppressed by the improved ground; and thus, a large compressive force was generated at the crown of the tunnel.

From the above, it is confirmed that the bending moments generated in the tunnel were suppressed in the area where the improved ground was placed, but that large cross-sectional forces were generated at the boundary between the improved ground and the unimproved ground. On the other hand, it was revealed that when the improved ground was placed, the self-weight acting on the tunnel increased and the compressive force generated in the lower part of the tunnel increased.

4   CONCLUSIONS

In this study, dynamic centrifugal model experiments under a gravitational acceleration of 50 *G* were conducted to clarify the dynamic behavior of shallow tunnels with pre-ground improvement. Two ground improvement patterns were investigated: (1) the ground around all the cross-sections of the tunnel was improved and (2) the ground around the crown of the tunnel and top section was improved. The results from these two patterns were compared with a case without ground improvement. The findings obtained from this research are as follows.

1) When the ground around the tunnel crown and top section was improved (Case-2), the response of the tunnel was amplified by the concentration of weight at the upper part of the tunnel. As a result, the different phases between the responses at the bottom and the top of the tunnel increased, causing large shear deformation of the tunnel at the boundary between the improved ground and the unimproved ground.

2) When the entire ground around the tunnel was improved (Case-3), the shear deformation of the tunnel could be suppressed by increments in the whole rigidity of the surrounding ground. Accordingly, it was found that the rocking motion of the tunnel would be dominant during earthquakes.

3) Although the improved ground preserved the shear deformation of the tunnel, large cross-sectional forces were generated at the boundary between the improved ground and the unimproved ground due to the difference in rigidity. Therefore, it can be said that the earthquake resistance of a shallow overburden tunnel can be improved when the entire ground around the tunnel is improved. On the other hand, when the ground around the tunnel crown and top section is improved, there is a possibility of damage to the tunnel at the boundary between the improved ground and the unimproved ground.

REFERENCES

Cui, Y., Kishida, K. & Kimura, M. 2018. Numerical study on the effect of pre-ground improvement method on control of ground subsidence occurring in shallow overburden NATM tunnels. *Proc. of the 7th China-Japan Geotechnical Symposium*, 31–37, Sanya, 16–18 March 2018.

Hashash, Y. M. A., Hook, J. J., Schmidt, B. & Yao, J. I. C. 2001. Seismic design and analysis of underground structure. *Tunnelling and Underground Space Technology*, 16, 247–293.

Jiang, Y.J., Wang, C.X. & Zhao, X.D. 2010. Damage assessment of tunnels caused by the 2004 Mid Niigata Prefecture Earthquake using Hayashi's quantification theory type II, *Natural Hazards*, 53 (3),425–441.

Kishida, K., Cui, Y., Nonomura, M., Iura, T. & Kimura, M. 2013. Discussion on the mechanism of ground improvement method at the excavation of shallow overburden tunnel in difficult ground, *Underground Space*, 1(2),94–107.

Konagai, K., Takatsu, S., Kanai, T., Fujita, T., Ikeda, T. & Johansson, J. 2009. Kizawa tunnel cracked on 23 October 2004 Mid-Niigata earthquake: An example of earthquake-induced damage to tunnels in active-folding zones. *Soil Dynamics and Earthquake Engineering*, 29 (2), 394–403.

Nonomura, M., Iura, T., Cui, Y., Kishida, K. & Kimura, M. 2013. Discussion on mechanical evaluation of ground pre-improvement method under shallow overburden and unconsolidated ground. *Japanese Geotechnical Journal*, 8(2),165–177. (in Japanese)

Shen, Y.S., Gao, B., Yang, X.M. & Tao, S.J. 2014. Seismic damage mechanism and dynamic deformation characteristic analysis of mountain tunnel after Wenchuan earthquake. *Engineering Geology*, 180, 85–98.

Wang, J. N. 1993. Seismic design of tunnels: A state-of-the-art approach. Monograph 7, Parsons Brinckerhoff Quade & Douglas, New York.

Wang, Z. Z., Gao, B., Jiang, Y.J. & Yuan, S. 2009. Investigation and assessment on mountain tunnels and geotechnical damage after the Wenchuan earthquake. *Science in China Series E: Technological Sciences*, 52 (2), 546–558.

Yashiro, K., Kojima, Y. & Shimizu, M. 2007. Historical earthquake damage to tunnels in Japan and case studies of railway tunnels in the 2004 Niigataken-Chuetsu earthquake. *Quarterly Report of RTRI*, 48(3),136–141.

Zhang, X. P., Jiang, Y. J. & Sugimoto, S. 2018. Seismic damage assessment of mountain tunnel: a case study on the Tawarayama tunnel due to the 2016 Kumamoto Earthquake. *Tunnelling and Underground Space Technology*, 71, 138–148.

*Tunnels and Underground Cities: Engineering and Innovation meet Archaeology,
Architecture and Art, Volume 4: Ground improvement in
underground constructions – Peila, Viggiani & Celestino (Eds)
© 2020 Taylor & Francis Group, London, ISBN 978-0-367-46868-2*

Excavation of a large diameter tunnel in sliding Carpathian flysch

F. Schiavone, F. Carriero, F. Bizzi & G. Orlati
Astaldi S.p.A., Warsaw, Poland

ABSTRACT: Along the "S7" Expressway, the execution of the Mały Luboń tunnel is ongoing. It is a large diameter tunnel, consisting in two tubes with a length of 2km, which approaches the Carpathians. The northern part of the tunnel develops through the Magura flysch unit, while the southern one through the Hieroglyphic flysch, which is highly tectonized and characterized by folds/overthrusts. During the excavation of the southern entrance, a landslide has been detected inducing the designer to produce a new technical solution, to be implemented during the ongoing construction activities, aimed to increase the slopes stability and to review the tunnel face consolidations, according to the A.DE.CO.-R.S. method. All the geological information and monitoring data are introduced, analyzed and implemented to lead to the solution adopted to complete the entrance and to excavate a large diameter tunnel in a sliding geological unit, for the first time through the western Carpathians.

1 INTRODUCTION

1.1 *Project Location and general overview*

The E77 is a part of the inter-European road system, which in Poland directly connects Gdańsk to Slovakia, passing through Warsaw, by means of the National road DK7 and the expressway S7.

Close to Slovakia – South of Kraków, between Lubień and Rabka-Zdrój cities, the work for the S7 extension is ongoing.

The alignment has a total length of 12km, divided in three sections. The second section, to be executed by Astaldi S.p.A., includes mainly two infrastructures: a concrete box viaduct with a total length of 310m, consisting of two decks and five spans, and a twin tube tunnel with a total length of 2km and a maximum excavated cross section of about 190m^2.

In terms of geometry, length and position, the Mały Luboń is the most important tunnel in Poland. It linearly develops through a North-South direction by means of two tubes which are transversally connected by pedestrian passages, every 170m approximately, while a vehicular cross-passage connects the tubes in the middle of the tunnel, where a lay-bys area is expected as well.

1.2 *Geological Structure*

The Mały Luboń tunnel approaches the western Carpathians, which is the second-longest mountain range in Europe.

Within the above mentioned area, the mining activity develops across the Carpathian Flysch Belt (Plašienka at al. 1997), which is a thin-skinned thrust belt, formed by nappes, consisting of flysch - alternating marine deposits of claystones, shales and sandstones which were moved tens of kilometers to the North.

Within this Belt, several units are divided according to their structural position, in the frame of mountain range, and can be recognized as follow:

- Magura nappes (Lower Cretaceous to Eocene) in northern Carpathians.
- Krosno-Silesia nappes (Lower Jurassic to Lower Miocene) in northern Carpathians.
- Moldavide nappes (Lower Cretaceous to Lower Miocene) in Romanian Carpathians.

At the smallest geological scale, almost all the Mały Luboń tunnel is included in the Magura flysch unit, where the percentage of sandstone lithology is predominant compared with the other ones. The thickness of the sandstone layers varies from decimeters to meter and folds/thrusts are observed at the biggest scale.

Close to the southern entrance, instead, the tunnel mining operations will be carried out through the Hieroglyphic flysch unit (Figure 1) that differs from the Magura one for the following reasons: the percentage of shale/claystones/clay lithologies is quite high compared with the sandstone one; the thickness of the sandstone layers stops to decimeters and layers are often discontinuous; fold/thrust lines can be observed at a smaller scale.

1.3 *Tunnel excavation technology*

For the first time in Poland, the A.DE.CO.-R.S. tunnel excavation method (Lunardi 2001) has been implemented in the S7 expressway project. Astaldi S.p.A. proposed the above-mentioned method in place of the N.A.T.M. one, which was expected by tender design.

The following key points have been considered for the modification of the excavation technology:

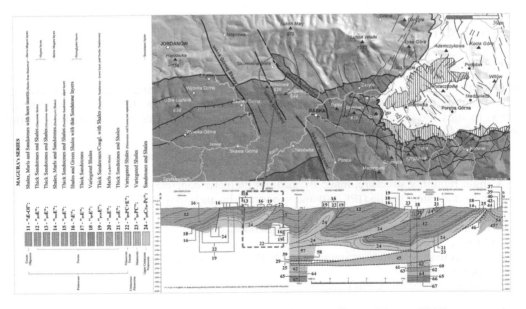

Figure 1. Geological structure of the project area: plan view according to Kłapyta (2015), cross-section according to Książkiewicz at al. (1991).

- To take into account the interaction between soil and temporary/final linings depending on the mutual stiffness.
- To take into account the time-reduction of the soil/rock parameters starting from the undisturbed configuration.
- To introduce the active design, especially in difficult geological conditions, by implementing tunnel core consolidations and crown supports, in order to perform a full-face excavation method.
- To introduce a proper monitoring system, strictly connected with the active design, aimed to limit as much as possible the plastic behavior of the soil around the tunnel, assuring an elastic equilibrium and then minimizing loads on temporary and final linings.

As a result of the above-mentioned requirements, with special reference to the southern entrance, a support class named "3" has been defined consisting on:

- Tunnel core consolidations, made by 46 fiberglass pipes (type 60/40 – L=15m) injected in drilled hole (Ø127mm) by cement mortar, under pressure.
- Tunnel crown protection, made by 45 forepoling steel pipes ($\varnothing_{ext}$=114.3mm – thk=8mm – L=15m) injected in drilled hole (Ø142mm) by cement mortar, under pressure.
- Temporary lining, made by a couple of steel ribs (type IPE180) with 1m longitudinal spacing.
- Maximum excavated cross-section equal to 190m^2, with maximum width equal to 18,31m and maximum height equal to 12,56m.
- Longitudinal distance between two consecutive consolidation cycles equal to almost 9m, with a consequent relative overlap equal to 6m.
- Maximum distance between tunnel face and final invert arch equal to almost 18m (2 excavated rounds).
- Maximum distance between tunnel face and final top-heading lining equal to almost 45m (5 excavated rounds).
- Final invert arch and final top-heading lining made by standard reinforced concrete.

2 THE SOUTH ENTRANCE

2.1 Brief description of the project

The South area of the Mały Luboń tunnel develops for almost 300m in length; the excavation have been supported by means of several earth-retaining systems. In order to provide an access to start the mining activities, the tunnel entrance has been designed on the basis of the following scheme (Figure 2):

- Three approach slopes with 2:3 (V:H) inclination and a maximum height equal to 8m, with intermediate berms, without any anchoring system.
- A main retaining structure, made by medium diameter concrete piles (Ø400mm) reinforced with a single steel profile (IPE300) and supported by anchoring nails by means of horizontal steel wailing beams.
- A drainage system represented by sub-horizontal drains, both along the approach slopes and the retaining wall.
- A watercourse regulation system behind both the approach slopes and the retaining wall's top beam.
- The total vertical excavation height is approximately 40m

2.2 Monitoring system

Following the designed monitoring plan, a certain set of instruments has been installed before to start any excavation:

- Optical targets along the crown of the slopes, every 20m, and along the middle of the slopes, every 10m;
- Optical targets on the top of the retaining wall and along it, every 5m;
- Vertical inclinometer pipes behind all the slopes – approximately in correspondence of the middle of the front slope and on the edges – extended till 5m below the final designed excavation level;
- Vertical piezometers close to inclinometer pipes – with same length of the above mentioned instruments;
- Load cells on ground anchors.

2.3 *Problems faced during execution and monitoring data*

During the excavation of the approach slopes, across June 2017, some local instabilities have been observed. These phenomena occurred in correspondence of some heavy rainfalls and, despite of the presence of sub-horizontal drains, only a few quantity of water has been caught and taken-out to the final destination.

After a light re-shaping of the above mentioned slopes, the excavation activities have been resumed and, in the meantime, a draining ditch system has been executed behind the slopes, aimed to interrupt the water flow in the upper ground layers and to rapidly deliver it far away from the area.

Between the end of August 2017 and the end of February 2018, the retaining wall and almost all anchors were completed, including drains. During this macro-phase, a big quiescent landslide has been detected, with an estimated slipping volume equal to 50000m^3 (refer to Figure 6, for the cumulative inclinometer pipe deformations and to Figure 7, for the horizontal optical target displacements).

An immediate action, by the Designer ("Studio Geotecnico Strutturale" – SGS S.r.l.) and Astaldi S.p.A., has been taken without stopping the construction activities in order to properly increase the global safety factor, despite the unexpected phenomenon.

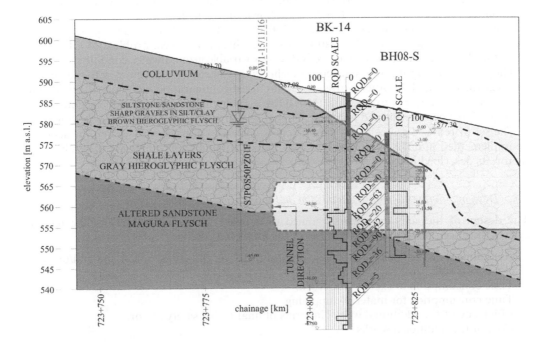

Figure 2.   South Entrance – Geological profile.

2.4 Back analysis

In order to properly assess the real stability conditions of the retaining system and then to design the actions to be taken to stop the sliding movement, a back-analysis has been numerically performed, starting from the characteristic geotechnical parameters, through a [c' - φ'] reduction process. The set of characteristic geotechnical parameters has been revised basing on additional geological investigations performed in the meanwhile.

For the numerical simulation, the code SSAP2010 (Slope Stability Analysis Program – Borselli, 2013) v. 4.9.0 has been used by implementing the Morgestern & Price (1965) criteria, based on the limit equilibrium method (LEM).

The calculation has been developed on several geometrical conditions, according to each real stage of the excavation, with the purpose to find the worst configuration for the slopes stability:

- Initial conditions (refer, later, to "MOD01");
- Excavation of the first slope (upper one - refer, later, to "MOD02");
- Excavation of the second slope (intermediate one - refer, later, to "MOD03");
- Excavation of the third slope (lower one - refer, later, to "MOD04");
- Installation of the retaining wall piles, lowering of the excavation and installation of anchors step-by-step (refer, later, to "MOD05"÷"MOD08").

For each arrangement, the ground water condition has been simulated in accordance with the monitoring evidences, collected since the beginning of the construction activities.

The first step of the study has been the assessment of the safety factor evolution, depending on the excavation stage and based on the characteristic values of the adopted geotechnical parameters. A global safety factor equal to 1,15 has been calculated for the configuration named "MOD04", while a factor equal to 1,21 has been determined in the final configuration named "MOD08", as shown in Figure 3.

The second step of the study has been the evaluation of the safety factor evolution, depending on the excavation stage, and based on a 25% reduction of the characteristic values of the applicable geotechnical parameters. A global safety factor equal to 0,92 has been calculated for the configuration named "MOD04", while a factor equal to 1,06 has been determined in the final configuration named "MOD08", as shown in Figure 4.

The apparent inconsistency between "MOD04" and "MOD08" results is, in part, connected with the absence of a stress-strain history, which is a typical problem of a LEM analysis. On the other hand, the difference is due to a variable ground water condition.

The results, obtained using reduced geotechnical parameter, are coherent with the phenomena described in §2.3. In other words, the first critical configuration has been detected almost at the end of the slopes excavation ("MOD04"), after that all the subsequent construction activities have been performed in a geotechnical system with a pseudo-plastic behavior.

The set of reduced geotechnical parameters has been used to properly design the actions to be implemented in order to strongly stop the movement and then to increase the global safety factor, as described in the §2.5.

2.5 Designed actions to stop the landslide and to improve the global stability

The design of the additional works, aimed to improve the global stability of the retaining system, has been performed on the basis of the following criteria:

- Availability of the construction machines, directly on the construction site and/or on the closest market.
- Time consumption to execute the work.
- Time consumption for material supplying.
- Efficiency of the additional works in terms of final global safety factor.
- Cost of the additional works.

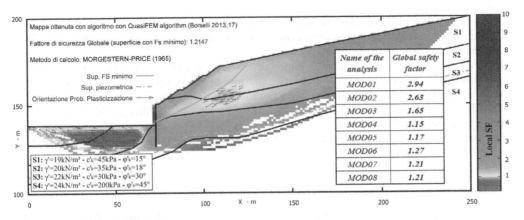

Figure 3. South Entrance – Global stability factor and evolution of the stability (table) for the final configuration, using characteristic geotechnical parameters (anchors are not plotted but they have been simulated in the model).

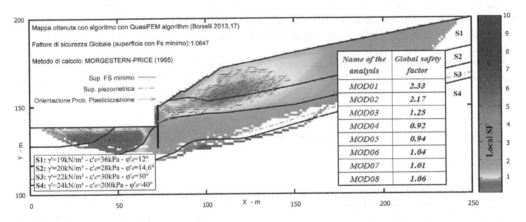

Figure 4. South Entrance – Global stability factor and evolution of the stability (table) for the final configuration, using design-reduced geotechnical parameters (anchors are not plotted but they have been simulated in the model).

On the light of the above, the best technical solution, among the many available, has been defined according to the following actions:

- Weight increase at the foot of the landslide, by means of a massive and reinforced concrete canopy tunnel, casted contiguously to the front and side of the retaining wall.
- Weight reduction on the top of the landslide, by means of a bigger excavation and wider berms, with consequent reduction of shear stresses along the slip surface.
- Insulation of the new approach slopes surface, by means of draining geotextile layer (below) and of PVC membrane (above).
- Execution of deeper drainage ditches behind the new approach slopes, aimed to avoid to let the water soaking the ground and then avoiding to reduce the shear resistance along the slip surface.

What above described, has been numerically analyzed through a slope stability analysis performed starting from the final configuration, already shown previously in Figure 3 and Figure 4:

- Numerical simulation of the concrete canopy tunnel (refer, later, to "MOD09");
- Numerical simulation of the ground weight reduction (refer, later, to "MOD10").

In order to take into account the real effect of the concrete canopy tunnel, the equivalent self-weight and the equivalent uniaxial compressive strength have been evaluated considering the voids inside the structure.

The result, shown in the Figure 5, highlights the beneficial effect given by the technical solution with respect to the global stability. The global safety factor, in fact, increased from 1,06 to 1,21 and this increment has been considered sufficient to ensure the safety of the system during a temporary design condition, up to the final backfilling of the tunnel entrance (final geometrical configuration).

The whole set of additional works has been started at the beginning of February 2018 and completed at the end of March 2018.

2.6 *Monitoring results after improvements*

The following pictures show the monthly monitoring results from September 2017 up to July 2018. The decrease in deformation speed is clearly evident, immediately after the end of February 2018. The behavior of inclinometer INO6 changed according to the modification of the slope geometry (weight reduction) and then of the pipe length. The deformations, highlighted in the above mentioned instrument, are mainly connected with the mining activity (started immediately after the implementation of the additional works as described in the §2.5 and §3.1). The inclinometer INO7 demonstrated the absence of additional sliding movement since the beginning of March 2018.

The optical targets, on the top of the soldier pile wall and along the approach slopes, shown as well that, since the beginning of March 2018, the deformation speed drastically reduced itself and after completely stopped starting from May 2018.

3 TUNNEL SUPPORT – RECALIBRATION AND MONITORING DATA

3.1 *Tunnel face consolidation*

Around the end of February 2018, the mining activities have been started in one of the two tubes (the left one – highlighted with red color in Figure 8) from the South entrance. As mentioned in §1.3, following the A.DE.CO.-R.S. method, a modification of the support class has been required in order to excavate the tunnel through the slip surface, with the appropriate safety factor.

Starting from the support class named "3", already described in §1.3, the following topics have been analyzed and the subsequent solutions have been implemented:

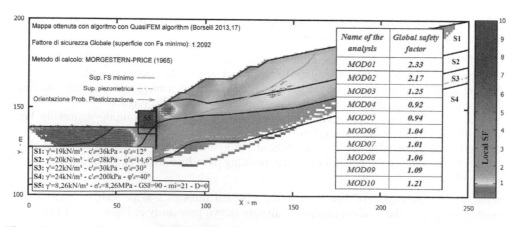

Figure 5. South Entrance – Global stability factor and evolution of the stability (table) for the final configuration, by implementing the additional works, using design-reduced geotechnical parameters (anchors are not plotted but they have been simulated in the model).

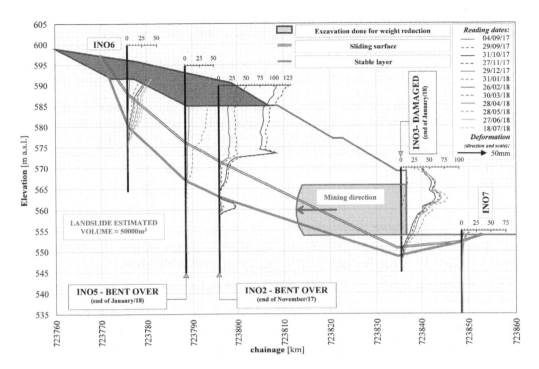

Figure 6.   South Entrance – Longitudinal profile – Cumulative inclinometer pipe displacements.

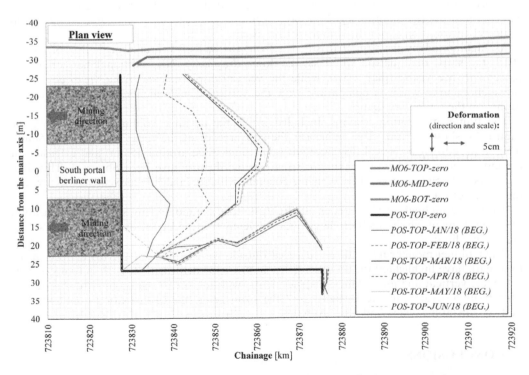

Figure 7.   South Entrance – Plan view – Horizontal optical target displacements at the top concrete beam (POS = South entrance portal, MO6 = other retaining wall structure, BEG. = beginning).

- Protection of the crown of the tunnel in order to reduce the ground surface settlements ahead of the tunnel face: obtained with an increase in the forepoling quantity from 45 to 61 tubes.
- Additional improvement of strength and strain parameters of the soil to be excavated, inside the landslide volume: obtained with an increase in the quantity of fiberglass pipes from 46 to 100.
- Stiffening of the tunnel contour in order to give additional support along the slip surface of the landslide: obtained with a reduction of the steel ribs spacing, from 1,00m to 0,75m, that induces a reduction of the longitudinal distance between two consecutive consolidation cycles, as well, from almost 9m to 7m. The immediate effect of this reduction is the increase of the overlap length between subsequent forepolings and subsequent fiberglass pipes.

3.2 *Evolution of monitoring data during mining activities*

In addition to the monitoring system for the South entrance, described in §2.2, the following instruments have been installed inside the tunnel, as expected by the project:

- One convergence section, for each excavated round (approximately every 9m), made by 5 optical targets.
- One extrusion section, at the end of each 2 rounds (approximately every 18m), after the consolidation activities and before to execute the final invert, made by 5 optical targets.
- One geotechnical section, every three rounds (approximately 27m), made by 3 couples of strain gauges and by a couple of load cells on the steel rib.
- One structural section, every six rounds (approximately 54m), made by 4 couples of strain gauges on the final reinforced concrete lining.

Hence, during the mining activities, started around the end of February 2018, both the external and internal monitoring data have been analyzed and daily evaluated to have continuous confirmation about the stability of the system.

As highlighted by the Figure 8, starting from the end of February/beginning of March 2018, the slope deformation speed slowed-down drastically. No additional horizontal displacements are visible (plan view), while the vertical displacements (front view) describe a subsidence curve which is compatible with the vertical displacements measured inside the tunnel on the temporary lining.

The tunnel monitoring sections, included inside the unstable volume (as defined in the Figure 6), shown a very small value for the convergence of the ropes (lower than 15mm), while a pseudo-rigid vertical displacement (maximum value around 45mm) and extrusion of the entire temporary lining (maximum value around 40mm) have been observed. This data indicates a kinematic mechanism which is coherent with the landslide movements occurred in the same period.

It must be highlighted that, under all circumstances, the monitoring shown a stabilization of the movement which has been, later on, completely stopped by the execution of the subsequent final linings.

Other tunnel monitoring sections, installed outside the unstable volume (as defined in the Figure 6), have highlighted the absence of any unpredicted deformation, both in terms of convergence, vertical and longitudinal displacements, with respect to the original tunnel design. This result provided, to the Designer and the Contractor, the confirmation about the fact that the unstable volume had been overcome.

4 CONCLUSIONS

In this Paper, the experience gained from the execution of the S7 Expressway during the excavation of the Mały Luboń tunnel, through the Carpathians, has been reported. An important

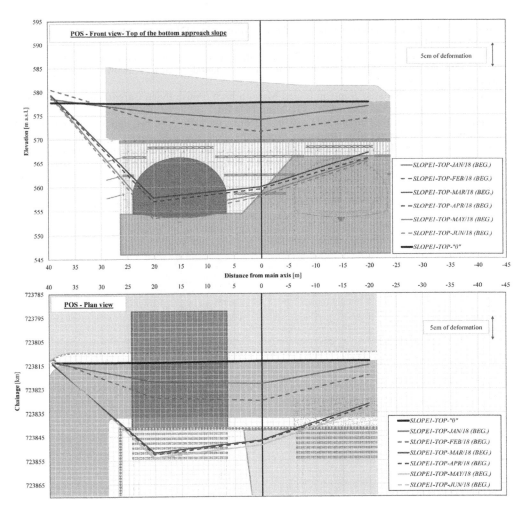

Figure 8.   South Entrance slopes – Front and Plan monitoring data – Optical target readings from January up to June 2018 (BEG. = beginning).

quiescent landslide would have hindered the commencement of the mining activities, but a proper and active action, by Astaldi and S.G.S., allowed to continue the construction sequence by means of the implementation of some important technical variations, aimed to increase the general safety factor of the tunnel and its entrance. The content of this document demonstrates the correctness and efficiency of the technical solution, which has been implemented while the construction activities were still ongoing, without losing time and keeping an appropriate level of safety.

REFERENCES

Plašienka, D., Grecula, P., Putiš, M., Kováč, M., Hovorka, D., 1997, Evolution and structure of the Western Carpathians: an overview. In Grecula, P., Hovorka, D., Putiš, M. (Eds.) Geological evolution of the Western Carpathians. Mineralia Slovaca – Monograph, Košice, pp. 1–24.

Kłapyta, P., 2015, Zarys budowy geologicznej Karpat w rejonie Rabki-Zdroju, pp. 329–336.

Książkiewicz, M., Rączkowski, W., Wójcik, A., 1991, Szczegółowa Mapa Geologiczna Polski: 1015 – Osielec (M-34-76-D). In Fortuna, J., Morawski, W. (Main Coordinators) and Zytko, K., Nescieruk, P. (Carpathian Region Coordinators) – Państwowy Instytut Geologiczny.

Lunardi, P., 2001, Progetto e costruzione di gallerie – Approccio ADECO-RS, Quarry and Construction, PEI-PARMA (Ed.).

Borselli, L., Greco, L., Petri, P., 2014, SSAP2010, Il software freeware per le verifiche di stabilità all'equilibrio limite (LEM) nei pendii naturali e artificiali, con metodi rigorosi e avanzati, Geologi e Territorio, n. 1/2014, pp. 22–32.

Tokarski, A. K., 1977, Rotated Joints in folded Magura flsch (Polish flysch Carpathians), Annales de la Société Geologique de Pologne, vol. XLVII-2, pp. 147–161.

Oszczypko, N., Jurewicz, E., Plašienka, D., 2010, Tectonics of the klippen belt and Magura nappe in the eastern part of the Pieniny Mts. (western Carpathians, Poland and Slovakia) – New approaches and results, Scientific annals, school of Geology, Aristotle University of Thessaloniki, Proceedings of the XIX CBGA Congress, Thessaloniki, Greece, Special Volume 100, pp. 221–229.

Aleksandrowski, P., 1980, Step-like tectonic lineation in the Magura flysch (western outer Capathians), Annales de la Société Geologique de Pologne, vol. L-3/4, pp. 329–339.

McCann, T., 2008, The Geology of the Central Europe volume 2: Mesozoic and Cenozoic, The Geological Society (Ed.).

Tunnels and Underground Cities: Engineering and Innovation meet Archaeology, Architecture and Art, Volume 4: Ground improvement in underground constructions – Peila, Viggiani & Celestino (Eds)
© 2020 Taylor & Francis Group, London, ISBN 978-0-367-46868-2

Design of ground freezing for cross passages and tunnel adits

J.A. Sopko
Moretrench, A Hayward Baker Company, Rockaway, NJ, USA

ABSTRACT: Ground freezing has been used extensively in the last ten years for the construction of transit tunnel cross passages and SEM tunnels, as well as short adits for utility tunnels. Quite often the ground freezing design requirements are more complex than in conventional shafts. The frozen earth structures are frequently located in urban areas where freezing forces against structures and utilities are common. Additional considerations must be given to frost heave and thaw consolidation. This paper discusses the design procedures related to frost action. Case histories are presented that indicate how frost effects can be controlled with heating pipes, coolant control, insulation methods, and in some cases additional structural bracing. A unique project is discussed where building columns required the installation of a jacking system to accommodate the heave and settlement.

1 INTRODUCTION

1.1 *Frost action in soils*

When water freezes, the volume increases by approximately nine percent. It is often assumed that the volumetric expansion of soil is as simple as estimating the pore volume and concluding that in a saturated soil this volume will expand by this percentage. Experience and laboratory testing have shown that the process is somewhat more complicated. The soil skeleton will expand when pressure in the ice exceeds the overburden pressure required to initiate separation of the soil skeleton. With sufficient ice pressure, the soil skeleton separates and a new ice lens forms (Andersland & Ladanyi 2004). Typically, the formation of ice lenses requires cyclic freezing and thawing as experienced in nature with seasonal temperature variation. This is not the case in artificial ground freezing. Additionally, ground freezing projects have shorter durations that do not permit the migration of groundwater to the freezing front.

The thawing process begins immediately when the ground freezing refrigeration system is turned off. In nature, the thawing process occurs from the ground surface down, or in in some cases, from a deeper stratum upward. This somewhat simplifies the mechanics of thaw consolidation. In artificial ground freezing, the thaw process begins radially from around each individual refrigeration pipe. As the ice melts and creates an excess of water, excess pore pressures can build up. Before this water can migrate from the frozen zone outward, there is a decrease in shear strength. Surface settlement can also occur depending on the magnitude of ice lenses. However, it must be remembered that the formation of ice lenses on short term artificial ground freezing projects is uncommon (Andersland & Ladanyi 2004).

1.2 *Mechanical properties of frozen soil*

When designing ground freezing projects, focus is typically given to the strength and thermal properties of the frozen soil. Constant strain rate and constant stress creep compression tests are used to determine the required size of the frozen mass. The thermal properties are evaluated to determine the duration and associated refrigeration load required to form the frozen mass.

A majority of frozen earth projects completed in last 30 years were shafts in remote areas or vacant city lots where frost heave and thaw consolidation were not considered. The design guidelines prepared by the International Symposium on Ground Freezing (ISGF) (Andersland et al. 1991) referred to two testing methods for evaluating the potential for frost susceptibility of soils by using the segregation potential or the CRREL (U.S. Army Cold Regions and Research Engineering Laboratory). These methods provided a qualitive approach to determine whether or not a soil would be subject to frost action but provided no values that could be used to evaluate the potential heave, settlement, or pressure on adjacent structures when artificial ground freezing is used.

These procedures were published well before numerical modeling was used in ground freezing design. Current models such as PLAXIS afford the capability of using the volumetric change in a soil upon freezing or thawing to act as an actual parameter in the input file. One simple method of determining the volumetric expansion is to simply measure the length and several circumferences of the specimen and compute the volume prior to and after freezing.

The volumetric expansion is considered while computing the stresses and deformations of frozen earth structures using PLAXIS. PLAXIS utilizes a system of staged construction in the computation process. One of the stages often labeled the freezing stage is the activation of the frozen soil zone that includes the volumetric expansion. PLAXIS permits the evaluation of the deformations during that phase as shown in Figure 1.

It should be noted that this method only accounts for the primary expansion upon freezing and does not account for the formation of ice lenses, which is not typically considered a factor in artificial ground freezing as previously noted. There are a few problems with the simplification of the described method of measuring frost expansion. Figure 2 illustrates the mechanics of heave during ground freezing.

This figure shows the vertical and lateral soil and hydrostatic pressures that resist heave. Different soil types exert varying pressures while frozen and can be measured in a laboratory. Vertical heave can only occur when the pressures generated by the expansion of the pore water through freezing exceed the horizontal and vertical stresses of the in-situ soil conditions.

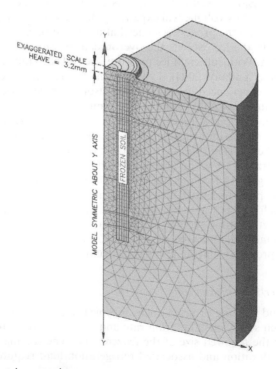

Figure 1.   Initial volumetric expansion.

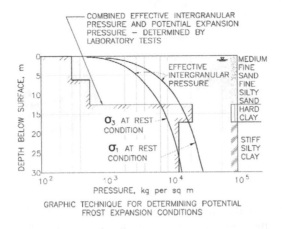

Figure 2. Frost pressures generated during freezing.

The simplified approach for measuring volumetric expansion does not account for the imposed lateral and vertical in-situ pressures. The author proposes a new method that is a modification of tests developed at Tufts University (Huang & Swan 2017) illustrated in Figure 3.

2 APPLICATIONS IN TUNNELING

2.1 *Frozen cross passages*

The advantage of this method is that it permits the application of both horizontal and vertical loads, providing a more realistic value of the anticipated in-situ frost heave and thaw consolidation.

Ground freezing is often the only method available to provide temporary earth support and groundwater control for cross passages that are not only very deep but exist in high permeability, cohesionless soils. A typical cross passage ranges in length from 3 to 8 m. Cross Passage 23 on the Seattle Northgate Link Tunnel is illustrated in Figure 4.

While the actual design of the frozen earth structure to support the cross passage excavation including structural and thermal analyses is not the topic of this paper, the frost pressures generated by the freezing process warrant discussion. The individual refrigeration pipes can be

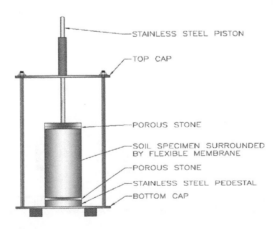

Figure 3. Heave/consolidation cell.

Figure 4. CP-23 frozen cross passage.

drilled and installed from within the tunnel(s) as shown in Figure 4 or from the ground surface. Regardless of the method used, the expansion of soil during the freezing process will exert considerable pressure on the segments of the two tunnels.

The three-dimensional mesh shown in Figure 5a is used to simulate the frozen zone. When that zone is activated in the model using a value of volumetric expansion from the laboratory tests, the resulting forces and displacements on the tunnels are computed as shown in Figure b. In this particular application, the volumetric expansion of the frozen soil was 3.1 percent, resulting in maximum forces of 4.28 kN/m^2 and a maximum displacement of the tunnel segments of 0.017 m. With the knowledge of these forces prior to freezing, sufficient bracing similar to that illustrated in Figure 4 can be installed.

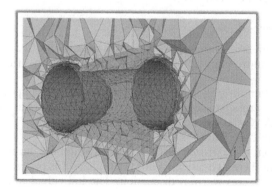

Figure 5a. Finite element mesh.

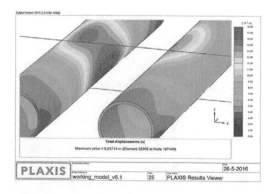

Figure 5b. Tunnel displacements.

2.2 Frozen tunnel adits

Tunnel adit construction is somewhat similar to cross passage installation but can often be considerably longer. An example of using ground freezing for adit construction at the First Street Tunnel in Washington, DC is presented in Figure 6.

In this case, ground freezing was used to make the connection from a micro tunneled adit to the main tunnel. The drilling and installation of refrigeration pipes was completed from the ground surface on a roadway laden with utilities including water and sewer lines. The frost susceptibly varies for different soil types. Sand and gravel soils typically exhibit minimal frost heave and thaw consolidation as compared with clay and silt soils (Andersland & Ladanyi 2004).

Review of the soils indicated they were not frost susceptible at this particular location. No mitigation techniques were implemented when the freezing process was initiated but a heave monitoring program was in place. During the initial freezing, unanticipated heaving was observed at the road surface. The contractor was faced with the situation of needing to maintain the freezing system to form the frozen mass at depth while at simultaneously limiting the growth of the frozen zone near the utilities. If it was known in advance that the non-frost susceptible soils would have behaved uncharacteristically and expanded while freezing, protection of the utilities could have been implemented in advance. The most expedient and effective method to limit the growth of the frozen mass and minimize heaving was to insulate the refrigeration pipes in the zone near the affected utilities. This process typically consists of the top 7 to 10 m of the pipe and is installed during the drilling and installation process.

Heave was observed at the road surface during the freezing process. This data was unfortunately not reported until the magnitude of the heave reached a near-threshold level. The heave values are shown in Figure 7.

As the magnitude of the heave continued to increase, it was necessary to act in order to prevent additional heaving and/or damage to the utilities. Since it was too late to install insulation on the refrigeration pipes, the contractor's only reasonable option was to turn off a few of the redundant refrigeration pipes and then drill/install heating pipes to prevent additional growth of the frozen zone and reduce the size of the already frozen area. Each vertical bar on the graph represents a various phase in controlling the heave. The first bar represents the time at which the contractor was informed that heave was occurring. Five days after notification and confirmation of the data, the redundant refrigeration pipes were turned off. Heave

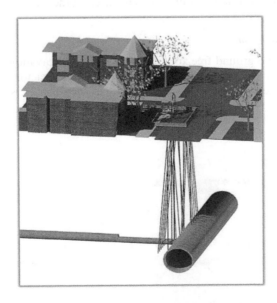

Figure 6. Frozen adit connection.

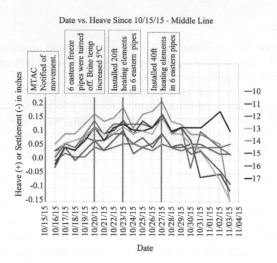

Figure 7.   Frozen adit connection.

subsided very briefly as the heating pipes were being drilled and installed. The heating pipes were initiated to a depth of 6 m three days later and heave was once again briefly reduced but continued. It was then necessary to heat the ground even deeper to a depth of 12 m. The deeper heating pipes were effective, as shown on the graph.

The heating pipes were essentially the same design of the refrigeration pipes: a steel pipe sealed at the bottom and filled with calcium chloride solution. An electric heating element was installed within each pipe and set to a temperature of approximately 35°C. A time-dependent heat transfer finite element model was used to evaluate the effects of the heat pipes and determine required energy to maintain the 35°C temperature.

This heave was measured at the road surface and not at the utilities. The mitigation using the heating pipes resulted in no measurable heave at the two utilities. After this connection, there were two other zones on the project that required freezing from the ground surface. Due to the unanticipated heave at this site, the refrigeration pipes were insulated on the subsequent projects. There was no measurable ground movement using the insulation.

2.3 Freezing near foundations

It is quite common where ground freezing is used to support excavations near existing building foundations. A large excavation was required within an aircraft factory in the Western United States. The line of refrigeration pipes was very close to existing pipe caps as shown in Figure 8.

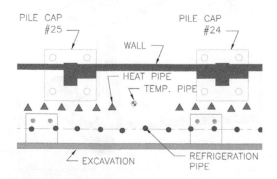

Figure 8.   Heating pipes to protect pile caps.

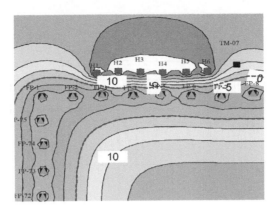

Figure 9.   Temperature contours with heat pipes activated.

It was necessary to evaluate the geometry, temperature, and electrical power load of the heating pipes in the design phase of the project. This was accomplished using a time dependent finite element heat transfer analysis. This analysis as presented in Figure 9 presents the temperature after 60 days of freezing. In this model, the heat pipes were considered a temperature boundary condition with the temperature fixed at 35°C. In the model as well during the actual construction, the heating pipes were activated approximately one week before the initiation of the freeze. The actual ground temperatures that were measured during the entire phase of construction were consistent with the modeled temperatures.

Heave at each of the pile caps was measured throughout the entire project. There was no heave during the freezing process nor any observed settlement during thawing. One significant lesson learned from the First Street Tunnel that was applied on this project was the timing of the heating process. On the First Street Tunnel, the heave was not anticipated given the soil conditions. It was not until the freezing was underway that the heave was observed and the heat pipe mitigation plan implemented. As shown in Figure 7, there was significant lag time from the start of the heating until a reduction in heave could be measured.

With that lesson in mind, the modeling confirmed in the field that when the heating pipes are turned on prior to freezing, the freeze will not advance into the warmed zone (provided sufficient heat energy is available).

2.4   Structural bracing to accommodate heave and settlement

In some cases, the frost heave and associated thaw consolidation cannot be prevented or even mitigated, as was the case with the MBTA's Silver Line in Boston, often referred to as the Russia Wharf Project. A binocular-shaped tunnel approximately 13 m wide by 18 m high was constructed under existing buildings in Boston (Lacy et al. 2004). As shown in Figure 10, the tunnel was constructed using NATM techniques under a building that was supported by pile foundations. Ground freezing was used to provide excavation support and groundwater control during construction.

The frozen ground would additionally serve as a medium to distribute the column loads as shown in Figure 10. The soil consisted of a fill layer of organic silts and clays over a silty marine clay. These soils are exceptionally prone to frost heave and thaw consolidation. The load distribution and NATM are shown in Figure 11. Laboratory testing confirmed there would be substantial heaving and settlement during the freezing and post-construction thawing. A jacking system as shown in Figure 12 was installed as a component of the load transfer from the columns to the frozen earth.

The jacks enabled the contractor to either lower or raise each building column to compensate for displacements during the various stages of construction. As anticipated,

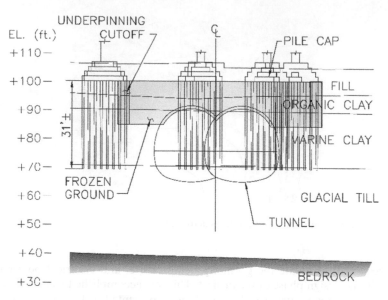

Figure 10. Russian Wharf NATM Tunnels.

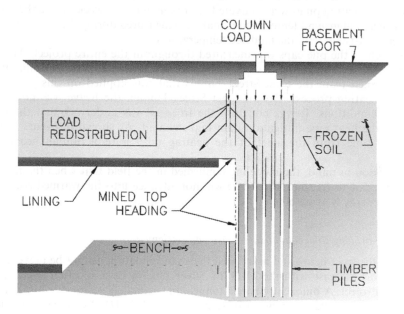

Figure 11. Load transfer to frozen earth.

the columns (measured at the pile caps) began to heave during the freezing. The heave was initially mitigated by restricting the coolant flow to each of the individual refrigeration pipes. This was only a temporary measure and could not accommodate the increased uplift forces caused by the expansion of the freezing ground. As the heave approached the pre-determined threshold limit of 0.003 m, the jacks were activated to lower the columns. Two adjustments of the jacks and continuous cycling of the refrigeration system and coolant flow were required to keep the movements below the threshold limit as shown in Figure 13.

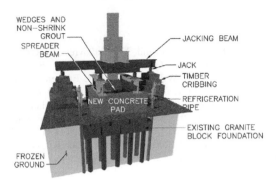

Figure 12.   Column jacking mechanism.

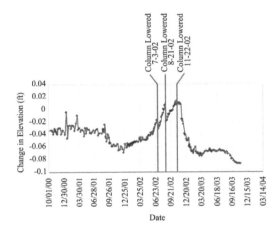

Figure 13.   Effects of jack adjustments.

3   CONCLUSIONS

Several aspects related to frost heave and thaw consolidation caused by artificial ground freezing have been discussed. A summary of these issues follows.

Finite element method programs such as PLAXIS can be used to evaluate heave deformations at the ground surface and settlement during thawing. In addition to deformation, these programs can also be used to evaluate the pressure against adjacent tunnels, utilities, and building foundations.

Direct measurement of a soil sample before and after freezing can provide reasonable parametric values for the input in the numeric models. A more accurate method using a triaxial cell and applying vertical loads and lateral pressures is currently being developed and tested.

Heave effects and frost pressures can be mitigated using different techniques. These techniques including deactivating redundant refrigeration pipes, reducing coolant flow or temperature, and installing and maintaining heating pipes.

In unique situations where frost-susceptible soils are present without a way to limit the growth of the frozen soil mass, it is possible to adjust the structure with bracing or a system of jacks to compensate for structural distortion.

REFERENCES

Andersland, O.B., Berggren, A.L., Fish, A., Fremond, M., Gallavresi, F., Gonze, P., Harris, J., Jessberger, H.L., Jordon, P., Klein, J., Maishman, D., Ohrai, T., Rebhan, D., Ryokai, K., Sego, D.C., Shuster, J.A., Thimus, J.F., Changsheng, W. & Williams, P. 1991. Frozen ground structures – basic principles of design. *Ground freezing 91; Proc. intern. symp., Beijing, 10–12 September 1991.*
Andersland, O.B. & Ladanyi, B. 2004. *Frozen ground engineering.* Hoboken: John Wiley & Sons.
Huang, Y. & Swan, C. 2017. Evaluating the behavior of a cohesive soil undergoing one cycle offreeze thaw. *Geotechnical frontiers; Proc. intern. symp., Orlando, 12–15 March 2017.* Reston: ASCE.
Lacy, H.S., Boscardin, M.D. & Becker, L.A. 2004. Performance of Russian wharf buildings during tunneling. Levent Ozdemir *(ed.)*, *North American tunneling conference; Proc., Atlanta, 17–22 April 2004.* Boca Raton: CRC Press.

*Tunnels and Underground Cities: Engineering and Innovation meet Archaeology,
Architecture and Art, Volume 4: Ground improvement in
underground constructions – Peila, Viggiani & Celestino (Eds)
© 2020 Taylor & Francis Group, London, ISBN 978-0-367-46868-2*

Factor of safety in ground freezing design

J.A. Sopko
Moretrench, A Hayward Baker Company, Rockaway, NJ, USA

ABSTRACT: Design of frozen earth structures for shafts, tunnels, and cross passages has evolved since the publication of guidelines in 2002 by the Working Committee of the International Symposium on Ground Freezing. This evolution is due to the development and implementation of numerical modeling methods, specifically the finite element method (FEM). The design was previously based on determining a thickness of the frozen earth structure with an average frozen temperature. The factor of safety was incorporated by dividing the unconfined compression strength of the frozen soil by two, and thus F.S. = 2. The finite element model allows more specific an analysis of a frozen structure cross section and adjusting strength parameters based on temperature and time-dependent strength reduction. Analyses results permit the evaluation of the internal stress regime of the frozen earth structure. This paper discusses the procedure and reports on key recent projects.

1 INTRODUCTION

1.1 *Factor of safety*

The factor of safety is the allowable stresses of a structural element divided by the actual stresses. In geotechnical engineering, the definition is more complex and somewhat undefined with frozen earth structures. The design engineer is often asked to identify the redundancy or factor of safety. In treating the frozen structure that provides temporary earth support and groundwater control as a structural element, focus is given to the allowable stresses within the structure to withstand the actual stresses imposed by surcharge loads, lateral or vertical earth pressures, and hydrostatic pressures. These allowable stresses are governed by two parameters: the strength of the frozen soil and dimensions of the frozen earth structure.

Published results of safety factors are documented as 1.0 (Harris 1995), 1.1 (Sanger 1968), and 3.0 (Sopko 1990). The author notes from his previous experience on over 50 frozen shafts that 2.0 is most often used in practice.

1.2 *Strength of the frozen soil*

Frozen soil exhibits time and temperature-dependent rheological behavior by deforming (creeping) with time under a constant applied stress, the rate of which is dependent on the temperature. Fine grained soils such as clays and silts are more subject to creep than sands and gravels. Reducing the temperature will reduce the rate of creep and may even prevent it. There are three phases of creep: primary, secondary and tertiary illustrated as I, II, and III in Figure 1.

The primary (I) phase is where the deformation rate decreases with time, the secondary (II) where deformation remains essential constant with time, and the tertiary (III) where deformation increases with time. The relationship between deformation rates and time is shown in Figure 2.

The deformation of frozen soil with time at a constant temperature and the rate of deformation increases with the applied stress to the soil. Figure 3 shows the relationship between time

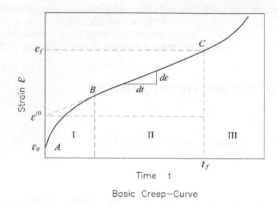

Figure 1.   Basic creep deformation curve.

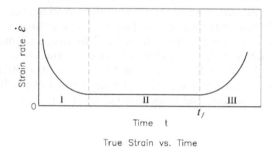

Figure 2.   Deformation rates versus time.

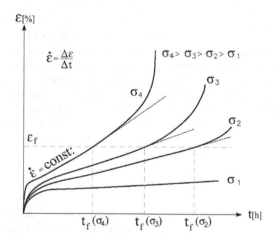

Figure 3.   Idealized multiple stress level creep curves.

and creep deformation based on four different stress levels depicting idealized creep curves. It is difficult to get curves of this consistency in the laboratory. At least three constant stress creep tests are typically conducted on samples to evaluate a design strength of the soil and ultimately determine the factor of safety. The constant stress creep test is conducted by applying a load to a frozen sample and measuring the deformation with time. The magnitudes of the theses stresses are typically 0.7, 0.5, and 0.3 of the unconfined compressive strength is

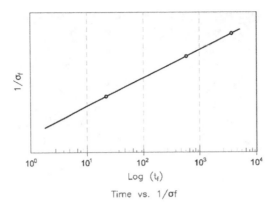

Figure 4. Time versus reciprocal of stress.

measured in a constant strain rate test. The results of the applied constant stress versus time is shown in Figure 3. It is noted that these tests must be conducted until the sample fails. A six percent strain is considered a failure. Sufficient time must be permitted to collect data that can be interpreted. This can take up to 1,000 hours in some cases.

It is necessary to begin with a frozen soil compressive strength in the design of frozen earth structures. As observed in the preceding figures, that strength is based on the length of time the structure will be exposed to combined pressures. This is unique to frozen earth. An excavation or tunnel that will be open for 30 days can be designed with a higher compressive strength than an excavation that will be open for 90 days or longer. The designer is then faced with evaluating a time-dependent compressive strength. There are two methods typically used. The first method, as proposed by the International Symposium on Ground Freezing (ISGF) (Andersland et al. 1991), uses the following equation:

$$q_f(t) = (\frac{\in f}{A * \tau^C})^{1/B} \qquad (1)$$

where $Q_f(t)$ = unconfined compressive strength at a given time; t = time; $\in_f$ = strain at the time of failure (Figure 3); and A, B, and C = creep parameters determined from the tests (Andersland et al. 1991).

Another method proposed by Sopko (1990) is shown in Figure 4. By plotting the time to failure versus reciprocal of applied stress, a relationship can be obtained to determine the unconfined compressive strength used in design.

It should be noted that the designer is often faced with less than ideal data and forced to use engineering judgement in evaluation of the laboratory test results. The import concept to note is that the compressive strength used in design is totally dependent on the required time the excavation is to remain open or unbraced. After determining this time, the strength can be evaluated by using Equation 1 or by finding the corresponding stress from the line generated line in Figure 4. It is also important to evaluate strength properties at different temperatures to be consistent with the proposed temperature of the frozen earth structure.

2 DESIGN OF FROZEN EARTH STRUCTURES

2.1 Frozen shafts – conventional design equations

Ground freezing is most often used to provide temporary earth support and ground water control for the excavation of deep shafts. The refrigeration pipes are typically drilled and installed around the perimeter of an excavation, similar to what is shown in Figure 5.

Figure 5. Typical frozen earth shaft design.

In the design phase, it is necessary to determine the required thickness of the frozen wall to safely support the excavation and resist lateral earth and hydrostatic pressures. In practice, the thickness in often defined as the -2°C boundary intrados and extrados. Several equations have been developed to determine the required thickness of a thick-walled cylinder subjected to external pressures as shown in section in Figure 6.

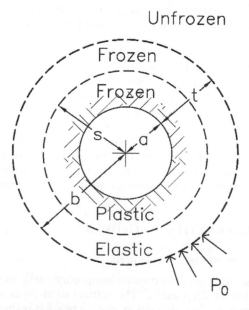

Figure 6. Definition of equation variables.

One of the more commonly used equations for determining the required frozen wall thickness shown in Equation 2 was proposed by Domke (1915).

$$t = a\left[0.29\left(\frac{P_0}{q}\right) + 2.30\left(\frac{P_0}{q}\right)^2\right] \tag{2}$$

This method considers the frozen soil intrados of the refrigeration pipes to be plastic while extrados elastic. It should be noted that this equation does not account for the time that a frozen earth structure will be excavated and subjected to loading. The time-dependent strength is acquired from either Equation 1 or extrapolated from a graph similar to Figure 4.

Equation 3 is based on work from Klein (1981) that was often used in the design of frozen shafts.

$$t = a\left[(0.29 + 1.42\sin\emptyset)\left(\frac{P_0}{q}\right) + (2.30 - 4.60\sin\emptyset)\left(\frac{P_0}{q}\right)^2\right] \tag{3}$$

wheres $= \sqrt{ab}$.

As with Domke's equation, the dependency is applied to the compressive strength. Klein incorporates the angle of internal friction in his equation. In most cases, the friction angle of frozen soil is assumed to be approximately equal to the angle for the unfrozen soil.

2.2 Factor of safety for conventional design equations

Prior to considering the incorporation of a factor of safety, the conservative nature of the assumptions in these equations must be addressed.

The equations are based on a uniformly loaded thick-wall cylinder. This is not the case with a frozen shaft. The load increases substantially with depth. Using a maximum or even average pressure is conservative.

An actual frozen earth shaft is essentially fixed or cantilevered at the base and typically tied into a strong or impermeable stratum. This is not the case with a free-standing cylinder. The cylinder does not transfer any load into the underlying strong stratum.

The equations assume a uniform frozen material. This is not the case as in practice, the core of the frozen earth wall near the refrigeration pipes is substantially colder and stronger than the interior and exterior zone of the wall. This stronger zone is not considered in the equations.

Frozen earth deforms when loaded, resulting in the plastic redistribution of stresses within the structure.

In addition to the inherent conservative nature, designers incorporate a factor of safety into the calculations, but the approaches vary. A simple approach would be increasing the calculated thickness of the frozen wall. For example, if the calculations yield a requirement for a 1 m wall, you could simply assign a factor of safety of 2 and use a 2 m wall. This is not practical in construction. Adding additional thickness to the frozen wall would increase the required freezing time. There are no documented references to using this approach.

As previously noted, the most common approach in design submittals that the author is familiar with is to assign a factor of safety of 2 by using half the tested time-dependent compressive strength. There is no published procedure citing this as a standard practice, but the author knows of no structural failures in using this approach.

Sanger (1968) suggests applying the factor of safety to the structural life or the time factor of strength. If the structure is to stay open for 100 days, assume 100 x 1.1 and use the decreased strength for 110 days from Equation 1 or Figure 4.

Sopko (1990) suggests applying the same approach but with a factor of safety of 3 for the following reasons: material properties are not always satisfactorily determined; refrigeration capacity may not always be available to reach the required frozen temperatures; the excavations may be open for longer than anticipated and sometimes in very warm climates where the ambient air temperatures could raise the temperature of the frozen wall and result in decreased strength; and impurities or dissolved salts may be contained in the groundwater and can weaken the frozen soil. Many improvements in standardized test procedures, refrigeration equipment, and excavation equipment have occurred since, rendering this suggestion obsolete.

Harris (1995) cites an interesting project in China where Equation 4 was used.

$$T = a[0.56\left(\frac{P_0}{q}\right) + 1.33\left(\frac{P_0}{q}\right)^2]$$ (4)

No factor of safety was applied here. Extreme creep deformation, basal heave, and broken refrigeration pipes occurred, suggesting a factor of safety of less than 1. This equation has not been accepted in the industry.

All of the cited references state the need and advantages using numerical methods, specifically the FEM, to analyze the stress and deformations with a frozen structure. The FEM permits analyses of shapes other than cylindrical and can be readily used in horizontal tunnel projects.

2.3 FEM design

Use of the FEM for design with programs such as PLAXIS has significantly changed the approach to the design and analysis of frozen earth structures. Figures 7a and b show typical PLAXIS models for actual projects.

These models permit the evaluation of the compressive hoop stresses defined by Figure 8. Evaluation of these stresses has permitted a new, straight forward approach to the determination and implementation of a factor of safety.

2.4 Suggested factor of safety method

The proposed standard method of applying a factor of safety in frozen earth shafts is based on the evaluation of the internal hoop stresses and comparing them to the time-dependent unconfined compression strength as shown in Equation 5.

$$\text{F.S.} = \text{Maximum Hoop Stress} / q_{(t)}$$ (5)

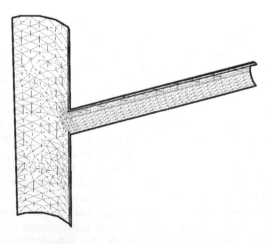

Figure 7a. Typical PLAXIS model for shafts and tunnels.

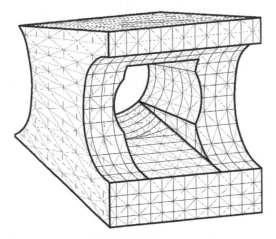

Figure 7b.   Typical PLAXIS model for shafts and tunnels.

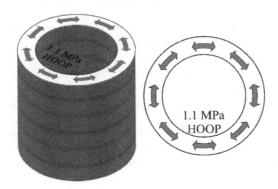

Figure 8.   Definition of hoop stresses for factor of safety computation.

The factor of safety can be increased by increasing the thickness of the frozen earth wall, re-evaluating the stresses, and then repeating equation 4.

This approach has been successfully used on several projects in North and South America. A factor of safety ranging from 1.2 to 2.0 has been used, resulting in safe structures and no measurable deformation. This variation is based on the confidence level of the laboratory testing and the number of tests run to confirm the strengths.

Established laboratory testing procedures must be adhered to. American Society for Testing and Materials (ASTM) D5520 (2011) and ASTM D7300 (2011) for the frozen compression tests are recommended. The samples must be representative of the field conditions and are of a high quality. The designer must be experienced with the FEM for frozen ground and have observed actual field performance. Appropriate shaft insulation and lining methods are required, as well as the implementation of a well-established quality control/assurance plan. A modification to this method has been used on two projects subsequently discussed.

Certain soils are highly susceptible to creep deformation. Situations have arisen where no deformation is permitted, making it difficult to assign a factor of safety. These cases require additional laboratory testing.

The constant stress creep compression tests were run at increasingly lower stress levels until a level was reached where there was simply no time-dependent deformation. The thickness of the frozen earth wall was increased to match those low levels and creep deformation was mitigated. This is not always practical but the only option in some cases.

The elastic modulus directly impacts the results of the FEM analysis. A very high value will result in lower deformations but much higher stresses, while lower values will result in lower stresses (thus affecting this approach to factor of safety determination) but increased deformation. Values for the elastic modulus can be obtained from the constant strain rate test. A time-dependent modulus was defined by Klein (1981) and is sometimes used.

3 CASE HISTORIES

3.1 Verglas, Rouyn-Noranda Quebec

The Quemont Mine located approximately 626 km northeast of Toronto required a large shaft excavation. The approach included a 61-m-diameter, 30-m-deep excavation support by ground freezing. One of the soil strata was a soft, water-bearing clay. The clay had a very high water content and would be susceptible to time-dependent creep deformation when frozen.

A series of constant strain rate and constant stress creep tests were conducted. Using the method previously described, it was determined that time-dependent deformation did not occur when stress levels in the laboratory were 1,000 kN/m^2 (Sopko et al. 2012) at -10°C.

Several iterations of the FEM stress analysis were conducted to determine the required frozen earth dimensions.

The analyses concluded that with a frozen wall thickness of 9.1 m, maximum internal stress would be on the order of 621 kN/m^2 as shown in Figure 9.

One of the problems with a project of this size was the need to insulate the exposed fact of the frozen wall when excavated. The insulation costs were prohibitive, requiring the contractor to complete the excavation from November 1 until March 31 when higher ambient temperatures would induce melting. This permitted five months of loading on the frozen earth structure.

Deformation of the frozen earth wall was monitored using three inclinometers drilled and installed within the frozen wall, as well as several monitoring points on the face of the wall. During the entire five months of excavation, there was no measurable movement, indicating the design approach was a success.

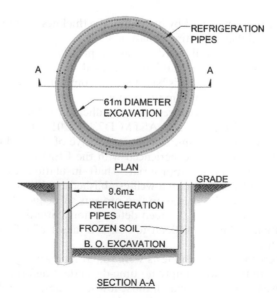

Figure 9.   Ground freezing for Verglas.

Figure 10.   Ground freezing system CP-23.

3.2   *Frozen cross passages, Northgate Link Tunnel, Seattle U.S.A.*

Seattle's Northgate Link project had 5.5 km of twin tunnels. There were 23 cross passages through water-bearing, unconsolidated soils. Ten of these cross passages were completed using ground freezing for temporary earth support and ground water control. Cross Passage 23 (CP-23) had been planned to use a conventional system of dewatering and bracing for elevation when unexpectedly unstable and running silt was encountered. Ground freezing was selected as the method for support as shown in Figure 10.

Unlike the previous project, a temporary shotcrete liner was installed at approximately 1 m intervals as the excavation was completed. This lining mitigated any potential time-dependent deformation. The design was based on the constant strain rate test conducted at -10°C, with a strain rate of 0.1 percent per minute). Results yielded a peak axial stress of 5.2 MPa and a tangent modulus of 1.5 MPa.

A three-dimensional finite element analysis was completed as shown in Figure 7. This analysis yielded a maximum stress of 0.148 MPa using the following equation for determining the factor of safety:

$$\text{F.S.} = 5.2 / .148 = 35 \tag{6}$$

The frozen earth wall had a thickness of 2 m governed by spacing of the individual refrigeration pipes at 1 m. A smaller thickness and more reasonable factor of safety would not be practical. It should be noted that if a time-dependent analysis was used, so would a much lower compression strength.

4   CONCLUSIONS

This paper has reviewed previously used methods of analysis and current state of the practice methods that have been highly successful. The convention methods incorporated a high level of conservatism in the assumptions and equations. The industry never established a standard for incorporating the factor of safety. Previous methods used high factor of safety values to compensate for unreliable equipment and non-standardized testing methods. Current equipment and test methods are very reliable and do not require consideration while determining the factor of safety. The method of relating the time-dependent compressive strength of a frozen soil to the internal hoop stresses of a frozen earth structure has proven to be safe and reliable.

REFERENCES

Andersland, O.B., Berggren, A.L., Fish, A., Fremond, M., Gallavresi, F., Gonze, P., Harris, J., Jessberger, H.L., Jordon, P., Klein, J., Maishman, D., Ohrai, T., Rebhan, D., Ryokai, K., Sego, D.C., Shuster, J.A., Thimus, J.F., Changsheng, W. & Williams, P. 1991. Frozen ground structures – basic principles of design. Ground freezing 91; Proc. intern. symp., Beijing, 10–12 September 1991.
ASTM (American Society for Testing and Materials). 2011. *D-5520 Laboratory determination of creep properties of frozen soil samples at constant rate of strain*. West Conshohocken: ASTM Press.
ASTM (American Society for Testing and Materials). 2011. *D-7300 Laboratory determination of strength properties of frozen soil samples by uniaxial compression*. West Conshohocken: ASTM Press.
Domke, O. 1915. On the stresses in a frozen cylinder of ground used for shaft sinking (in German). *Glückauf*: 51–47.
Harris, J.S. 1995. *Ground freezing in practice*. London: Thomas Telford Services Ltd.
Klein, J. 1981. Dimensioning of ice-walls around deep freeze shafts sunk through sand formations under consideration of time (in German). *Glückauf-Forschung* 43: 112–120.
Sanger, J.S. 1968. Ground freezing in construction. *Journal of the Soil Mechanics and Foundations Division* 94(1): 131–158.
Sopko, J.S. 1990. New design method for frozen earth structures with reinforcement. *Unpublished Ph.D. Thesis*. East Lansing: Michigan State University.
Sopko, J.S. Braun, B. & Chamberland, R., 2012. *Largest frozen ground excavation in North America for excavation of crown pillar*; Proc. third intern. conference on shaft design and construction, London, 24–26 April 2012.

Tunnels and Underground Cities: Engineering and Innovation meet Archaeology,
Architecture and Art, Volume 4: Ground improvement in
underground constructions – Peila, Viggiani & Celestino (Eds)
© 2020 Taylor & Francis Group, London, ISBN 978-0-367-46868-2

Stockholm bypass project – passage under the Lake Mälaren

B. Stille
Department of Civil Infrastructure, AECOM Nordic AB, Stockholm, Sweden

F. Johansson & F. Ríos Bayona
Department of Civil Infrastructure, AECOM Nordic AB, Stockholm, Sweden
Department of Civil and Architectural Engineering, Division of Soil and Rock Mechanics, KTH, Royal
Institute of Technology, Stockholm, Sweden

R. Batres Estrada
Department of Civil Infrastructure, AECOM Nordic AB, Stockholm, Sweden

M. Roslin
Swedish Transport Administration, Stockholm, Sweden

ABSTRACT: In the last years, the Swedish Transport Administration has been working on improving and expanding road communications in Sweden. The Stockholm Bypass Project, one of the biggest projects in Swedish history, consists of a 21 km long highway that goes around the city from north to south. In order to reduce the environmental impact, 17 km of the total length will be excavated underground passing through several regional fault zones and subsea passages. One of the most difficult technical challenges in this project is the passage under the Lake Mälaren and the regional fault zone in the Fiskar fjord. This paper presents the utilized methodology to design the temporary rock support and to manage the risks and uncertainties for the excavation through the fault zone, which mainly originate from the limited information about the rock conditions and the relatively large width of the tunnels.

1 INTRODUCTION

In the process of improving road communications in Sweden, the Swedish Transport Administration has been working in the last years to expand the infrastructure in the country. The Stockholm Bypass, one of the biggest infrastructure projects in Swedish history, consists of a 21 km long highway that goes around the city of Stockholm from north to south with an estimated cost of 2.5 billion €. In order to preserve the natural and cultural values of this Nordic city, 17 km of the total length will be built in two tunnels (one per direction) of approximately 17 m width each.

The general good quality bedrock in the Stockholm area provides excellent geological conditions for underground construction, which explains the large number of tunnels and underground facilities that previously has been built under the city. However, the Stockholm Bypass passes through several regional fault zones; together with the fact that these regional faults zones are passed under water, constitutes a challenge for engineers. One of the most difficult technical challenges in the Stockholm Bypass Project is the subsea passage under Lake Mälaren and the regional fault zone under the Fiskar fjord. Despite the vast amount of rock tunnels that exists under the city of Stockholm, none of them has been excavated through this fault zone previously.

The uncertainties and risks associated with this fault zone in the Fiskar fjord originate from limited information about the rock cover, the rock quality in the fault zone, possible large water leakages and existing *in-situ* stresses together with the relatively large width of the tunnels. Consequently, many of these risks must be managed under the tunnel excavation.

This paper presents the utilized methodology for the design of the temporary rock support in the tunnels, and how the different risks will be managed under the tunnel excavation through the regional fault zone. The design of this subsea passage under the Lake Mälaren has been carried out by a company consortium between ÅF/AECOM.

The work presented in this paper starts with a description of the expected geology in the fault zone. Thereafter, based on the geological assessment and other general conditions, the potential failure mechanisms are identified and the technical solution together with a verification based on numerical calculations is presented. A control program is proposed to verify the selected design and technical solutions under the tunnel construction.

2 GENERAL DESIGN METHODOLOGY

Due to the complex geological conditions expected during the construction of the subsea passage under the Fiskar fjord between Sätra and the island of Kungshatt, the main principles of the observational method are applied. The followed strategy for designing the temporary rock support of the main tunnels follows the flow chart presented in Figure 1.

3 GENERAL CONDITIONS

3.1 *Geology and rock mass quality*

The bedrock at the location of the Stockholm Bypass was formed during the Svecofennian orogen (Stålhös, 1969; Gaál & Gorbatschev, 1987). The rock mass in the area where the subsea passage under the Lake Mälaren will be excavated consists mainly of gneiss of sedimentary origin with intrusions of gneiss-granite and granite (SGU Sveriges Geologiska Undersökning, 2018). The gneiss encountered in the passage is partly heavily fractured, weathered and altered with chlorite, clay and graphite. In addition, the rock mass is crossed with dykes of pegmatites and metabasites of varying degrees of alteration. An illustration of the geological profile in the subsea passage under the Lake Mälaren between Sätra and the island of Kungshatt is displayed in Figure 2.

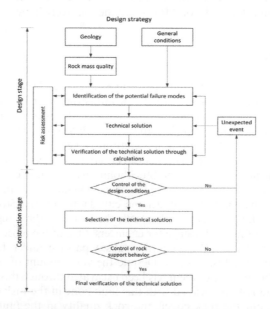

Figure 1.   Design strategy for the temporary rock support in the subsea passage under the Fiskar fjord in the Lake Mälaren between Sätra and the island of Kungshatt.

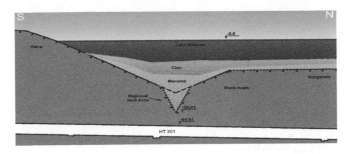

Figure 2. Illustration of the geological profile in the subsea passage under the Lake Mälaren between Sätra (left) and the island of Kungshatt (right).

The information about the geological conditions in the subsea passage under the Lake Mälaren presented in this paper comes from several drill cores performed in the area. For instance, a horizontal drill core performed at the front of one of the main tunnels nearby the Lake Mälaren showed that the fault zone in the Fiskar fjord may have several meters of core losses indicating poor rock mass quality conditions, see Figure 3. The encountered material consisted of cataclasite rock and gauge material. This consolidates the idea that the regional fault zone under the Lake Mälaren is composed of a weak and highly fragmented material in the core of the fault zone.

Geologically, the regional fault zone in the Fiskar fjord can be described as a strike-slip fault. The southern part of the region has even rotated over the northern part following a dip-slip movement. This fault zone has simultaneously been investigated in other infrastructure projects such as the new metro line, the Citylink and a new waste water treatment plant. The results of the investigations indicate similar geological conditions along the fault zone showing a complex fault matrix with a possible duplex geological formation. This means that there might be additional secondary geological structures against the main fault zone under the Lake Mälaren, see Figure 4.

In the subsea passage between Sätra and the island of Kungshatt the fault zone is assumed to be approximately 200 m wide with a central core, approximately 30 to 50 m wide. The fault zone consists of minor parallel geological structures oriented in the E-W direction following the main fault orientation. In addition, it might be possible to meet both sub-horizontal and sub-parallel geological structures with respect to the tunnel direction due to the relative movements between the blocks forming the fault. The results from the performed investigations indicate that the fault zone has a dip of about 70 degrees towards the south.

Two different rock mass qualities are mainly expected for the central core and the transition zones in the identified fault zone in the subsea passage between Sätra and Kungshatt. The rock mass in the central core of the fault zone is expected to be highly fractured and weathered with an estimated RMR_{base} index between 29 and 40 (the base index indicates that water and

Figure 3. Horizontal drill core through the regional fault zone showing several meters of continuous core loss (pieces of wood).

1571

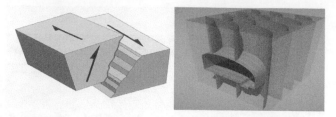

Figure 4. Strike-slip fault with rotation of the southern part over the northern part (Left); possible duplex formation in the fault zone (Right).

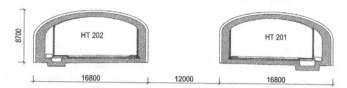

Figure 5. Geometry for the main tunnels in the subsea passage between Sätra and Kungshatt.

orientation of the joints with respect to the tunnel direction are not considered). Local areas with lower rock mass quality may be found according to the results from the drill cores (see Figure 3). On the other hand, the rock mass quality in the transition zones of the fault zone are expected to be of a slightly better quality with an estimated RMR_{base} index between 40 and 59.

3.2 Tunnel geometry and rock cover

The main tunnels that will be excavated in the subsea passage between Sätra and Kungshatt under the Lake Mälaren in the Stockholm Bypass project are approximately 17 m wide and 8.7 m high. The distance between both tunnels is approximately 12 m (see Figure 5). In addition, two exploratory tunnels parallel to the main tunnels, approximately 6 m in width, are planned to be excavated first. These exploratory tunnels will give additional information about the quality of the rock mass in the fault zone and reduce the total production time for the tunnel excavation process at the other side of the subsea passage.

The investigations performed to identify the depth of the bedrock in the subsea passage between Sätra and Kungshatt with respect to the tunnel location indicate that the rock cover can vary between 10 and 20 m in the deepest point. This uncertainty originates from the distance between the performed investigations. The rock surface in the deepest part of the subsea passage is covered with approximately 23 m of moraine, 8 m of clay and 16 m of water.

4 TECHNICAL SOLUTION

4.1 Grouting

The poor quality rock mass together with a relatively low rock cover and the presence of an "infinite" amount of water increase the risk of having stability problems due to possible flowing ground conditions during the tunnel excavation. A combination of cement grouting and an observational methodology was the adopted solution to control water hazards in these rock conditions.

The grouting of the rock mass in the tunnels in the subsea passage between Sätra and Kungshatt will be performed with two different rounds in order to reduce the rock mass conductivity to a defined target value of 10^{-8} m/s. The effectivity of each performed grouting round is verified under tunnel excavation by measuring water inflow in a number of drilled control holes. If the rock mass conductivity target is not reached after the second round, a third round may be performed until the defined value for the rock mass conductivity is reached (10^{-8} m/s).

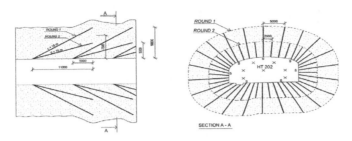

Figure 6. Cross section of the designed grouting procedure to manage water hazards in the regional fault zone in the subsea passage between Sätra and Kungshatt parallel (to the left) and perpendicular (to the right) to the tunnel alignment.

Figure 6 displays the main rock mass grouting procedure designed for the excavation of the tunnels through the regional fault zone in the parallel and perpendicular direction to the tunnel alignment respectively.

4.2 Temporary rock support

Failures in the rock mass due to high *in-situ* stresses that exceeds the local rock mass strength might occur during tunnel excavation through the fault zone. In this case, the installation of the rock support in early stages to improve both tunnel crown and tunnel face stability is considered important.

The designed temporary rock support for the tunnels in the subsea passage is as follows:

- Pipe umbrella with a length of 15 m, an overlap of 5 m and drilled at c/c 500 mm spacing. The steel pipes have a diameter of 140 mm and a thickness of 10 mm. Figure 7 displays a cross section of the pipe umbrella procedure.
- To avoid stability problems at the tunnel face, self-drilling rock bolts installed at c/c 1.5 m are considered necessary. After analyzing the expected rock conditions through the fault zone in the subsea passage, an overlap of 5 m for the self-drilling rock bolts was considered in the design process.
- 200 mm thick shotcrete and 5 m long rock bolts $\phi 25$ mm with a distance c/c 1 m perpendicular to the tunnel direction and c/c 2m in the parallel direction. In order to improve the resistance of the rock support, two additional steel meshes are installed in the second and third shotcrete layers. Figure 8 displays a cross section with the main support elements during this procedure.

4.3 Excavation sequence

The expected poor rock mass quality conditions in the fault zone, together with the relatively large width of the tunnels may present difficulties during the tunnel excavation. For this reason, the excavation of the main tunnels will be divided into a gallery and bench excavation. In addition, tunnel advance will be performed in stages of 2 m. Figure 9 displays the proposed division for the tunnel excavation of the main tunnels through the fault zone.

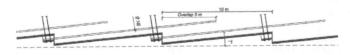

Figure 7. Cross section of the proposed pipe umbrella support for the subsea passage.

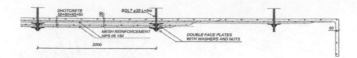

Figure 8.   Cross section showing the designed procedure for the installation of shotcrete and rock bolts.

Figure 9.   Division of the excavation sequences in the gallery and the bench for the main tunnels.

5   CALCULATIONS

The computer software Plaxis (Plaxis, 2017) was used to verify the proposed technical solution presented in section 4. Numerical calculations were performed to analyze the soil-structure interaction during the tunnel excavation through the regional fault zone in the subsea passage between Sätra and Kungshatt. The expected rock mass in the center and transition parts of the regional fault zone was modeled with linear elastic perfectly plastic material behavior with a Mohr-Coulomb failure criterion. Table 1 displays the utilized rock mass properties in the numerical calculations.

The soil-structure interaction of the proposed pipe umbrella solution was performed by analyzing a two-dimensional cross-section parallel to the tunnel direction. In this numerical model the varying rock cover in the subsea passage and the expected rock mass qualities in the core and transition parts of the fault zone were taken into account. In order to get correct deformations in the two-dimensional numerical model, spring elements were installed along the tunnel walls. The stiffness of these spring elements was calibrated through performing numerical calculations for a cross section perpendicular to the main tunnels.

Two different load cases with the estimated lowest and highest *in-situ* stresses in the regional fault zone were analyzed to capture the pipe umbrella behavior. The results from the numerical models in Plaxis showed deformations between 20–40 mm in the pipe umbrella with the highest *in-situ* stresses and the development of plastic hinges in some parts of the steel pipes, see Figure 10.

In order to analyze the mechanical behavior of the rock-support system, consisting of shotcrete and rock bolts, two-dimensional numerical models were performed in Plaxis. The calculations were performed in one cross section with lowest rock cover in the subsea passage with the expected rock qualities in the core and transition parts of the regional fault zone.

Table 1.   Mechanical properties for the rock mass in the regional fault zone utilized in the numerical calculations.

Parameter	Transition part	Center part
Cohesion [MPa]	1.0	0.6
Friction angle [°]	46	35
Dilation angle [°]	0	0
Young's modulus [GPa]	3.0	0.7
Tensile strength [MPa]	0.01	0
Poisson's ratio [-]	0.25	0.25

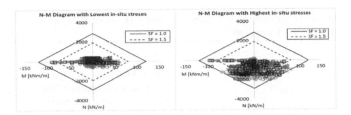

Figure 10. N-M diagrams with the calculated loads for the pipe umbrella solution during tunnel excavation through the fault zone in Plaxis with the estimated lowest (Left) and highest (Right) *in-situ* stresses in the subsea passage.

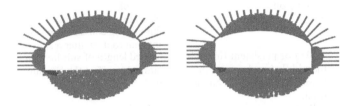

Figure 11. Areas with rock mass plasticity around the tunnels in the core part of the fault zone for the load case with the estimated highest *in-situ* stresses in the subsea passage.

In line with the performed analysis for the design of the umbrella pipe solution, a sensitivity analysis was performed to study the mechanical behavior of the rock support during the tunnel excavation with the estimated lowest and highest *in-situ* stresses in the subsea passage. The results showed a tunnel deformation up to 45 mm and plasticity of the rock mass around the tunnels in the core part of the analyzed fault zone, see Figure 11.

The results from the performed numerical calculations showed acceptable mechanical behavior of the temporary rock support for the tunnels in the subsea passage between Sätra and Kungshatt.

6 CONTROL PROGRAM UNDER TUNNEL EXCAVATION

Since the design of the temporary rock support is performed prior to the excavation of the tunnel for predefined rock mass qualities in the fault zone, it is necessary to verify through observations both that the design conditions are valid and that the support behaves as expected. In order to obtain this and follow the main principles of the observational method, two different control programs were established.

6.1 *Control of the design conditions*

In order to verify that the rock conditions assumed in the design of the temporary rock support are met during excavation, a control program with several parameters representing the design conditions has been established.

The parameters to be observed during the tunnel excavation are rock mass quality including fracture characteristics, rock cover, water leakage and pressure. The rock mass quality is expressed with the RMR-index. Verification of the rock cover will be performed by probe drilling that will be carried out during the tunnel excavation, while water pressure will be measured in the performed drilled holes. Permissible values for the parameters in the control program are presented in Table 2.

Table 2. Permissible values for control parameters to verify the validity of the conditions assumed in the design of the temporary rock support.

Control parameter	Limit value	Comment
RMR_{base}:		The RMR-index is used as control parameter for the rock mass quality. The average value for 10 m is used for the assessment
Transition part of the fault zone	$RMR_{base} \geq 40$	
Center part of the fault zone	$40 > RMR_{base} \geq 29$	
Rock mass and rock support properties	See section 6.2	Control of the response of the interaction system (rock mass/rock support) is performed by measuring the tunnel deformation
Water inflow	4 l/min 0.2 Lugeon	Water leakage or water measurement in every drilled hole
JRC	6	Relatively planar joints
Ja	8	Joint infilling consisting of consolidated clay. No joint-wall contact after a shear deformation of 10 cm
Length of discontinuities	> 10 m or persistent through the middle bench over a distance of 4 m from the tunnel front.	Mapped length of sub-horizontal fractures which is judged to affect the bearing capacity of the bench.
Rock cover	> 10 m	Drill hole

6.2 *Control of rock support behavior*

The rock mass together with the tunnel support constitute parts of an interactive system, where the stresses in the rock mass and the stiffness of the support together determine the behavior of the system. A single property is difficult to verify and separate from other ones. Initial stresses, rock mass quality and support determine the deformations in the tunnel; and in the end the strains in the support, which are the critical parameter. The control of the behavior of the support is therefore performed by measurements of the deformations.

The measurements of the deformations will mainly be carried out with optical convergence measurements in predefined sections of the tunnel. The measurement points are installed 6 m from the tunnel front. Based on previous calculations, it is assumed that approximately 50% of the total deformations have occurred at the time of installation. Acceptable limits for support are based on those results and a reduction is made for the deformations that have occurred prior to the installation. When the tunnel is excavated with a gallery and bench, the measurements are initiated when the gallery is excavated.

Deformation measurements will also be performed in two sections along the tunnel direction based on a combination of inclinometers and extensometers installed in the umbrella pipes. The measurements are performed as an indirect method to estimate initial stresses and rock mass (deformation) properties, since these parameters govern the load taken by the umbrella pipes. These measurements are considered to be a necessary parameter in case additional support may be needed.

For the control of the deformation, three predefined limits have been established: observational limit, alarm limit 1 and alarm limit 2. The observational limit refers to a limit that follows the predictions in the design. If alarm limit 1 is exceeded, predefined measures are executed. In case alarm limit 2 is exceeded, the excavation should be stopped and the deformations analyzed.

The measurements are continuously carried out until deformations have ceased to increase and are stable for a time of at least three weeks.

In addition to the measurements of the deformations, visual inspection is carried out of the shotcrete support in order to observe if any cracking occurs – and if that is the case, how it progresses. The control program and observational values for the temporary rock support behavior are presented in Table 3.

Table 3. Control program and observational values for the temporary rock support behavior.

Control parameter	Limit value	Comment
Visual observation of cracks in the shotcrete installed at the tunnel front	Observational limit: No cracks appear Alarm limit 1: 3 m Alarm limit 2: 6 m	The observational limit is based on the design assumption that no cracks are expected in the installed shotcrete under stable conditions. For alarm limit 1, the level of deformation implies that relatively large cracks have started to occur in the shotcrete. Alarm limit 2 is set to account for the continuously growing cracks in the shotcrete. This may indicate creeping behavior and unstable conditions during the excavation.
Visual observation in the tunnel	Identification of cracks in the shotcrete/Mapping	Cracks in the shotcrete will be mapped up to 20 m back from the tunnel front. In case the observational or alarm limits stated below are exceeded, additional mapping must be performed
Inclinometer measurement in the transition part of the fault zone	Observational limit: 7 mm Alarm limit 1: 12 mm Alarm limit 2: 17 mm	The observational limit is based on the assumption that 30% of the total deformations has occurred when the inclinometer is installed. Inclination changes are converted to deformations. If alarm limit 1 is reached, the yield strength of the pipe umbrella support is considered to be exceeded. If alarm limit 2 is reached, a plastic hinge is assumed to be formed in the pipe umbrella support.
Inclinometer measurement in the center part of the fault zone	Observational limit: 20 mm Alarm limit 1: 30 mm Alarm limit 2: 40 mm	The observational limit is based on the assumption that 30% of the total deformations has occurred when the inclinometer is installed and approximately 5% of the deformations occurs 10 m in front of the tunnel front. Inclination changes are converted to deformations. A plastic hinge formation is observed in the calculations when the observational limit is reached. Alarm limit 1 and alarm limit 2 are set after the value for the convergence measurement limit.
Deformation measurement and extensometer	Observational limit: 10 mm Alarm limit 1: 15 mm Alarm limit 2: 20 mm	The observational limit is based on a stress level in the pipe umbrella support equal to the 70% of the yield strength (characteristic value). If alarm limit 1 is reached, the yield strength of the pipe umbrella support is considered to be exceeded at some extent. If alarm limit 2 is reached, a plastic hinge is assumed to be formed in the pipe umbrella support.
Convergence measurement in the transition part of the fault zone	Observational limit: 5 mm Alarm limit 1: 10 mm Alarm limit 2: 15 mm	The observational limit is based on the assumption that approximately 50% of the total calculated deformation has occurred when the measurement point is installed (6 m from the tunnel front). If alarm limit 1 is reached, there is a risk for cracks in the shotcrete. Alarm limit 2 is set at a level which may indicate that the rock mass behavior differs from the estimated variation in the performed assessment.

(*Continued*)

Table 3. (*Continued*)

| Convergence measurement in the center part of the fault zone | Observational limit: 20 mm
Alarm limit 1: 30 mm
Alarm limit 2: 40 mm | The observational limit is based on the assumption that approximately 50% of the total calculated deformation has occurred when the measurement point is installed (6 m from the tunnel front).
If alarm limit 1 is reached, cracking in the shotcrete has occurred.
Alarm limit 2 is set at a level which may indicate that the rock mass behavior differs from the estimated variation in the performed assessment in combination with load variation. |

7 CONCLUDING REMARKS

In this paper the methodology used for the design of the temporary rock support for the tunnels in the subsea passage between Sätra and Kungshatt in the Stockholm Bypass Project was presented. The presence of a poor quality rock mass together with limited rock cover and the access to an unlimited amount of water constitute the major risks for the excavation. A technical solution consisting of cement grouting, pipe umbrella, rock bolts and shotcrete reinforced with wire-mesh was proposed as support. In addition, the control program is divided into two parts and follows the main principles of the observational method. The first part contains limit values for control parameters in order to verify that the rock conditions assumed in the design of the temporary rock support are met during excavation, while the second part contains predefined limits for acceptable deformations of the rock-support system. Based on this methodology, a safe passage through the fault zone can be accomplished.

Despite the extensive measurements described in this paper, there are remaining risks with low probability of occurrence during the tunnel excavation. Many of these risks are related to the geological and geometrical hazards in the subsea passage such as swelling clay and rock cover. These risks are considered acceptable and manageable under tunnel construction but may lead to long-term stability problems. For this reason, a permanent concrete lining will be built in the main tunnels in parts of the subsea passage as a permanent solution.

REFERENCES

Gaál, G. & Gorbatschev, R. 1987. An outline of the Precambrian Evolution of the Baltic Shield. Precambrian Research 35: 15–52.

Plaxis. 2017. Plaxis 2D Reference manual.

SGU Sveriges Geologiska Undersökning (2018). https://www.sgu.se/produkter/kartor/kartgeneratorn/

Stålhös, G. 1969. Beskrivning till Stockholmstraktens Berggrund. SGU BA 24.

*Tunnels and Underground Cities: Engineering and Innovation meet Archaeology,
Architecture and Art, Volume 4: Ground improvement in
underground constructions – Peila, Viggiani & Celestino (Eds)
© 2020 Taylor & Francis Group, London, ISBN 978-0-367-46868-2*

Deep intervention shaft excavation in Kuala Lumpur limestone formation with pre-tunnelling construction method

Y.C. Tan, K.S. Koo & D.I. Ting
G&P Geotechnics Sdn Bhd, Kuala Lumpur, Malaysia

ABSTRACT: In Malaysia, Klang Valley Mass Rapid Transit Line 2 (KVMRT2) is consisting of 38.7km elevated tracks and 13.5km underground tunnels with several shafts. For intervention shaft 2 (IVS2), deep soil mixing (DSM) block was designed to provide excavation support in overburden alluvial soil, followed with vertical rock excavation until 57.5m deep in Kuala Lumpur limestone formation. Ground treatment in limestone was established to minimize groundwater drawdown during deep excavation to prevent excessive ground settlement. A mined adit was constructed to connect from the shaft at mid-depth to south-bound tunnel. Meanwhile, north-bound tunnel at bottom of the shaft was constructed by tunnel boring machine cutting through concrete-filled block inside the shaft compartment. This paper covers the challenges in design and construction of the deep excavation works for the shaft, ground improvement scheme for the bored tunnelling near the shaft, temporary works for mined adit and unconventional construction sequence of bored-through tunnel.

1 INTRODUCTION

It is estimated that the demand for travel in the Malaysia's Klang Valley to reach 18 million trips per day by year 2020. Hence, a public transportation network capable of ferrying large numbers of passengers efficiently in the form of rail-based mode is needed. The first line of the Klang Valley Mass Rapid Transit project (also known as KVMRT1) begins from Sungai Buloh at the north-west, runs through the city centre of Malaysia's capital city (i.e. Kuala Lumpur) and ends in Kajang, which is located at the south-east of the city. This Sungai Buloh-Kajang line began operations in two phases respectively on 16 December 2016 and 17 July 2017. The 51-km length transit line is estimated to serve a daily ridership of about 400,000 passengers in the city.

Meanwhile, the second line of the MRT project (KVMRT2) was given the approval by the federal government in October 2015 in order to complement the KVMRT1 line to serve a corridor with a population of around 2 million people. While this transit line is currently being constructed, it will have a length of 52.2km stretching from Sungai Buloh, through the centre of Kuala Lumpur and Serdang until Putrajaya, which is Malaysia's federal administrative centre. This on-going KVMRT2 line will have a distance of 13.5km running through underground tunnel as compared with only 9.5km of the KVMRT1 line. The tunnels will be built using tunnel boring machines and/or the cut-and-over method depending on the depth of the tunnels.

For the underground sections of KVMRT2, total of six escape shafts and intervention shafts are provided for the purpose of emergency escape and ventilation. Among them, Intervention Shaft 2 (also known as IVS2) is the largest in size and the deepest. It is a circular shaft with maximum diameter of 19.7m between two sides of the excavation faces. The maximum depth of the shaft is up to 57.5m until final excavation level below existing ground level (i.e. RL42m). Twin bored tunnel lines bounding for North and South respectively are running at different elevations while interfacing with the IVS2. The south-bound tunnel is located at higher elevation and is connected to the shaft via a mined adit. Meanwhile, north-bound tunnel is directly crossing through the bottom section of the shaft.

2 GENERAL SITE CONDITIONS

2.1 Site Location

The site for IVS2 shaft is located at the center of a roundabout (known as Bulatan Kampung Pandan), which is an intersection between Jalan Tun Razak and Jalan Kampung Pandan, located in the Kuala Lumpur city center. A landmark building of Berjaya Times Square is just 1.3km away from the site. Location of the site with existing buildings and roads is shown in Figure 1.

The shaft is located in close proximity to KVMRT Line 1 tunnel as well as its Escape Shaft 3 building, Stormwater and Roadway Tunnels (SMART) motorway and North Junction Box, and several existing buildings (including IKEA Cheras). Besides that, at the time of designing, it is known that future vehicular viaduct of Duta Ulu-Kelang Expressway (DUKE) will be running adjacent to IVS2, potentially with some piers located close-by. All these structures and some other services/utilities are posing major constraints to the design and construction of the shaft.

It is very crucial to ensure that the temporary retaining structure system for the excavation of the shaft would be designed to limit ground movements and does not cause damaging impacts to those structures during the entire construction period.

2.2 Ground Condition

Based on available boreholes within the close vicinity of the shaft, it is found that the site is underlain by Alluvium overlying the Kuala Lumpur Limestone as shown in Figures 2 & 3. The Alluvium is very loose sand/very soft to soft silt with SPT-N not more than 30 overlying the limestone bedrock. The depth of bedrock varies from 4.5m to 8.0m below ground level around the shaft area. The limestone bedrock consists of highly fractured rock to strong rock with Rock Quality Designation (RQD) ranges from 0% to 93%. The interpreted groundwater level is 2m below ground level based on available water standpipe readings. Summary of geotechnical design parameters for Intervention Shaft 2 is presented in Table 1.

Kuala Lumpur Limestone is well known for its highly erratic karstic features, such as irregular bedrock profiles, variable weathering condition, cavities, pinnacle zone and slime zone. Challenges in such karstic limestone formation are faced by the designers in designing an appropriate temporary retaining structure system to facilitate the underground excavation works.

Figure 1.   Location of Intervention Shaft 2 (IVS2).

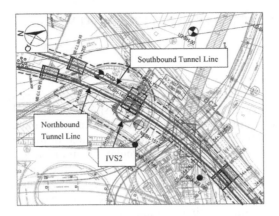

Figure 2.   Boreholes layout plan.

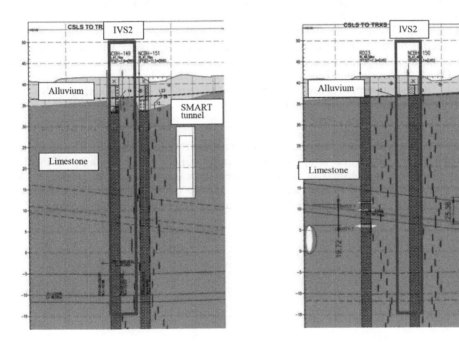

Figure 3.   Borehole and geological profile along Northbound tunnel (left) and Southbound tunnel (right).

Table 1.   Geotechnical Design Parameters.

Geological Formation	Alluvium	Limestone
Depth below ground	0m – 8m	>8m
Average SPT-N	5	-
Unit Weight	19 kN/m^3	24 kN/m^3
Loading Stiffness	15,000 kPa	1.0E6 kPa
Unloading Stiffness	45,000 kPa	-
Effective Cohesion	1 kPa	400 kPa
Effective Friction Angle	29°	32°
Permeability	1.0E5 m/sec	1.0E6 m/sec
At Rest Lateral Earth Pressure	0.52	0.50

3 TEMPORARY RETAINING STRUCTURE SYSTEM

3.1 Deep Soil Mixing Wall

3.1.1 Design Considerations

The Intervention Shaft 2 consists of a circular shaft. The advantage of circular shaft is there will be unobstructed excavation area and working space which results in faster overall construction, especially for deep underground shafts. Given such shape of shaft, the proposed temporary retaining structure system is therefore configured in the same circular form. Referring to Tan et al. (2016), the design of this circular retaining structure is also based on the hoop stress concept in which the induced hoop stress due to earth pressures shall not exceed the allowable compressive stress of the material of the retaining structure.

Deep soil mixing (DSM) wall is adopted where the entire block of soil will be mixed with cement grout to form an interlocking columns block, rather than a conventional retaining wall with or without lateral supports (e.g. steel struts, ground anchors and etc.) as shown in Figure 4. The width of DSM block is 4m all around based on design assumption of 8m thick overburden subsoil above bedrock. Typical arrangement of the DSM interlocking columns arrangement is shown in Figure 5.

The DSM technique involves the process of mixing soil with cement slurry by using a mechanical tool, which is drilled into the ground. The cement-mixed soil block will have enhanced compressive strength, reduced permeability and increased stiffness as compared to the original soil. As the DSM wall is designed as a circular block based on hoop stress concept, steel reinforcements and lateral supports are not required. Furthermore, rock socketing is also not needed as compared to conventional retaining wall.

3.1.2 Monitoring and Performance

The execution practice and quality control of DSM works are in accordance to British Standard BS EN 14679:2005 (Execution of special geotechnical works – Deep Mixing). An inspection and test plan were established in which the frequency of strength tests on the DSM samples extracted from the ground is set as four cores per $1000m^3$ of DSM area. Core samples were collected from the center of the DSM columns and at intersection between columns for

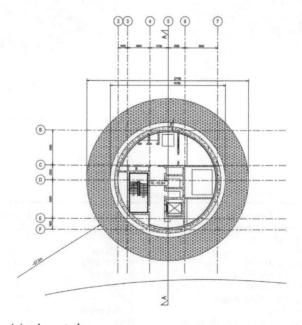

Figure 4.   Deep soil mixing layout plan.

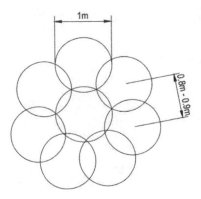

Figure 5.   Typical DSM columns arrangement.

unconfined compressive strength (UCS) tests. Generally, UCS strength of all test samples are more than the designed allowable compressive strength of 1.0MPa after 28 days.

Monitoring scheme is also implemented to monitor the actual performance of the DSM wall in terms of wall movement and settlement. For this purpose, four inclinometers were installed within the DSM block zone and thirty-two settlement markers were placed on the ground surface along two section lines across the circular shaft. Maximum wall movement monitored throughout the excavation works was less than 4mm. The deflection profile of one of the inclinometer is shown in Figure 6. Meanwhile, the maximum ground settlement over the same period only recorded less than 6mm. The measured deflection profiles and ground settlement readings could imply a very stiff circular DSM block given the hoop action and counter-balancing effect. The overview of the shaft excavation works is shown in Figure 7.

3.2   Rock Slope Strengthening Works

Bedrock is anticipated at about 8m below ground level and thereafter rock excavation works will be carried out for another 50m until the final excavation of 57.5m deep. Rock excavation

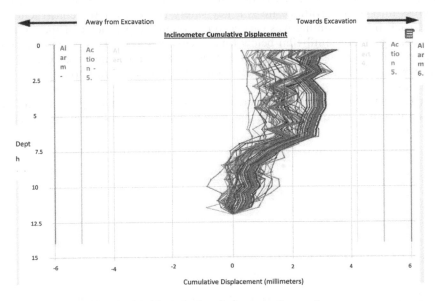

Figure 6.   Deflection profile of DSM block during shaft excavation works.

Figure 7.   Overview of Shaft Excavation Works.

by controlled blasting method is the most practical way to use. Excavation in rock will be carried out in stages of every 2 to 3m height.

After exposing the finished rock face, on-site rock mapping will be conducted by suitably qualified and experienced engineering geologist. The rock mapping data which includes dip angle and dip direction of visible joints/faults will feed into Stereonet analysis to determine any potential planar, toppling or wedge failures. Subsequently, stability analyses will be performed to calculate the factor of safety of the identified failure modes and the stabilization works using mainly steel reinforcement rock bolts will be designed.

Glass fibre reinforced polymer (GFRP) type rock bolts will be used to create a soft-eye where tunnel bored-through is anticipated. GFRP reinforcement will be designed in accordance with ACI 440. Shotcrete with steel wire mesh will be applied to the full surface of the exposed rock face to obtain a uniform surface for water proofing for permanent reinforced concrete works in later stage and also to prevent any localized rock instability.

3.3   Grouting Works

In consideration of the potential cavities and solution channels in limestone formation, water ingress into the excavation site is a key safety concern for the construction works. Therefore, rock fissure grouting was carried out around the perimeter of the circular shaft from the rock head level until 10m below final excavation level. This is aimed to minimize the risk of groundwater ingress and groundwater drawdown during rock excavation which also to prevent excessive ground settlement that could cause distress to adjacent buildings and structures.

In order to form an effective curtain surrounding the shaft, primary grout points were carried out at a distance of 6m away from shaft excavation surface with grouting spacing of 4m

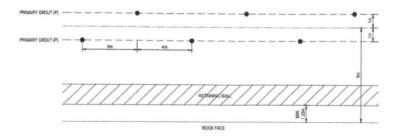

Figure 8.   Typical arrangement of primary fissure grout points.

staggered all-round the shaft perimeter. Secondary grout points were added at middle spacing between the primary grouted points if high grout intake of more than 5,000 liter for a prescribed pressure at a given depth or cavities are detected. Tertiary holes and further grout points could be added if needed. Schematic arrangement of the curtain grouting is shown in Figure 8.

4 GROUND IMPROVEMENT SCHEME FOR ADITS

As mentioned in Section 1 of this paper, there is an adit that connects the shaft to the South bound Tunnel. The adit is inside the limestone rock. Before the mined adit excavation works, fissure grouting works were carried out 5m all around the adit to create a grouted block.

Meanwhile, for TBM break-in at the receiving locations of the shaft, ground improvement/grouting is also carried out for a minimum length equivalent to the length of the TBM shield plus one ring of tunnel lining. The respective ground improvement zones are shown in Figure 9.

The main purpose of the ground improvement zones are (1) to provide the ground with sufficient strength (2) to reduce the permeability of the ground over the adit length (3) to allow TBM to reduce face pressure from full face pressure of balancing earth and hydrostatic pressure to atmospheric pressure without causing instability of ground and excessive water inflow into the receiving shaft.

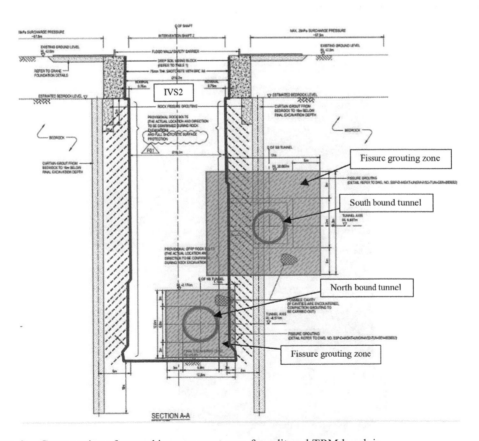

Figure 9.   Cross section of ground improvement zone for adit and TBM break-in.

5 TEMPORARY CONSTRUCTION WORKS FOR TUNNEL/ADIT

5.1 Adit connecting to South Bound Tunnel

The adit is in a horseshoe shape with a clear width between the side walls of 8m. The internal height of the adit close to the shaft is 8m and the height will increase to 10.85m (while the crown remains at a constant level) after the first 3m length of the adit from the shaft side.

Generally, the mined tunnel will be constructed using the Sprayed Concrete Lining (SCL) approach which is characterized by installing temporary support in the form of rock bolts and Steel Fibre Reinforced Shotcrete (SFRS) during excavation. When excavation is completed, a permanent support, in the form of a cast in place concrete lining, is installed.

With the objective of expediting the construction of permanent shaft structure and tunneling schedule, it was decided to complete the adit structure first when vertical excavation in shaft reaches this adit level prior to arrival of tunnel boring machine (TBM) at South Bound line. The length of the mined adit is until the entire bored tunnel section and then allow for TBM bore-through at the 'soft-eye' on the receiving adit wall.

Besides that, it is crucial to create a temporary wall in between the shaft and the bore-through area to cater for the tunnel bore through impact (50kPa pressure). This temporary wall is going to connect to the permanent lining of the adit and to be demolished once the TBM operation is completed. The compartment behind the temporary wall, where tunnel is coming through, is filled up with mass concrete (minimum 5 MPa compressive strength). After all these are in place, then South Bound tunnel drive can bore through into the adit. Figure 10 shows the general arrangement of the adit. Meanwhile, the overview of the adit construction works is shown in Figure 11.

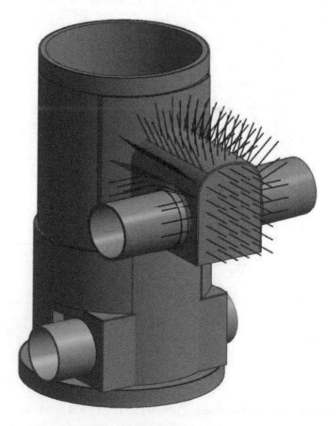

Figure 10.   General arrangement of adit and TBM break-in area.

Figure 11.   Horseshoe shape adit in construction.

5.2   *Bore-through North Bound Tunnel*

The construction sequence for the shaft connection to North Bound Tunnel is planned such that shaft is excavated until the final level in precedence over the TBM arrival. After the final excavation level of the shaft is reached, permanent base slab, permanent upper slab, all external walls and tunnel eye are then constructed.

Similar to the adit connecting to South Bound Tunnel, a temporary reinforced concrete wall is to be constructed with connection to the permanent base slab and upper slab creating a room for TBM bore through. This room is filled up with mass concrete and back grouting is carried out to fill any voids in the crown of the room. After that, North Bound tunnel drive can bore through the tunnel eye coming into the concrete-filled room inside the shaft. The temporary wall and mass concrete will be excavated after completion of the tunnelling operation.

6   CONCLUSION

Deep soil mixing wall was successfully used as temporary earth retaining scheme for shallow overburden subsoil up to 8m depth above limestone bedrock. Ground improvement scheme with rock fissure grouting works were carried out around the perimeter of the circular shaft to prevent excessive groundwater ingress into excavation pit as well as to control groundwater drawdown at the surrounding area of shaft. Rock excavation was carried out with stages of control blasting and follow by rock condition mapping by engineering geologist up to maximum excavation depth of 57.5m below ground. Necessary rock slope strengthening measures will be assigned at every stage of excavation with rock mapping data and condition assessment before allow to next stage of excavation work. Considering overall tunneling schedule, post-tunneling construction method was adopted. The permanent structure of connection adit will be completed in advance before TBM arrive by providing required bore-through facility without affecting tunneling schedule.

REFERENCES

BS EN 14679. 2005. Execution of special geotechnical works – deep mixing
Tan Y.C., Chow C.M., Koo K.S. & Nazir R. 2016. Challenges in design and construction of deep excavation for KVMRT in Kuala Lumpur limestone formation. YGEC, Kuala Lumpur.

Yew Y.W. & Tan Y.C. 2015 Excavation support for TBM retrieval shaft using deep soil mixing technique, Kuala Lumpur. *International Conference and Exhibition on Tunnelling and Underground Space 2015*, Grand Dorsett, Petaling Jaya, Selangor, Malaysia.

Koo K.S. 2013. Design and construction of excavation works for Klang Valley mass rapid transit underground station at Cochrane, Kuala Lumpur, Malaysia. *5th International Young Geotechnical Engineering Conference*, Paris, France.

Raju V.R. & Yee Y.W. 2006. Grouting in limestone for SMART tunnel project in Kuala Lumpur. *International Conference and Exhibition on Tunnelling and Trenchless Technology*, Subang, Selangor, Malaysia.

Tunnels and Underground Cities: Engineering and Innovation meet Archaeology, Architecture and Art, Volume 4: Ground improvement in underground constructions – Peila, Viggiani & Celestino (Eds)
© 2020 Taylor & Francis Group, London, ISBN 978-0-367-46868-2

Expansion of existing TBM tunnel by ground freezing method

Y. Tanaka & N. Takamatsu
Tokyu Construction Co., Ltd., Tokyo, Japan

ABSTRACT: This is a case report on Sumidagawa sewer trunk line 3rd project, the construction project expanding the middle of existing Sumidagawa sewer trunk line by non-open cut method in Japan. This expanded section required an outer diameter of 9.5m in order to receive the TBM which started from Senjusekiya pumping station. Sumidagawa sewer trunk line 3rd project contains works involving large-scaled ground freezing, removal of existing segments, excavation of frozen soil, assembly of expanded type segments and enforced thawing of frozen soil. We completed all of these works safely under operating conditions of high water and earth pressure at 40m below ground.

1 CONSTRUCTION OVERVIEW

1.1 *Sumidagawa Sewer Trunk Line Project*

Sumidagawa Sewer Trunk Line Project is a flood control project aimed at enhancing the rainwater discharge capacity of the Senju area in Adachi-ku, Tokyo, Japan between the Arakawa and Sumidagawa rivers. Sumidagawa Sewer Trunk Line (SSTL) will have capacity to accommodate rainwater accumulated from 292.85 ha, which is about 70% of the Senju area. This project consists of trunk line construction works, pumping station building and connecting works, all of which lie between the trunk line and the pumping station. Figure 1 indicates the construction site of this project. Meanwhile, this paper specifically concentrates on the works relating to SSTL 3rd project.

1.2 *SSTL 3rd project*

The main work of SSTL 3rd project is the expansion of the existing TBM tunnel, which was constructed in SSTL 1st and 2nd project. The inner diameter of the existing tunnel is 4.75 m to drain a large amount of rainwater in the Senju area. On the other hand, a new tunnel supplied by SSTL 4th project was excavated connecting Senju Sekiya pumping station to the above tunnel by use of a TBM of 6.5 m in diameter and causing difficulties in penetration to the existing tunnel directly. Likewise, the thick overburden and some important infrastructures prevented the construction of an arrival shaft for the TBM at the middle of the existing tunnel. Due to this condition, we decided to adopt an unprecedented plan where the existing tunnel was expanded from 5.5 m to 9.5 m in outer diameter without adopting an open cut method. In order to implement this, we selected large-scaled ground freezing as an auxiliary method. The total volume of freezing soil on this site was 3,700 m^3, which makes it one of the largest scale applications among domestic sewage projects. Figure 2 and Figure 3 show the longitudinal section and the cross section of the expanded section in SSTL 3rd project, respectively. Likewise, the workflow of SSTL 3rd project is shown in Figure 4.

1.3 *Site Condition*

SSTL was constructed at a depth of 40 m in the Senju area. Presently above this tunnel exist prefectural Bokutei-dori Street, railway facilities of Keisei Main line and many underground

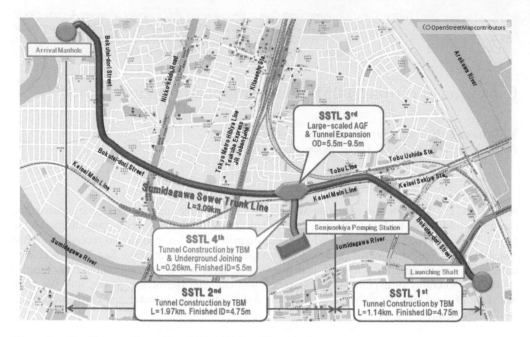

Figure 1.   Sumidagawa Sewer Trunk Line Project (Plan view).

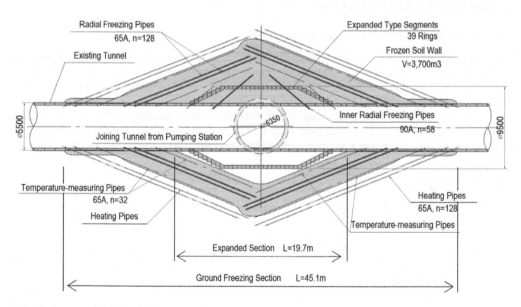

Figure 2.   Expanded section in SSTL 3rd project (Longitudinal section).

installations (as shown in Figure 1). Geologically, the area is composed of a soft layer with a thickness of 20 m from the top and having a cohesive soil with value N of 0. Also an alternative layer of fine sand and silt having a thickness of 15 m, along with a 2m layer of gravel and diluvial formations, consisting of large amounts of silt (as shown in Figure 3).

Generally, freezing expansion is likely to occur in the layer of silt or clay in comparison with the sand or gravel layer. The planned freezing zone for SSTL 3rd project contained a substantial amount of cohesive soil, causing high freezing-expansion of the soil. Therefore, it was necessary to predict the influence of frost heaving and thawing subsidence on the above structures around the existing tunnel in advance.

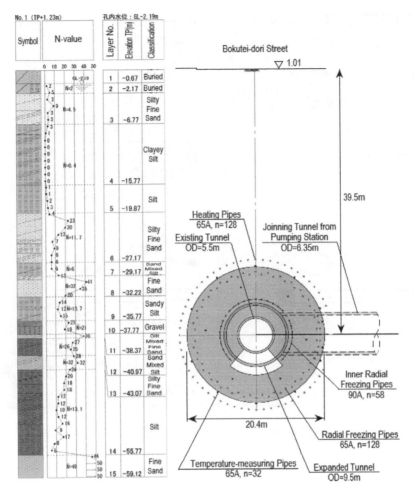

Figure 3. Expanded section in SSTL 3rd project (Cross section).

2 SCHEME OF EXECUTION

In the ground freezing work, anti-freeze liquid called "brine" at - 30 °C circulates in freezing pipes buried into the ground, depriving heat from the ground to form frozen soil. Here, Figure 5 shows the schematic diagram of these frozen soils that gradually combine to form a glowing frozen soil wall. The frozen soil wall acts as a load-bearing wall as well as a water-stopping wall impending high earth pressure and water pressure.

Prior examination of the ground freezing work is very important in order to obtain an ample performance of frozen soil in SSTL 3rd project.

2.1 Investigation of Groundwater

Empirically, in the case of the brine type freezing method, it is impossible to form the required frozen wall when the groundwater flow velocity is 2 m/day or more.

At this site, we measured the flow velocity in the immediate vicinity of the expanded section and confirmed that the groundwater flow velocity is 0.7 m/day or less in the full year. This means, no countermeasure controlling the flow velocity was necessary for the freezing method.

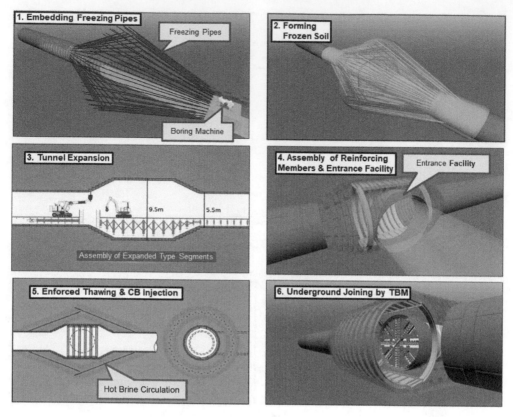

Figure 4. Workflow of SSTL 3rd project.

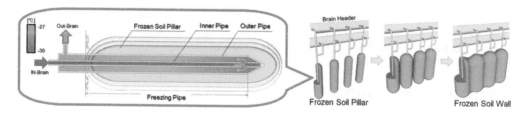

Figure 5. Schematic diagram about the formation of the frozen soil wall.

2.2 Performance Requirement for Frozen Soil

According to the site plan, the existing steel segments were supposed to be removed partially after forming the frozen soil around the expanded section. Therefore, the frozen soil wall created by the freezing method, temporarily supports all of the earth pressure and water pressure.

Based on this situation, we designed the frozen soil wall as conjugated rigid uniform rings divided into 1 m widths. From Figure 6, the calculation model adopted for this design can be understood. Furthermore, the design strength of the frozen soil can be known from Table 1, which has been estimated using the geological information of SSTL 3rd project. Considered values taken under the circumstances of having salinity and an average temperature of 0 % and -12 °C respectively in the frozen soil.

As a result of the overall review, we applied a thickness of 3.4 m to the frozen soil wall.

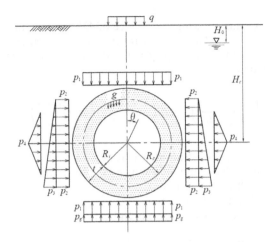

Figure 6.   Calculation model of the frozen soil wall.

Table 1.   Design strength of the frozen soil.

Compression Strength σ_{uc} (kN/m^2)	3,450
Bending Strength σ_{ub} (kN/m^2)	2,050
Shear Strength $\sigma_{u\tau}$ (kN/m^2)	1,700

2.3   Arrangement of Freezing Facilities

We determined the arrangement of the freezing pipes to take into consideration the above thickness of frozen soil. Initially, to cover the expansion part of 19.7 m in length, 186 No. radial freezing pipes were enclosed. Secondly, 704 No. sticking freezing pipes winded around the root of radial freezing pipes to block groundwater completely from the interface between the existing segments and the frozen soil. Furthermore, temperature-measuring pipes for the inspection were installed along the radial freezing pipes of the growth of frozen soil and heating pipes were used for suppressing the excessive growth of the frozen soil (as shown in Figure 2-3).

2.4   Adoption of Enforced Thawing

Formerly, it was predicted that if the frozen ground of 3,700 m^3 in SSTL 3rd project was naturally thawed, it would take about 1 year and 8 months to return to its original state. This concluded that natural thawing would be inappropriate in terms of construction time and cost, and thus "forced thawing" was adopted which promoted the thawing of frozen ground by circulating hot water of about 60 °C in freezing pipes.

2.5   Prediction of Influence Caused by Ground Freezing

Approximately 80% of the freezing zone planned for SSTL 3rd project was cohesive soil and therefore the soil easily expanded on freezing. We predicted that the frost heaving and thawing subsidence caused by the ground freezing work would likely be substantial in this site.

We therefore proceed to calculate the frost heaving and thawing subsidence around the expanded section by using equation (1). Additionally, in order to examine the necessity and extent of countermeasures, we analyzed the deformations of surrounding structures by 3D FEM (as seen in Figure 8).

As a result, we were able to estimate that frost heaving of up to 90 mm may occur at 40 m below the ground and also of approximately 10mm at the center of the railway on the ground.

On the other hand, a maximum thawing subsidence of 436 mm was predicted in the event that no countermeasures were taken. Therefore, a total volume of cement-bentonite (CB) of 536 m³ was planned to be injected into the gaps generated in the thawed ground as a countermeasure against thawing subsidence.

$$G(x,y) = \frac{\eta}{4}\int_{h_1}^{h_2}\left\{\mathrm{erf}\left(\frac{1+x}{ah}\right) + \mathrm{erf}\left(\frac{1-x}{ah}\right)\right\}\cdot\left\{\mathrm{erf}\left(\frac{w+y}{ah}\right) + \mathrm{erf}\left(\frac{w-y}{ah}\right)\right\}dh \qquad (1)$$

Here,

$G(x,y)$: Frost Heaving or Thawing Subsidence at (x,y)
η: Rate of Frost Heaving or Thawing Subsidence (as shown in Figure 7)
l: Half of X-direction Width of Frozen Soil
w: Half of Y-direction Width of Frozen Soil
a: Effect Extent of Frost Heaving or Thawing Subsidence
(a=1 in dispersion angle = 45)
h_1, h_2: Depth at Top and Bottom of Frozen Soil
$\mathrm{erf}(x)$: Gaussian Error Function

$$\mathrm{erf}(x) = \frac{2}{\sqrt{\pi}}\int_0^x e^{-\lambda^2}\, d\lambda$$

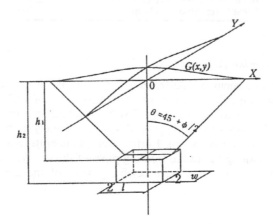

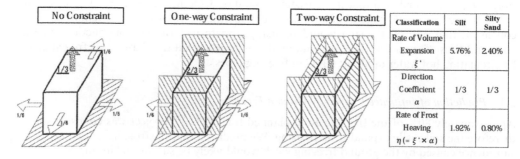

Classification	Silt	Silty Sand
Rate of Volume Expansion ξ'	5.76%	2.40%
Direction Coefficient α	1/3	1/3
Rate of Frost Heaving $\eta\,(=\xi'\times\alpha)$	1.92%	0.80%

Figure 7.   Coefficient for each direction of the frozen expansion in different constraint conditions.

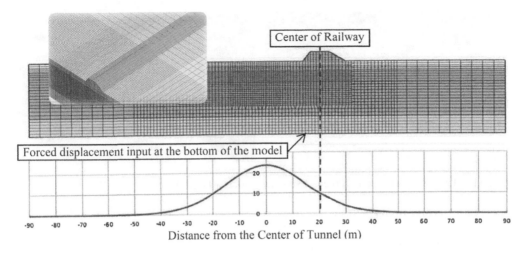

Distance from the Center of Tunnel (m)

Figure 8.   Prediction of the uplift on the ground due to frost heaving by 3D FEM.

2.6 *Safety Measure*

In order to prevent the ground from collapse due to the melting frozen soil wall, it was necessary to maintain the temperature below 10 °C inside the tunnel during the expansion work. Thus, thermal simulation as its counteragent was executed for each operation process based on the data relating to the exhaust gas volume and temperature of each of the construction machinery. The arrangement of the machinery as well as their operation time was also examined.

Meanwhile, it was crucial to stop the spreading damage as soon as possible, in case the frozen soil wall tended to break. Therefore, as a countermeasure, we installed two bulkheads with a thickness of 1,400 mm at the both ends of the expanded section as shown in Figure 9 and Figure 10. This was done to prevent the large scale collapse of the ground due to closing of the bulkheads and injecting water inside.

Furthermore, monitoring the behavior of the surrounding structure and ground at all times was required in order that immediate preventive measures could be taken if necessary. We then, installed almost 1,500 measuring instruments around the expanded section and constantly monitored their measurement values. The values were then unitarily managed and computer modelled as 3D shapes which changed colors according to their values (as can be seen in Figure 11).

Figure 9.   Bulkhead at the expanded section.

Figure 10.   Water storage tank for injection.

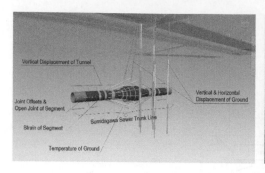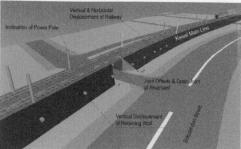

Figure 11.   Unified management of the measurement values using 3D model.

3   RESULT AND ANALYSIS

3.1   *Freezing Operation*

Within Figure 12 the internal state of the expanded section in forming frozen soil is clearly shown. Soon after the commencement of the freezing work, a longitudinal deformation of the existing tunnel began to appear. This deformation continued to increase and eventually reached up to 160 mm in upheaval, focusing around the center of the expanded section within two months.

The main cause of this unexpected deformation was due to the increase in the formation of frozen soil above the planned amount. Separating the formed frozen soil inside and outside of the radial freezing pipes, the thickness of the former was almost the same as the original plan due to the effect of the heating pipes. However, it grew excessively where heating pipes were not installed inside the radial freezing pipes. Following this result, we obtained the predicted values close to the measured values after calculating again the frost heaving displacement around the expanded section and taking into consideration the excess frozen soil.

Contrarily, the frost heaving on the ground was up to 8 mm, which was approximately in agreement with the assumptions obtained in advance from the 3D FEM analysis. At the completion of the frozen soil formation, the deformations of various surrounding structures were gradually increasing, but thereafter there was no major change.

3.2   *Tunnel Expansion*

The expansion work of a 19.7 m length of tunnel took 8 months to complete without any collapsing of the frozen soil wall by some improvement plans. As a result, Figure 13 shows that there were almost no cracks or melting of the frozen soil appearing during the excavation or removal and also there was no problem with the assembly of segments (as can be seen in Figure 14).

Figure 12.   Internal state of the tunnel in forming the frozen soil.

Figure 13. Excavation of the frozen soil.

Figure 14. Assembly of the expanded type segments.

Figure 15. Internal state of the tunnelin thawing the frozen soil.

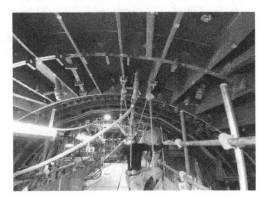

Figure 16. CB injection at the top of the tunnel.

3.3 Enforced Thawing

Figure 15 shows the internal state of the expanded section in the thawing frozen soil. During the forced thawing process, a very small amount of water leakage in the welded part of the entrance facility was confirmed, but overall there were no major problems encountered. The maximum thawing subsidence of the expanded section was 47 mm from the start of the enforced thawing and this had no influence on the new segments.

In the prior examination, it was predicted that the maximum subsidence that would occur without countermeasure would be 436 mm. However, during construction, the actual extent that was observed was only one-twentieth of the predicted value. This situation was only possible due to the CB injection in the gaps filled by thawing (as seen in Figure 16) together with the dissipation of the freezing expansion pressure by excavation of the frozen soil near the existing tunnel.

4 CONCLUSION

From the commencement of the operation of freezing facilities in May 2016, it took 5 months for the frozen soil formation, 8 months for the tunnel expansion, 5 months for the forced thawing, and 10 months for other accompanying works. Therefore, an overall period of 2 years and 4 months was required to complete all of the major engineering challenges involving demanding problems and difficulties. We were able to complete the main engineering challenges without major problems and the work did not have any effect on the surrounding structures under high water pressure or under high earth pressure at 40 m underground. This successful outcome is due to the utilization of the combined knowledge and experience of the experts from within and outside the company.

We completed the enforced thawing by the end of August 2018 and are now preparing for the underground joining being done by the TBM, which started from Senjusekiya pumping station in SSTL 4th project.

However, we expect that the underground joining in SSTL 4th project is likely to be one of the most difficult sections of work on the SSTL. We will continue to strive to maintain and improve safe construction of this work together with good quality, until the overall successful completion of SSTL 4th project.

REFERENCES

Ueda, Y., Ohrai, T. & Tamura, T. 2007. Experimental Study on the Moduli of Deformation of Frozen Soil. *Journal of Japan Society of Civil Engineers*. Ser. C, vol. 63: 577–589.

Ueda, Y., Ohrai, T. & Tamura, T. 2007. Three-Dimensional Ground Deformation Analysis with Frost Heave Ratios of Soil. *Journal of Japan Society of Civil Engineers*. Ser. C, vol. 63: 835–847.

Matsuura, M., Kobayashi, O., Matsuda, Y. & Kosaka, H. 2007. Fundamental Study on the Underground Joining of a Large Depth Sewer Shield Tunnel. *Journal of Tunnel Engineering*. vol. 17: 43–54.

Kushida, Y., Tomida, K., & Yamamoto, M. 2008. The Outline Design Technique for Frozen Soil Tunnel. *Proceedings of Tunnel Engineering*. vol. 18: 315–326

Tunnels and Underground Cities: Engineering and Innovation meet Archaeology, Architecture and Art, Volume 4: Ground improvement in underground constructions – Peila, Viggiani & Celestino (Eds)
© 2020 Taylor & Francis Group, London, ISBN 978-0-367-46868-2

3D Numerical modeling of artificial ground freezing in mining engineering

H. Tounsi, A. Rouabhi & M. Tijani
MINES ParisTech, PSL Research University, Centre de Géosciences, Fontainebleau, France

F. Guérin
ORANO, Paris La Défense, France

ABSTRACT: For decades, Artificial Ground Freezing (AGF) technique has been used for waterproofing and strengthening soils and soft rocks. However, the resulting pore pressure build up may endanger the stability of adjacent structures located inside and outside the frozen area. Consequently, the use of this technique requires the development of numerical tools that not only allow to predict the extent of the frozen zone around the cooling sources, but also to anticipate the ground movements and the site stability problems. On the basis of the Theory of Porous Media, a thermo-hydro-mechanical coupled model is derived and validated against field data collected in the Cigar Lake underground mine, comprising temperature and tunnel displacement measurements. Three-dimensional finite-element simulations of the Cigar Lake mine have allowed a better understanding of the freezing process and its impacts, in a challenging mining environment.

1 INTRODUCTION

The artificial ground freezing technique is becoming increasingly widespread in civil and mining engineering, especially in areas with complex hydrogeological conditions (see for e.g. Pimentel et al. 2012, Vitel et al. 2016). It prevents the risk of water inflow into the underground working space, and also improves, temporarily, the mechanical properties of the ground (Andersland 2004). However, the expansion of pore water during freezing and the migration of unfrozen water result in a hydraulic pressure rise that can cause damage to neighboring structures and buildings, even if they are located outside the frozen area. Consequently, to safely use this technique, it is imperative to predict the evolution of the frozen area and the stability of the site by using accurate and reliable numerical tools.

A lot of effort has already been put into the development of models that take into account coupling between thermal, hydraulic and mechanical mechanisms (e.g. Neaupane & Yamabe 2001, Nishimura et al. 2008, Zhou & Li 2012, Zhou & Meschke 2013, Lai et al. 2014, Zhang & Michalowski 2015). Most often, these models aim to gain a better understanding of the frost heave problem in permafrost regions and are validated against classic one-dimensional laboratory frost heave tests (Lai et al. 2014, Zhang & Michalowski 2015). Validation against in situ field data corresponding to shallow underground structures driven in zones of degrading permafrost is rarely carried out in literature (Nishimura et al. 2008). In addition, the use of ground freezing, until recently, was based more on observation and practical experience gained from previous projects than on theoretical modeling. The monitoring of temperature and displacement, when implemented, was of poor quality and not regular. Consequently, it could not be used to feed numerical models for future projects. Moreover, the majority of existing papers dealing with the use of artificial ground freezing in urban underground construction projects presents only temperature data and a coupled thermo-hydraulic numerical analysis (Pimentel et al. 2012, Yan et al. 2017 and Hu et al. 2018). The mechanical impact of freezing is rarely investigated, and the ground movements and the displacements in nearby structures are very seldom recorded or published (Russo et al.2015).

In this paper, a thermo-hydro-mechanical coupled model is presented and validated against *in situ* temperature and displacement measurements collected in the Cigar Lake mine. It is one of the world's largest Uranium deposits where the artificial ground freezing technique has been successfully used to mitigate water inflow into underground workings (the orebody area and the 480 m deep mine tunnels) and to consolidate the ground preventing thus instabilities (Bishop et al. 2016). The focus of this paper is on the use of artificial ground freezing technique to allow a safe extraction of the ore material and on the freezing induced settlement of mine tunnels located 15 m below the vertical freezing pipes.

The theoretical formulation is derived within the framework of the Theory of Porous Media (Coussy 2005) and based on the liquid saturation degree function for modeling phase-change fronts and the concept of effective stress to describe stress partitioning in both frozen and unfrozen ground.

2 THEORETICAL FRAMEWORK

A porous medium, initially fully saturated by pure water, and subjected to negative temperatures is considered. It is then constituted of three phases α: soil particles ($\alpha = \sigma$), pure liquid water ($\alpha = \lambda$) and ice ($\alpha = \gamma$). Each phase is characterized with a density ρ_α, a volume fraction n_α (with $\Sigma_\alpha n_\alpha = 1$), a saturation degree $S_\alpha = n_\alpha/n$ (with $S_\lambda + S_\gamma = 1$), with n the porosity of the medium, and an apparent volumetric mass density $\rho^\alpha = n_\alpha \rho_\alpha$. We assume that ice follows the movement of the solid skeleton phase σ and behaves like a fluid phase endowed with its own pressure p_γ and temperature T (Rouabhi et al. 2018). The latter is common to all phases (local thermal equilibrium). Moreover, the porous medium is assumed to be isotropic.

2.1 Balance equations

The THM behavior is governed by the following balance equations of mass, energy and momentum where the notation $\dot{\varphi}$ is used to denote the time derivative of a quantity φ following the motion of the solid skeleton:

$$\dot{\rho}^\lambda + \dot{\rho}^\gamma + \left(\rho^\lambda + \rho^\gamma\right)\dot{\varepsilon}_v + \nabla.(\rho_\lambda V) = 0 \tag{1}$$

$$\left(\sum_\alpha \rho^\alpha C_{p\alpha}\right)\dot{T} + \rho_\lambda C_{p\lambda} V.\nabla T = -\nabla.\psi - (\dot{\rho}^\gamma + \rho^\gamma \dot{\varepsilon}_v)\Delta h \tag{2}$$

$$\nabla.\sigma + \rho g = 0 \tag{3}$$

where $\varepsilon_v = \text{tr}(\varepsilon)$ is the volumetric component of the strain tensor ε, V is the filtration velocity vector of the liquid phase, $C_{p\alpha}$ is the heat capacity at constant pressure of phase α, ψ is the surface density of the amount of heat exchanged by conduction, $\Delta h = h_\gamma - h_\lambda$ is the latent heat of water phase change, namely the difference in enthalpy between the pure liquid water and ice, σ is the total stress tensor, $\rho = \Sigma_\alpha \rho_\alpha$ is the mass density of the porous medium and g the gravitational acceleration vector.

The mechanical equilibrium at the water-ice interfaces is expressed through a relation between the capillary pressure p_c (the difference between the pressure p_γ of the phase γ and the pressure p_λ of the phase λ: $p_c = p_\gamma - p_\lambda$), the temperature T and the liquid saturation degree S_λ, as follows:

$$S_\lambda = S(p_c, T) \tag{4}$$

where S is an empirical function that may be deduced from laboratory tests (Watanabe & Mizoguchi 2002) or via a back analysis of *in situ* measurements (Vitel et al. 2016).

2.2 Constitutive equations

The system of balance equations must be supplemented by the generalized laws of Darcy and Fick to express the secondary unknowns V and ψ, respectively, as follows:

$$V = -\frac{k_r(S_\lambda)k_0}{\eta_\lambda}(\nabla p_\lambda - \rho_\lambda g), \ \psi = -\Lambda \nabla T \qquad (5)$$

where η_λ is the dynamic viscosity of the liquid water, k_0 is the intrinsic permeability of the porous medium, and k_r is the liquid phase relative permeability that varies with the liquid saturation degree to reproduce the hindering effect of the formation of ice on the movement of the remaining liquid water in the pores. Its expression is derived from the van Genuchten model, as follows:

$$k_r = \sqrt{S_\lambda}\left(1 - \left(1 - S_\lambda^{1/m}\right)^m\right)^2 \qquad (6)$$

Λ denotes the thermal conductivity of the porous medium calculated using a geometric mean of the thermal conductivities of the three phases (Côté & Konrad 2005), as follows:

$$\Lambda = \Lambda_\sigma^{1-n}\Lambda_\lambda^{nS_\lambda}\Lambda_\gamma^{n(1-S_\lambda)} \qquad (7)$$

The saturation degree S_λ is given by the following van Genuchten model:

$$S_\lambda(p_c) = \left(1 + (p_c/P)^{1/(1-m)}\right)^{-m} \qquad (8)$$

where m and P are two material parameters.

In order to determine partial porosities, an evolution equation of the global porosity is generally used:

$$\dot{n} = (B - n)\left(\dot{\varepsilon}_v - 3A\dot{T} + \frac{(1 - B)S_\lambda}{K}\dot{p}_\lambda + \frac{(1 - B)(1 - S_\lambda)}{K}\dot{p}_\gamma\right) \qquad (9)$$

where B is the Biot coefficient, A is the drained linear thermal expansion coefficient and K the drained bulk modulus.

In addition, equation (3) should be supplemented by the solid skeleton phase effective stress law, as follows:

$$\dot{\sigma} + B\dot{p}\mathbf{1} = \mathbf{D} : (\dot{\varepsilon} - A\dot{T}\mathbf{1}) \qquad (10)$$

with

$$\dot{p} = S_\lambda\dot{p}_\lambda + (1 - S_\lambda)\dot{p}_\gamma \qquad (11)$$

$\mathbf{D}$ is the drained Hooke's elasticity tensor characterized by a Poisson's ratio v and an equivalent Young's modulus E of the mixture (Zhou & Meschke 2013):

$$E = S_\lambda E_0 + (1 - S_\lambda)E_f \qquad (12)$$

where E_0 and E_f are the Young's moduli of the material in the unfrozen state and the fully frozen state, respectively.

Finally, the chemical equilibrium between liquid water and ice phases enables the determination of the expressions of capillary pressure p_c and enthalpy variation Δh. If a linear approximation is accepted, we obtain:

$$p_c = \rho_y L_0 + \langle T/T_0 - 1 \rangle \tag{13}$$

$$\Delta h = L_0 + L'_0(T - T_0) \tag{14}$$

with $T_0 = 273.15$ K the coexistence temperature at reference pressure $p_0 = 0.1$ MPa, $L_0 = 333427$ J/kg and $L'_0 = 2458.25$ J/kg/K. $\langle . \rangle$ are the Macauley brackets such as p_c is positive when T $<$ T_0 and equal to zero otherwise.

To establish the expressions (13) and (14), the densities and the heat capacities of the water phases were considered constant, which allowed simplifying the model's formulation (Rouabhi et al. 2018).

3 NUMERICAL MODELING

All numerical simulations were carried out using the software COMSOL where the partial differential equations presented in Section 2 were implemented.

3.1 Presentation of the Cigar Lake mine

At the Cigar Lake mine, the area that is subjected to freezing is located between the surface and 15 m below the orebody lying between 410 m and 450 m depth. The refrigerant, cooled to -30 °C, is sent underground through a network of 465 m length and 57.15 mm radius vertical freezing pipes, prior to beginning uranium extraction and until the end of mining operations. The mining method that is used is the jet boring system. It is a non-entry approach, where mining is carried out from tunnels driven 25 m to 30 m beneath the deposit, in the basement rock. Boreholes for ground temperature measurement were installed. Each borehole was equipped with several sensors at different depths, in the vicinity of the orebody zone. The mine tunnels were also equipped with displacement sensors anchored in the rock behind the liner.

A representative tunnel (named tunnel 797) regularly monitored and located below the vicinity of a borehole for temperature measurement (ST797-21) has been chosen in order to validate the numerical model with respect to field measurements. Data corresponding to 1100 days of ground temperature and displacement measurements was collected.

The methodology we adopted consisted first of calibrating the liquid saturation degree for all the geological layers using temperature in situ measurements and the real freezing pipe layout. Subsequently, a regular grid of freezing pipes was assumed, and the THM simulations were carried out using a simplified geometry (shown in Section 3.3).

3.2 Calibration of the liquid saturation degree

The water saturation degree function, hard to determine experimentally, was estimated through a back analysis of temperature in situ measurements recorded by the closest temperature borehole to tunnel 797: borehole ST791-21. TH simulations were carried out using a 3D model that comprises six horizontal and homogeneous layers, crossed by the real freezing pipes layout in the vicinity of the borehole ST791-21.

Table 1 presents, for each geologic unit, the values of the TH simulation constants in addition to the calibrated parameters m and P of the liquid saturation degree function. It shows that the geology in this area varies considerably with depth. The simulated results, displayed in Figure 1, are in good agreement with the in situ measurements at two different and

Table 1. Thermal and hydraulic parameters for each geologic unit.

Layer depth [m]	Λ_σ [W m^{-1} K^{-1}]	n	m	P [MPa]	k_0 [m^2]
[0;-405]	3.4	0.22	0.55	1	4.32×10^{-15}
[-405;-425]	3.4	0.18	0.6	1.2	2.89×10^{-12}
[-425;-435]	3.1	0.24	0.62	1.3	5.78×10^{-17}
[-435;-455]	2.7	0.27	0.52	2.6	1.16×10^{-15}
[-455;-465]	1.8	0.2	0.3	0.7	2.31×10^{-15}
[-465;-520]	1.8	0.2	0.3	0.7	6.94×10^{-16}

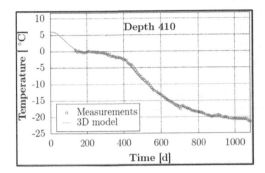 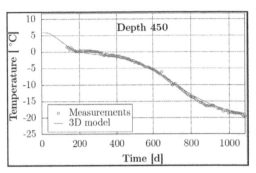

Figure 1. Comparison between measured and simulated temperatures at depths: 410 m and 450 m.

representative depths: 410 m and 450 m. Indeed, both the latent heat induced plateau around 0 °C and the temperature decrease over the long term, following the plateau, are well reproduced.

3.3 THM numerical simulations

The geometry that was used in the THM simulations is the blue block shown in Figure 2. The latter displays a regular freeze pipe grid with a spacing of 6 m, the studied mine tunnel and the existing symmetries. An XZ cross section of the geometry is shown in Figure 3, including the main model's dimensions.

The initial temperature was set equal to 6 °C and the pressure equal to the hydrostatic pressure. Regarding the boundary conditions, a constant far field temperature condition, equal to the initial temperature, was applied at all the outer vertical boundaries and at the bottom. The Newton's law of cooling was used at the pipe walls: $\psi.n = H (T - T_c)$, with H the forced convective heat transfer coefficient, n the outward unit vector normal to the boundary of the freeze pipe and T_c the coolant temperature (T_c = -30 °C). The displacement was constrained in the horizontal direction at the lateral boundary and at the pipes walls and completely constrained at the bottom. The excavation took place after 150 days of active freezing. A 60% stress release ratio was applied to the tunnel's wall, and the pressure, initially equal to the hydrostatic pressure was canceled after the excavation. Along all the external boundaries the hydraulic flux was equal to zero.

In addition to the thermal and hydraulic parameters given in Table 1, Table 2 presents the remaining constants used in the calculation.

The capability of the proposed THM model to reproduce the mechanical impact of freezing is assessed in Figure 4, where a good agreement between measured and predicted vertical displacements in the crown (target 1) and the shoulder (target 3) of section N° 26 of the studied tunnel is shown. It should be noted that this section has the advantage of being excavated just before the phase change temperature was reached above in the frozen area (see Figure 1).

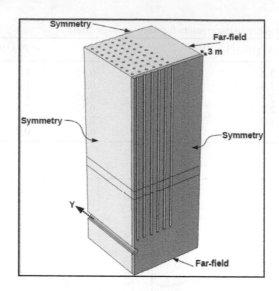

Figure 2. Simplified geometry (not to scale).

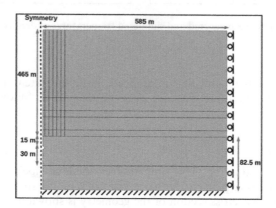

Figure 3. Vertical cross section of the computational domain (not to scale).

Table 2. Constants for numerical simulations.

ρ_σ [kg m^{-3}]	ρ_λ	ρ_γ	$C_{p\sigma}$ [J kg^{-1} K^{-1}]	$C_{p\lambda}$	$C_{p\gamma}$
2500	999.84	916.72	900	4219.44	2096.71
Λ_λ [W m^{-1} K^{-1}]	Λ_γ	H [W m^{-2} K^{-1}]	A [°C^{-1}]	B	
0.6	2.3	12	10× 10^{-6}	1	
Geological unit		E_0 [MPa]	E_f [MPa]	v	
Upper material	[0;-465 m]	250	1125	0.4	
Basement rock		3500	15750	0.25	

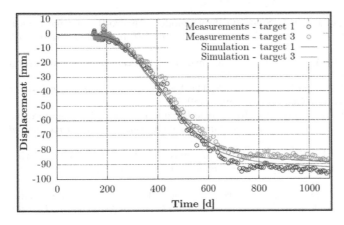
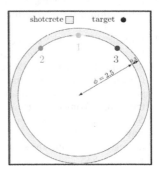

Figure 4.   3D THM model: simulated and measured vertical displacements in section 26.

Consequently, it allowed us to approximately isolate the freezing induced settlements from those due to tunnel's excavation. If we assume that the vertical displacements stabilize at a constant value when the distance behind the tunnel face has exceeded five times the radius of the tunnel, which was the case in sections not exposed to the effects of freezing, all the remaining settlement is due to the upper freezing operation. Figure 4 shows a short plateau at the beginning corresponding to the excavation activity and then a significant increase of displacement. Another change of slope occurs around 600 days in full accordance with the temperature change of slope (see Figure 1).

4   CONCLUSION

In this paper, a coupled thermo-hydro-mechanical model is briefly presented and applied to the modeling of the ground freezing activity in the Cigar Lake mine. The model has allowed us to predict the evolution of the extent of the frozen area, and to evaluate the potential impacts of freezing on the ground movements and on the displacements in the vicinity of an underground mine tunnel.

However, several improvements could be investigated. In particular, the use of an advanced rheological model for the unfrozen rockmass, where the tunnel is driven, and for the upper frozen mass. In addition, the effects of salinity on the freezing behavior of the porous medium deserve careful consideration in order to increase the predictive capabilities of the model. Indeed, if the pore water in the Cigar Lake mine is considered as totally pure, the saturating fluid in other hydro-geological contexts may present high salinity levels inducing a shift in the coexistence temperature.

REFERENCES

Andersland, O. B., & Ladanyi, B. 2004. *Frozen ground engineering*. John Wiley & Sons.
Bishop, C., Mainville, A., & Yesnik, L. 2016 Cigar Lake Operation - Northern Saskatchewan, Canada (Technical report, Cameco Corporation).
Côté, J., & Konrad, J. M. 2005. A generalized thermal conductivity model for soils and construction materials. *Canadian Geotechnical Journal* 42(2): 443–458.
Coussy, O. 2005. Poromechanics of freezing materials. *Journal of the Mechanics and Physics of Solids* 53(8): 1689–1718.
Hu, J., Liu, Y., Li, Y., & Yao, K. 2018. Artificial Ground Freezing In Tunnelling Through Aquifer Soil Layers: a Case Study in Nanjing Metro Line 2. *KSCE Journal of Civil Engineering*: 1–7.

Lai, Y., Pei, W., Zhang, M., & Zhou, J. (2014). Study on theory model of hydro-thermal–mechanical interaction process in saturated freezing silty soil. *International Journal of Heat and Mass Transfer* 78: 805–819.

Neaupane, K. M., & Yamabe, T. 2001. A fully coupled thermo-hydro-mechanical nonlinear model for a frozen medium. *Computers and Geotechnics* 28(8): 613–637.

Nishimura, S., Gens, A., Olivella, S., & Jardine, R. J. 2008. THM-coupled finite element analysis of frozen soil: formulation and application. Géotechnique 59(3): 159–171.

Pimentel, E., Papakonstantinou, S. & Anagnostou, G. 2012. Numerical interpretation of temperature distributions from three ground freezing applications in urban tunnelling, *Tunnelling and Underground Space Technology* 28: 57–69.

Rouabhi, A., Jahangir, E., & Tounsi, H. 2018. Modeling heat and mass transfer during ground freezing taking into account the salinity of the saturating fluid. *International Journal of Heat and Mass Transfer* 120: 523–533.

Russo, G., Corbo, A., Cavuoto, F., & Autuori, S. 2015. Artificial Ground Freezing to excavate a tunnel in sandy soil. Measurements and back analysis. *Tunnelling and Underground Space Technology* 50: 226–238.

Vitel, M., Rouabhi, A., Tijani, M., & Guérin, F. 2016. Thermo-hydraulic modeling of artificial ground freezing: Application to an underground mine in fractured sandstone. *Computers and Geotechnics* 75: 80–92.

Watanabe, K., & Mizoguchi, M. 2002. Amount of unfrozen water in frozen porous media saturated with solution. *Cold Regions Science and Technology* 34(2): 103–110.

Yan, Q., Xu, Y., Yang, W., & Geng, P. 2017. Nonlinear transient analysis of temperature fields in an AGF project used for a cross-passage tunnel in the Suzhou Metro. *KSCE Journal of Civil Engineering*: 1–11.

Zhang, Y., & Michalowski, R. L. 2015. Thermal-hydro-mechanical analysis of frost heave and thaw settlement. *Journal of Geotechnical and Geoenvironmental Engineering* 141(7): 04015027.

Zhou, J., & Li, D. 2012. Numerical analysis of coupled water, heat and stress in saturated freezing soil. *Cold Regions Science and Technology* 72: 43–49.

Zhou, M. M., & Meschke, G. 2013. A three phase thermo-hydro-mechanical finite element model for freezing soils. International journal for numerical and analytical methods in geomechanics 37(18): 3173–3193.

Tunnels and Underground Cities: Engineering and Innovation meet Archaeology,
Architecture and Art, Volume 4: Ground improvement in
underground constructions – Peila, Viggiani & Celestino (Eds)
© 2020 Taylor & Francis Group, London, ISBN 978-0-367-46868-2

Mechanical behavior of shallow embedded and large cross-section tunnel supporting using Freeze-Sealing Pipe-Roof method

D.M. Zhang
Key Laboratory of Geotechnical and Underground Engineering of Education, Tongji University, Shanghai, China

J. Pang
Department of Geotechnical Engineering, College of Civil Engineering, Tongji University, Shanghai, China

ABSTRACT: Gongbei Tunnel is one section of Hong Kong-Zhuhai-Macao Bridge connecting Zhuhai city and Macao, and undercrossing the Gongbei Port. The tunnel is shallow embedded with a large cross-section of 345 m^2 built in saturated soft soil. To guarantee the construction safety and avoid the damage of the surroundings, the Freeze-Sealing Pipe Roof method, which is an innovative combination technology with pipe jacking roof for per-support and ground freezing for waterproof, is applied in the excavation of Gongbei Tunnel. The frost heaving of the soil surround the tunnel would increase the contact pressure between the soil and primary support. The primary support and the secondary lining have different mechanical response in two different contact modes: complete contact and incomplete contact. The loading transfer between the primary support and the secondary lining is more efficient in the complete contact mode.

1 INTRODUCTION

The Hong Kong-Zhuhai-Macao Bridge is the first world-class sea-crossing project in China. Gongbei Tunnel is one of the eight important components of the Hong Kong-Zhuhai-Macao Bridge, connecting the Zhuhai city and Macao. The tunnel was excavated beneath the largest China Mainland port – Gongbei Port. The section under Gongbei Port is constructed by sub-surface excavation method. The remaining sections are constructed by cut and cover method as shown in Figure 1 (He et al., 2013). The total length of underground excavation tunnel is 255 m, including a transition curve of 88 m and a circular curve of 167 m in length as shown in Figure 2. The minimum radius of plane curvature of Gongbei Tunnel is about 900 m, the buried depth is 4~5 m, and the maximum excavation cross-sectional area is about 345 m^2 (Zhang et al., 2013; Li et al., 2014). The underground excavated section of Gongbei Tunnel has double-decker and two-way highway with a design speed of 80 km/h, three lanes for each direction (Zhang et al., 2016).

The underground excavated section of Gongbei Tunnel is located at the junction of Macao and Zhuhai, a very environmentally sensitive location with poor geological conditions, and thus is extremely difficult to be constructed. The tunnel is constructed in layered soft soils with high groundwater pressure. In addition, there are a large number of buildings and auxiliary structures around the tunnel, such as Gongbei port and Macao inspection building, entry and exit gallery, Zhuhai railway station, water pipelines, power pipelines and communication pipelines. Besides, there are many piles of Gongbei Port and Macao inspection building and its auxiliary buildings foundation as shown in Figure 3.

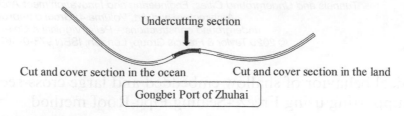

Figure 1. Gongbei Tunnel at the Zhuhai link of the Hong Kong-Zhuhai-Macao Bridge.

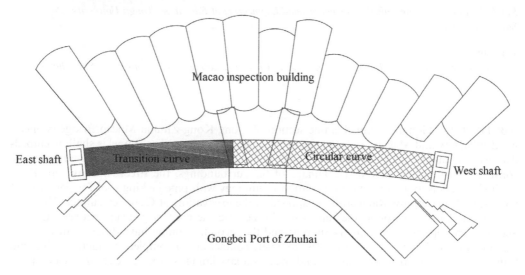

Figure 2. The plan and location of Gongbei Tunnel.

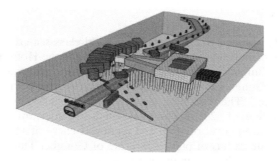

Figure 3. The 3D graph of Gongbei Tunnel and surrounding buildings.

2 CONSTRUCTION SCHEME

2.1 *Pipe roof design*

The underground excavated section of Gongbei Tunnel is constructed through the "Freeze-Sealing Pipe Roof (FSPR) method". According to the engineering requirements, the normal traffic of the Gongbei Port cannot be affected during the construction process, and the settlement of the ground cannot be greater than 30mm. Gongbei Tunnel passes through the sensitive area of Gongbei Port using pipe jacking roof. Two working shafts are arranged at both ends of the tunnel. Then the pipe roof is pushed into the soft soils from two shafts to form the pre-supporting system of the tunnel. And the excavation is carried out from two

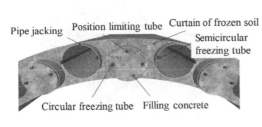

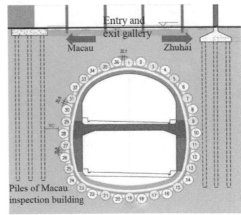

| (a) Freeze tube setting method | (b) The section of tunnel |

Figure 4. The curved freezing-sealing pipe roof of Gongbei Tunnel (a) Freeze tube setting method (b) The section of tunnel.

shafts under the protection of the pipe roof (Pan et al., 2015). Based on engineering experience, the water proof between the pipes mainly adopts the method of grouting after locking the bolts. However, the longitudinal axis of the tunnel has a large curvature, and it is difficult to ensure that the water stop lock between the two pipe jacks is successfully connected. Therefore, the sealing effect of this method is poor, and it cannot meet the sealing requirements between the pipe jacking under the geological conditions of the Gongbei Tunnel. In order to form the water-proof structure of Gongbei excavation tunnel, freezing-sealing pipe roof method is designed to combine curved pipe jacking technology and ground freezing method (Hu et al., 2014).

The pipe roof is alternately arranged with "solid pipe" and "hollow pipe" and the freezing tubes are arranged inside the steel pipes as shown in Figure 4a. Considering the weakening effect of the thermal disturbance caused by tunnel excavation on the curtain of frozen soil, and it is necessary to strictly control the volume of the frozen soil to limit surface frost heaving. Three special tubes are arranged inside the steel pipe: circular freezing tube, semicircular freezing tube and heating salt water position limiting tube. Through the freezing action of circular freezing tubes, a curtain of frozen soil is formed between the two pipes. In the stage of excavation and lining construction, the working of semicircular freezing tubes is used to resist the weakening of frozen soil. Through the working of the position limiting tubes, the frost heaving effect on the outside of the solid pipe is controlled (Zhang et al., 2015).

The whole pipe roof is made up of 36 steel pipes with the diameter of 1620mm. The upper 17 steel pipes are 20 mm thick, jacked from east shaft to west shaft. And the bottom 19 steel pipes are 24 mm, jacked from the opposite direction. The minimum clearance between each adjacent pipe is ranging from 355 to 358 mm as shown in Figure 4b (Xie et al., 2016). Each segment of steel pipe is 4 m in length, connecting with F-type socket. After the completion of the pipe jacking operation, sub-regional freezing is performed to form a waterproof curtain.

2.2 Excavation sequence

The underground section of Gongbei Tunnel is excavated with the bench cut method. The whole cross-section is divided into 5 benches and 14 blocks as presented in Figure 5a. And the excavation cross-sectional area is a huge oval, which is about 18.9 m in width and 20.6 m in height. The procedure of the excavation and lining application of Gongbei Tunnel are followed:

The excavation section is divided into five benches of A, B, C, D, and E, respectively, from top to bottom. The Bench A is divided into two partitions, and each of the other four benches

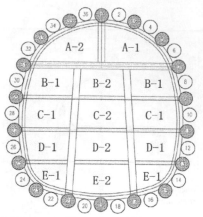

(a) Layout of tunnel partitions

(b) Excavation and supporting of bench cut method

Figure 5. The excavation method of Gongbei Tunnel (a) Layout of tunnel partitions (b) Excavation and supporting of bench cut method.

is divided into three partitions. For the Bench A, firstly, the working face soil of areas A-1 and A-2 is reinforced through horizontal grouting along the longitudinal direction. After grouting reinforcement is completed, the area A-1 is first excavated. Then the area A-2 is excavated 5 m behind the A-1 working face.

The Bench B is 20 m away from the Bench A in the longitudinal direction. The Bench C is 15 m away from the Bench B and the Bench D is also 15 m away from the Bench C. The Bench E is 10 m away from the Bench D. For the second bench, the working face soil of area B-1 is reinforced through horizontal grouting along the longitudinal direction, and then the area B-2 is reinforced while excavating area B-1. The area B-2 is excavated 5 m behind the areas on both sides. When excavating 40~80 cm in each area, the primary support and temporary support structures are promptly applied. The secondary lining is applied inside each area, 10 m away from the excavation face. The excavation sequence of Bench C, D, E and the construction sequence of the primary support, temporary support, and secondary lining are similar to the Bench B as shown in Figure 5b.

2.3 Design of supporting system

Figure 6 shows the three lining structures of the underground excavated section of Gongbei Tunnel. For the first lining, sprayed concrete over a steel reinforcing mesh and steel sets is used as primary support with a thickness of 30 cm. The secondary lining is carried out with C35 form-work concrete with a built-in grid steel frame. The third lining is finished with C45 reinforced concrete designed to withstand all water and earth pressure. The frozen soil could seal water leakage between the pipe roofs (Liu et al., 2018). To make the pipe roof stiffer, every other jacked pipe (hollow pipes) is filled with C35 micro-expansive self-compacting concrete after third lining is completed (Cheng et al., 2012). 25b section steels are used for temporary support, which are fabricated in a factory and installed on site. Both ends of the temporary support are connected with the jacked pipes filled with concrete to make the supporting system more stable.

The design parameters are shown in Table 1.

2.4 Monitoring point arrangement

The Gongbei Tunnel is constructed by "Freeze-Sealing Pipe Roof (FSPR) method", which is first used in tunnel engineering, the contact pressure between the soil and primary support and the internal forces of lining are not clear in this construction condition. In order to ensure the safety of construction and explore the variation laws of contact pressure and

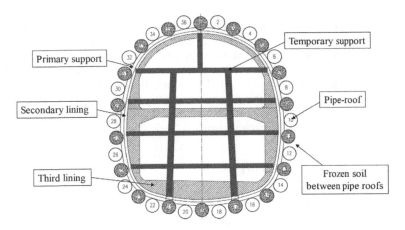

Figure 6. The cross-section layout of Gongbei Tunnel.

Table 1. List of parameters of supporting system in the underground excavated section.

Supporting type		Design parameters
Primary support	Steel mesh	A8 steel bars arranged in one layer at intervals of 20 × 20 cm
	Section steel	22b I-steel arranged with spacing of 0.4 m, welded with pipe roof, and vertically connected with 16 I-steels.
	Sprayed concrete	C30 Sprayed concrete, 30 cm thick.
	Temporary support	25b section steel with a spacing of 0.8m, and vertically connected with square steel 80 × 60 × 6 mm, temporary vertical support arranged with small A50 tube at intervals of 40 cm.
Secondary lining	Grid steel frame	Section, 22 cm height and 20 cm width, arranged at longitudinal spacing of 35-45 cm.
	Concrete	C35 formwork concrete
Third lining	Reinforced concrete	C45 reinforced concrete, 60-219 cm thick

internal forces of lining, the contact pressure between the soil and primary support, the moments of primary support and secondary lining are monitored separately.

Figure 7a shows the monitoring point arrangement of contact pressure between the soil and primary support. The contact pressure is measured by placing the pressure cell in the soil

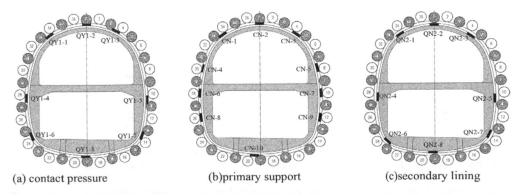

(a) contact pressure   (b)primary support   (c)secondary lining

Figure 7. Monitoring point layout of underground excavated section (a) contact pressure (b) primary support (c) secondary lining.

surrounding the tunnel. And there are a total of 10 monitoring sections along the longitudinal direction of the tunnel, 8 measuring points in each section. Figure 7b shows the moment monitoring point arrangement of primary support. The moment of primary support is measured by the strain gauge embedded in concrete. And there are a total of 10 monitoring sections along the longitudinal direction of the tunnel, 10 measuring points in each section. Figure 7c shows the moment monitoring point arrangement of secondary lining. The moment of secondary lining is also measured by the strain gauge. And there are 10 monitoring sections in total, 8 measuring point in each section.

3 ANALYSIS OF INTERNAL FORCES

3.1 *Variation of contact pressure between the soil and primary support*

There are three typical cross-sections from east to west along the longitudinal direction of Gongbei Tunnel as shown in Figure 8. Cross-sections K2+390 and K2+600 are located in the sensitive area closing the two shafts on both sides of the tunnel. Cross-section K2+520 is in the middle of the tunnel along the longitudinal direction where the freezing effect is not so good. And the field data show that the contact pressure variation between the soil and primary support. The positive value indicates increasing pressure and the negative value means pressure decreasing. The altitude of Gongbei Tunnel is higher in the east and lower in the west in the longitudinal direction, resulting in the groundwater to flow from east to west. Therefore, the groundwater level is lower in the east section. And owing to the influence of the freezing effect, the soil at the bottom of the tunnel with large water content has a large amount of expansion. And it produced a large pressure on the bottom of the tunnel as shown in Figure 8a. The groundwater level is getting higher gradually near the west shaft. The water content of the soil near the middle of the tunnel increases significantly, and the amount of soil frost heaving also increased near the middle of the tunnel. Therefore, the closer to west shaft, the greater contact pressure variation between the soil and primary support at the middle of the tunnel as shown in Figures 8b and 8c.

The three continuous lines in Figure 9 indicate the contact pressure variation at the crown of the three typical cross-sections in Figure 8, and the three dashed lines in Figure 9 indicate the crown displacement in the same time. According to the measured data of the crown displacement, the crown of the tunnel sunk in the underground excavated sections near two shafts (dashed lines of K2+390 and K2+600). With the sinking of the crown, the contact pressure between the soil and primary support should decrease. However,

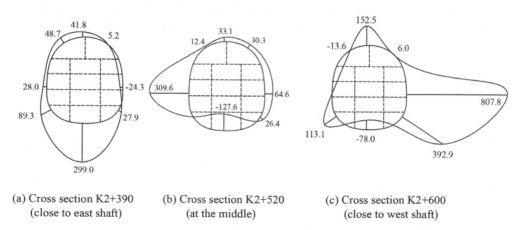

(a) Cross section K2+390   (b) Cross section K2+520   (c) Cross section K2+600
(close to east shaft)   (at the middle)   (close to west shaft)

Figure 8.   Contact pressure variation between the soil and primary support (unit: kPa) (a) Cross section K2+390 (b) Cross section K2+520 (c) Cross section K2+600 (close to east shaft) (at the middle) (close to west shaft).

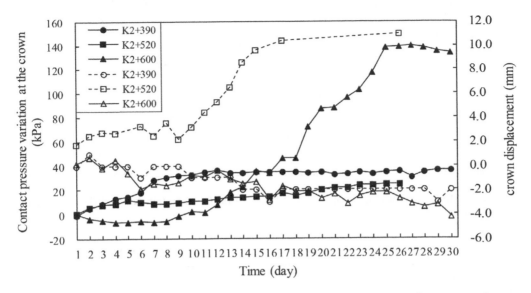

Figure 9.   Contact pressure variation between the soil and primary support at the crown and crown displacement.

the actual situation is the contact pressure has been increasing (continuous lines of K2+390 and K2+600). Especially in the cross section K2+600, with the sinking of the crown, the contact pressure has even increased to 138.6 kPa. This is a phenomenon caused by the frost heaving effect. During the freezing of the soil, the volume of the soil expands, and giving the pressure to primary support. The sinking of the crown is not sufficient to fully release the stress generated by the expansion of the soil (Li et al., 2018). The crown of the tunnel moved up in the middle underground excavated section of Gongbei Tunnel (dashed line of K2+520). As a result, the pressure between the soil and supporting structure increased. Meanwhile the soil has been expanding due to freezing. Eventually, these two reasons lead to a continuous increase in contact pressure between the soil and primary support (continuous line of K2+520).

3.2   *Moment of primary support*

Figure 10 shows the maximum moment envelope of primary support. The positive value indicates that the outer side of primary support is compressed, and the negative value means that the inner side of primary support is compressed. When the contact pressure increase between the soil and primary support is small at the position of tunnel crown as shown in Figures 8a and 8b, the inner side of primary support is generally compressed at the corresponding position as shown in Figures 10a and 10b. When the contact pressure increase is large enough, such as up to 152.5 kPa as shown in Figure 8c. This pressure forces primary support at the crown to move down and primary support on both sides of the crown just has a lesser downward movement. In this case, it will cause a large pressure on the outer side of primary support at the crown as shown in Figure 10c. In the middle of the tunnel cross-section, the outer side of primary support is generally compressed.

For cross-sections K2+390 and K2+520 in Figure 11, the inner side of primary support is compressed at the crown. After installing the secondary lining, the primary support has different mechanical response characteristic in two different contact modes: complete contact and incomplete contact. In the complete contact mode, the moment of primary support is significantly reduced after installing the secondary lining (cross-section K2+390). In this case, the secondary lining is fully contact with the primary support and the secondary lining shares a portion of the external load. In the incomplete contact mode, the moment of primary

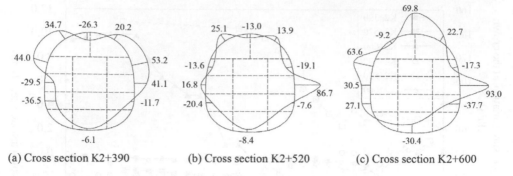

(a) Cross section K2+390   (b) Cross section K2+520   (c) Cross section K2+600

Figure 10.   Moment of primary support (unit: kN • m) (a) Cross section K2+390 (b) Cross section K2+520 (c) Cross section K2+600.

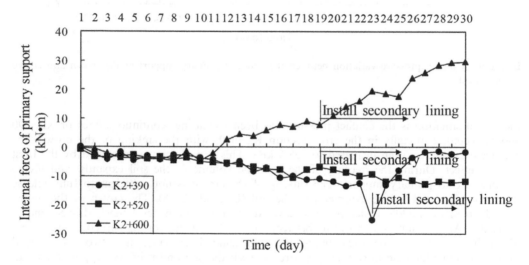

Figure 11.   Moment of primary support at the crown.

support is almost unaffected after installing the secondary lining (cross-section K2+520). In this case, the secondary lining is not fully contact with the primary support. So it cannot effectively share the load with the primary support. For cross-section K2+600, the moment curve of primary support is still developing along the original trend. And it should belong to the incomplete contact mode. Therefore, the loading transfer between the primary support and the secondary lining is more efficient in the complete contact mode.

3.3   Moment of secondary lining

Figure 12 shows the maximum moment envelope of secondary lining at three typical cross-sections. The positive value indicates that the outer side of secondary lining is compressed, and the negative value means that the inner side of secondary lining is compressed. The moment of the secondary lining is obviously larger than the primary support. It indicates that the secondary lining shares more external load than the primary support. In cross section K2+390, the maximum moment in the lower left corner of the secondary lining is up to 746.1 kN • m as shown in Figure 12a. It may cause excessive tensile stress on the outer side of secondary lining concrete and the concrete is possibly going to crack.

The inner side of secondary lining is mostly compressed at the position of tunnel crown. The cross-sections K2+520 and K2+600 are in the incomplete contact mode at the beginning,

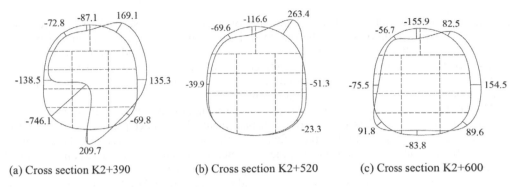

(a) Cross section K2+390   (b) Cross section K2+520   (c) Cross section K2+600

Figure 12.   Moment of secondary lining (unit: kN • m) (a) Cross section K2+390 (b) Cross section K2+520 (c) Cross section K2+600.

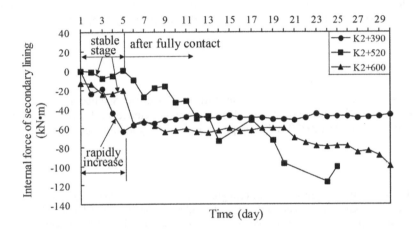

Figure 13.   Moment of secondary lining at the crown.

and the secondary lining at the crown is not fully contact with the primary support. Therefore, the moment of the secondary lining enters a stable stage first. Then with the relative movement between the secondary lining and primary support, these two structures gradually come into full contact with each other. After fully contacted, the secondary lining begins to share the external load, and the moment of the secondary lining is significantly increased. And the cross section K2+600 belongs to the complete contact mode. The secondary lining is fully contact with the primary support from the beginning. So the moment of the secondary lining increases rapidly at the beginning. Then with the rebalancing of the stress field, the moment of the secondary lining gradually remains stable after a period of time as shown in Figure 13.

4   CONCLUSIONS

An innovative "Freeze-Sealing Pipe Roof (FSPR) method" was used for a shallow-buried and very large sectional tunnel. This project is constructed under high underground water pressure and sensitive surrounding environment condition. There are many technical challenges such as controlling the jacking precision of the long pipe roof, controlling the precise thickness of the frozen soil curtain, and the internal force variation law of the tunnel lining is indistinct. In this case, the experience of the design and construction and the internal force characteristics of tunnel lining show that:

1. The pipe jacking roof is an effective pre-support method in soft soils. And the advantage of the freezing method is that the frozen soil can form a waterproof curtain to prevent water leakage during excavation. Freeze-Sealing Pipe Roof method combines the advantages of these two methods. It can effectively resolve the problem of deformation control and water leakage in super shallow-buried underground excavation under poor geological condition and sensitive environmental condition.

2. The contact pressure between the soil and primary support is significantly affected by the freezing effect. Regardless of the crown displacement, the contact pressure is increased in most instances. This shows that the influence of the pressing force caused by the expansion of the frozen soil is significant. During the construction, the amount of soil expansion should be strictly controlled to prevent the primary support damage by excessive expansion force.

3. The moments of primary support and secondary lining are related to the contact mode between them. If the secondary lining and primary support can be in full contact in time, the secondary lining can timely share the external load. Then the moment of the primary support will reduce or maintain stable, preventing the primary support from being damaged. Therefore, in the construction of secondary lining, we should try to make the secondary lining fully in contact with the primary support.

ACKNOWLEDGEMENTS

This study is financially supported by the National Natural Science Foundation of China (Grants No. 41772295 and No. 51478344), Shanghai Science and Technology Committee Project (Grant No. 17DZ1203201).The support is gratefully acknowledged.

REFERENCES

Cheng, Y., Liu J.G. 2012. Design scheme of Gongbei Tunnel. *Highway Tunnel* (3): 34–38. (in Chinese)

He, X.L., Cheng, Y., Guo, X.H., Tuo, Y.F. 2013. Gongbei Tunnel design of HongKong-Zhuhai-Macau Bridge connector project. *Soil Engineering and Foundation* 27 (1): 21–24. (in Chinese)

Hu, X.D., Ren, H., Chen, J., Cheng, Y., Zhang, J. 2014. Model test study of the active freezing scheme for the combined pipe-roof and freezing method. *Modern Tunnelling Technology* 51 (5): 92–98. (in Chinese)

Li, T.A. 2018. Study on deformation characteristics and construction control of shallow buried soft rock tunnel. *Highway Engineering* 43 (2): 155–161. (in Chinese)

Li, J., Liu, Z.G., Zhang, P. 2014. The hugest curved jacking pipe roof tunnel of Hong Kong-Zhuhai-Macao bridge. *In: International Conference on Pipelines and Trenchless Technology 2014 (ICPTT). Xiamen, China*: 422–431.

Liu, J.G., Ma, B.S. & Cheng, Y. 2018. Design of the Gongbei Tunnel using a very large cross-section pipe-roof and soil freezing method. *Tunnelling and Underground Space Technology* 72: 28–40.

Pan, J.L., Gao, H.D., Shi, P.X. 2015. A study of combined pipe-roof scheme optimization for the bored section of the Gongbei Tunnel. *Modern Tunnelling Technology* 52 (3): 55–62. (in Chinese)

Xie, H.M., Wang, X.Y., Wang, W.Z., Li, X., Zhang, P., Ma, B.S. 2016. Experimental study on pipe's mechanical characteristics of steel curved pipe jacking. *Geological Science and Techology Information* 35: 79–82. (in Chinese)

Zhang, P.P., Ma, B.S., Zhao, W. 2013. The largest curved pipe roofing tunnel project in the world. *In: Pipelines 2013, Fort Worth, Texas*: 953–963.

Zhang, P., Ma, B.S., Zeng, C., Xie, H.M., Li, X. & Wang, D.W. 2016. Key techniques for the largest curved pipe jacking roof to date: A case study of Gongbei Tunnel. *Tunnelling and Underground Space Technology* 59: 134–145.

Zhang, J., Hu, X.D., Ren, H. 2015. Case study on control of freezing effect by installing limiting tubes in freezing-sealing pipe roof in Gongbei Tunnel. *Tunnel Construction* 35 (11): 1157–1163. (in Chinese)

Tunnels and Underground Cities: Engineering and Innovation meet Archaeology,
Architecture and Art, Volume 4: Ground improvement in
underground constructions – Peila, Viggiani & Celestino (Eds)
© 2020 Taylor & Francis Group, London, ISBN 978-0-367-46868-2

Frost heave control during excavation of Gongbei tunnel with freeze-sealing pipe roof as pre-support

T. Zhou, P.X. Shi & J. Zhang
School of Rail Transportation, Soochow University, Jiangsu, China

J.L. Pan
China Railway 18 Bureau Group Co., Ltd., Tianjin, China

ABSTRACT: The Gongbei tunnel is one of the most technically challenging components of the Hong Kong-Zhuhai-Macau Bridge project. It was excavated using a pipe roof in combination with artificial ground freezing as pre-support. This paper presents the ground frost heave techniques during Gongbei tunnel excavation. Based on the thermal-mechanical coupling theorem, the development of the frost heave during ground freezing is simulated using COMSOL FE software. The effects of frost heave control using different techniques including pre-grouting, heating pipes to limit frozen ring thickness, and combination of the pre-grouting and heating pipes are analyzed and compared. The simulation results show that pre-grouting is critical to frost heave control by reducing the permeability of the soils and inhibiting the migration of moisture. The heating pipes are an effective supplement to frost heave control by limiting the thickness of frozen ring. The simulated ground surface heave is compared favorably with the filed measurement.

1   INTRODUCTION

In recent years, the construction scale of tunnels in China has been continuously expanded. The construction length and section dimension of tunnels are increasing, and the construction environment is becoming more and more complicated. According to The International Tunnel Association, a super-large section tunnel is defined as one with a clearance area of more than 100 m^2 (Luo, Y.B. 2017). In China, The excavation section area of many tunnels has exceeded 300 m^2. For example, Gongbei Tunnel is excavated with the section area of 336.8 m^2, taking cross-section with upper-and-lower layers and two-way six lanes. It is a component of Zhuhai Link in Hong Kong-Zhuhai-Macao Bridge and crosses a narrow strip between the customs clearance port. The minimum depth of the tunnel is 4~5 m.

It is recognized as a difficult problem in international tunnel community that excavating shallow and super-large-section tunnels which section area exceeding 300 m^2 under complex geological and construction environment. Traditional methods cannot adapt to the project and the innovation of construction technology is required. A new idea for the construction of super-large-section tunnels under complex conditions is provided when constructing Gongbei Tunnel. 36 steel pipe with a diameter of 1.62 m are used to form a pipe roof as the main support structure. And meanwhile, the soil around the pipe roof is frozen artificially to form a curtain for cutting off water. Because of the sensitive surrounding environment, the frost heave control is the key technology during the construction of Gongbei Tunnel.

The frost heave caused by artificial freezing mainly consists of in-situ phase change frost heave and segregation frost heave (Bronfenbrener, L. & Bronfenbrener, R. 2010). The segregation frost heave is the major one (He, P. et al. 2001). In-situ phase change frost heave is caused by the icing of water existing in the soil and the segregation frost heave is caused by the icing of migration water. According to the study of Harlan, R.L. 1973, the freezing process of the soil is a complex process of hydrothermal coupling and the amount of segregation frost heave mainly depends on the change of formation temperature field and the migration of water in the soil. Therefore, controlling the temperature field of the soil and inhibiting the migration of water are two major methods of controlling the frost heave of the soil. Zhou, J.S. et al. 2006. adopt the intermittent freezing mode which could control the depth of freezing to break the steady state of the formation temperature field. The result shows that soil samples have 80% decrease in the frost heave, from 0.72 cm (which in a continuous freezing mode) to 0.14 cm after frozen for 60 h. Zhang, M.Y. et al. 2017. study the frost heaving characteristics of the cold zone with water supply and large soil permeability coefficient. It is found that when the permeability coefficient of soil is large, taking the method of controlling the temperature field will still cause a large amount of segregation frost heave. In order to control the frost heave of soil with large permeability coefficient, it is necessary to reduce the permeability coefficient of the soil to inhibit the migration of water. Hu, X.D. 2009. finds that when the soil permeability coefficient decreases from 1.7×10^{-6} cm/s to 6×10^{-7} cm/s, the frozen-heave factor reduces from 4.5% to 2%.

Soil pre-grouting can reduce the permeability coefficient and frozen-heave factor, increase strength and control the frost heave and the melting and sinking of soils. After grouting in the undisturbed zone of the Yuanliangshan tunnel in Yuhuai Railway, Zhang, M.Q. et al. 2006. find that the porosity of soil decreases from 82.8% to 18.3~44.3% and the water content decreases from 120.3% to 15.7~29.1%. Zhang, Q.S. et al. 2015. find that the uniaxial compressive strength of soil samples taken from a fault in the tunnel increases by 181%, from 0.17 MPa to 0.48 MPa after grouted with cement paste which water cement ratio is 1:1.

The technic of freeze-sealing pipe roof used in Gongbei Tunnel is the first attempt in China. According to previous studies, it can be concluded that a single method like pre-grouting is often adopted to do researches on the control of frost heave at present. The study on the method of combining pre-grouting and heating pipes to control the frost heave is less. Therefore, based on the thermal-mechanical coupling theorem, the displacement of frost heave in surface is simulated by finite element in this paper. The results when adopting different methods like pre-grouting and taking heating pipes are compared and analyzed. What's more, the mechanisms of frost heave control are discussed through comparing and analyzing the distribution of surface displacement of simulation and filed measurement.

2 ENGINEERING BACKGROUND

The Gongbei Tunnel is a key-controlling project of the Zhuhai Link in the Hong Kong-Zhuhai-Macao Bridge. It is located in Xiangzhou District of Zhuhai City and adjacent to Macau. The excavation section of Gongbei tunnel is considered as one having the largest section dimension in the world and passes through Gongbei Port, which is known as the biggest land port of China. Figure 1 shows the plan view of excavation section. It has a length of 255 m along an alignment with a curvature of 886~906 m. The tunnel is buried in the water-rich soft soils with sea-land

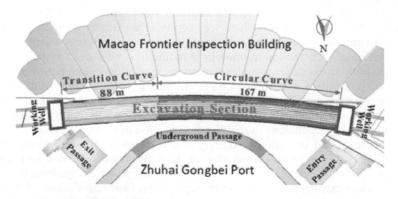

Figure 1. Plane view of Gongbei tunnel.

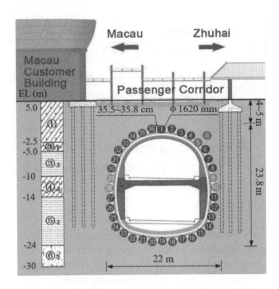

Figure 2. Cross-section of Gongbei tunnel.

phases, which geological conditions are extremely poor. As is shown in the Figure 2, the strata which tunnel passes through from top to bottom are respectively fill (①), muddy silt (③-1), sand (③-3), muddy silty clay (④-3), sand (⑤-2), coarse sand (⑥-2), silt (⑦-1).

Because of the special geographical position and sensitive political environment, a method called "pipe roof + artificial freezing" is adopted to control the deformation of surface. The pipe roof acts as a support for load, and the curtain formed by frozen soil is of benefit to inhibiting the water between the pipes. 36 steel pipes with a diameter of 1.62 m make up the pipe roof and each steel pipe consists of 64 pipe joints with a length of 4 m. A spacing between two pipes is 35.5~35.8 cm (Shi, P.X. et al. 2017). As is shown in Figure 3, the cross section of pipe roof is divided into five zones of A, B1, B2, B3 and C to conduct the construction of freezing.

Figure 4 shows that a circular freezing pipe and a heating pipe are arranged in the odd-numbered steel pipe, and a irregular shape freezing pipe is arranged in the even-numbered steel

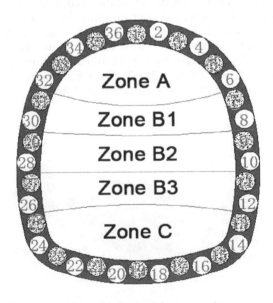

Figure 3. Zonation of artificial ground freezing in tunnel cross-section.

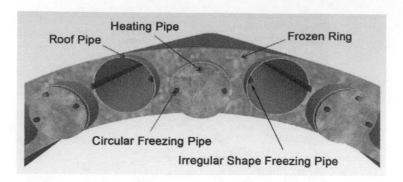

Figure 4.   Arrangement of freezing pipes.

pipe. During the process of freezing, the circular freezing pipe is used for active freezing, the irregular shape freezing pipe is used for maintenance of freezing, and the hot water is circulated in the heating pipe to control the thickness of the frozen soil. As is shown in Figure 5, three frozen sections are divided along the longitudinal direction of the tunnel. The circular freezing pipes are activated all together, and the irregular shape freezing pipes are activated by different areas and periods.

The pre-grouting of soil around the pipe roof and activating heating pipes are the two main methods of frost heave control in this project. After the pipe jacking, the soil outside the pipe is grouted through the holes reserved inside the pipe roof so as to compact the soil and then the permeability coefficient of soils and the influence of the peripheral hydraulic force on the frozen area are reduced. As a result, the frost heave caused by artificial freezing is limited.

The grouting material is made of ultra-fine cement and bentonite, which water-cement ratio (W/C) is 1:1, and amount of bentonite is 3%. The method of borehole inspection is used to inspect the effect of grouting. After drill sampling, the site hydrostatic test is carried out and the permeability coefficient of grouting soil is obtained from the unit water absorption. Then taking the samples for a laboratory geotechnical test and freezing test, the density, water content, strength and frozen-heave factor of grouting soils are obtained. The thermo-physical parameters and frozen-heave factor of the formation before and after grouting are shown in Tables 1-3, which is obtained from the geological exploration report of excavation section of the Gongbei tunnel.

As shown in Table 1-3, some indexes of the strata, such as the porosity, the frozen-heave factor and elastic modulus, are improved after grouting. From Table 1, the porosity of the strata decreases by 13% from 0.26~0.67 to 0.22~0.60, and the permeability coefficient decreases from 10-2~10-7 cm/s to 10-5~10-7 cm/s. The elastic modulus increase from 1.6~15.0 MPa to 20.1~32.7 MPa, which increase by 118~1156% after grouting. The water content of the strata reduces from 10.2%~38.1% to 9.2%~33.6%, while the frost-heaving ratio decreases by 12% from 3.82%~6.63% to 3.53%~5.46% according to Table 3.

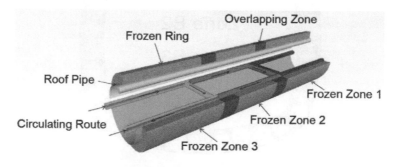

Figure 5.   Longitudinal section of artificial ground freezing.

Table 1. Soil parameters before and after grouting.

No.	Thickness m	Unit Weight kN/m^3		Porosity		Permeability coefficient cm/s		Poisson ratio		Water content	
		B	A	B	A	B	A	B	A	B	A
①	7.5	18.5	20.7	0.67	0.6	3.44×10^{-5}	1.72×10^{-5}	0.35	0.23	16.1	12.8
③-1	2.5	15.4	17.2	0.67	0.6	2.29×10^{-7}	1.14×10^{-7}	0.27	0.21	38.1	33.6
③-3	5.0	20.0	22.4	0.31	0.27	8.55×10^{-4}	1.70×10^{-5}	0.25	0.22	13.5	12.2
④-3	4.0	17.9	20.0	0.55	0.5	3.48×10^{-7}	1.74×10^{-7}	0.30	0.21	38.0	33.3
⑤-2	10.0	20.5	22.9	0.26	0.22	4.61×10^{-3}	1.70×10^{-5}	0.28	0.23	13.5	12.2
⑥-2	2.5	19.5	21.8	0.32	0.28	3.31×10^{-2}	1.70×10^{-5}	0.23	0.23	10.2	9.2
⑦-1	11.5	18.0	20.2	0.66	0.59	6.69×10^{-7}	3.48×10^{-7}	0.23	0.21	33.0	26.4

Note: B represent soil parameters before grouting; A represent soil parameters after grouting.

Table 2. Thermo-physical properties of soil layers.

No.	Soil Stratum	Specific heat J/kg/K		Thermal conductivity W/m/K				
		Before frozen	After frozen	-20°C	-10°C	-0.5°C	10°C	20°C
①	Fill	1420	1360	2.292	1.974	1.765	1.633	1.295
③-1	Muddy Silt	1410	1340	2.248	2.053	1.795	1.421	1.245
③-3	Sand	1400	1330	2.245	2.049	1.763	1.413	1.224
④-3	Muddy Silt Clay	1510	1450	2.391	2.011	1.885	1.734	1.409
⑤-2	Sand	1420	1360	2.358	2.073	2.008	1.748	1.479
⑥-2	Coarse Sand	1430	1360	2.338	2.051	1.878	1.681	1.375
⑦-1	Silty	1580	1470	2.317	2.091	1.896	1.684	1.541

Table 3. Frozen-heave factor of soils.

No.	Soil Stratum	Frozen-heave factor %	
Before frozen	After frozen		
①	Fill	6.63	5.46
③-1	Muddy Silt	6.59	5.23
③-3	Sand	4.25	3.53
④-3	Muddy Silt Clay	6.32	5.02
⑤-2	Sand	4.13	3.61
⑥-2	Coarse Sand	3.82	2.88
⑦-1	Silty	5.36	4.37

3 NUMERICAL SIMULATION

A two-dimensional finite element model of the tunnel cross section is established by using COMSOL finite element software. The thermal coupling equations are solved through the coefficient-type differential equation module so that the ice content θi is calculated. The input parameters include soil weight, porosity, permeability coefficient water content, specific heat capacity and thermal conductivity, which are listed in Table 1 and Table 2. The ice content θi is then substituted into the solid mechanics module as the criterion of frost heave so that the frost heave volume of soil is calculated. The input parameters include soil weight, elastic modulus, Poisson's ratio and frozen-heave factor, which are detailed in Table 1 and Table 3. On this basis, the simulation of pre-grouting of the soil surrounding pipe roof and the control on frost heave by opening the heating pipes can be realized through changing thermo-physical parameters and the frozen-heave factor of strata, and the working state of heating pipes.

3.1 Model establishment

The geological strata is divided into seven layers, and the groundwater level is 1 m below the ground surface. As shown in Figure 6, the model size is set to 150 m×50 m according to the area affected by tunnel construction. The soil around pipes jacking, except for 14#-21# pipes jacking at the bottom, is grouted to form a grouting reinforcement ring. The whole model is divided freely utilizing triangular meshes, and the minimum element size is set to 1 cm, so that a total of 69954 elements are established with the average mesh quality of 0.92.

3.2 Boundary conditions

The soil temperature considering the effect of water migration is calculated by the coefficient differential equation PDE module. The boundary temperature of ground surface is set to 21°C, and that of circular freezing pipes, irregular shape freezing pipes, heating pipes and empty pipes are -28°C, -24°C, 8°C and -10°C, respectively. Surface displacement is calculated by using solid mechanics module, in which the horizontal displacement of the lateral boundary is limited; the pipe jacking and the bottom are fixed, while the upper boundary is free.

3.3 Results analysis

In order to analyze the effect of frost heave control using different measures including heating pipes to limit frozen ring thickness, pre-grouting, and the combination of the pre-grouting and heating pipes, this paper simulates the surface displacement when three different measures are carried out respectively to control the frost heave of the soil. Figure 7 shows the simulation results of surface displacement after the soil is frozen for 30, 50, 70 days using three different measures.

It can be seen from Figure 7 that with the extension of freezing time, the thickness of the frozen ring is continuously enlarged and the surface displacement increases gradually and

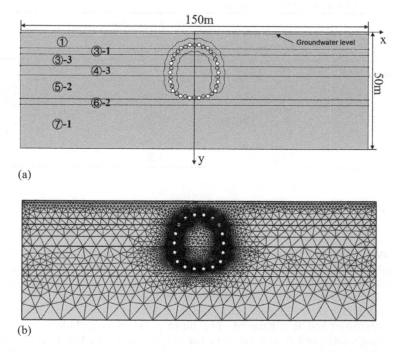

Figure 6. Schematic diagram of the finite element model. (a) the calculation model. (b) the mesh model

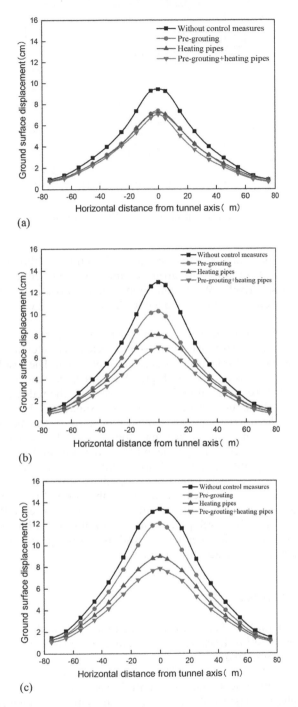

Figure 7. Ground surface displacement of different frost heave control measures. (a) 30d. (b) 50d. (c) 70d

reaches a maximum near the central axis of the tunnel. The maximum surface displacements without measures of frost heave control are 9.4, 12.9 and 13.5 cm respectively for different freezing time of 30, 50, 70 d. It is obvious that the curves of surface displacement with three control measures are below the one without any control measures. This indicates that all three control measures play an active role in inhibiting the frost heave. After freezing for 70 days,

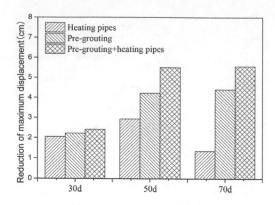

Figure 8.   Reduction of maximum surface displacement of different frost heave control measures.

the maximum surface displacements of four situations including combination of the pre-grouting and heating pipes, heating pipes, pre-grouting and absence of control measures are 13.5, 12.3, 9.0 and 7.7 cm respectively. When combining the pre-grouting and heating pipes, the maximum surface displacement reduces by 5.8 cm, showing the best effect on controlling frost heave of soil.

In order to find the effect of three measures on frost heave control intuitively, the maximum in displacement curve without frost heave control is selected as a reference value, and then the maximum of the other three displacement curves is subtracted. The displacement difference obtained is drawn into tables as is shown in Figure 8 for the later comparative analysis.

It can be seen from Figure 8 that after 30 days of freezing, the reduction of maximum surface displacement of the three measures is similar, approximately 2.2 cm. Besides, after freezing for 50 and 70 days, the maximum displacement reductions of the measure of pre-grouting is greater than that of the measure of heating pipes, which shows that grouting is critical to frost heave control by reducing the permeability of the soils and inhibiting the migration of moisture. The effect of grouting on frost heave control is more visible in the later stage of freezing. In addition, it can be easily found that regardless of freezing time, the maximum of displacement reduction shows up at the measure of combination of the pre-grouting and heating pipes, which means the surface displacement is the smallest when taking the measure of combination. This is because the development rate of frozen soil is still relatively fast when the measure of pre-grouting is used independently. And as a consequence, the volume of frozen soil expands and increases the surface displacement.

4   FIELD MEASUREMENT

After freezing for 30, 50 and 70 days with the frozen mode of double-pipe including circular freezing pipe and irregular shape freezing pipe, the surface displacement of field measurements and the simulations are compared as shown in Figure 9.

As shown in Figure 9, after freezing for 30, 50 and 70 days, the maximum surface displacements of field measurement show up near the central axis of the tunnel, which fixed at 6.7, 7.3 and 7.7 cm respectively. The results of the simulation are similar to the measured data, which is 7.0, 7.5 and 7.8 cm respectively. In addition, with the extension of freezing time, the growth of maximum surface displacement approaches to 0, indicating that the effect of frost heave control can keep well in the late freezing period. The reasons can be concluded in two aspects. One is that pre-grouting reduces the permeability of the soil and improving the mechanical properties. Meanwhile, the frozen-heave factor of the soil and the segregation frost heave caused by migration of water in the soil is reduced as well. The other is heating pipes reduce the development rate of frozen soils. Both of them are effective to control the surface displacement.

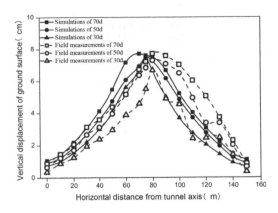

Figure 9. Comparison of ground surface displacement between measured and numerical results.

5 CONCLUSIONS

This paper takes the excavation section of Gongbei Tunnel as an example, simulate the displacement of soil during artificial freezing, compare and analyze the effect of three measures including pre-grouting, heating pipes and combination of pre-grouting and heating pipes on frost heave control. Through comparing the results of simulation and field measurements, the mechanism of frost heave control is explained from the distribution of surface displacement. The conclusions are as follows:

1. The results of simulation based on thermal-mechanical coupling theorem are in good agreement with the field measurements. It shows that the frost heave of the soil caused by artificial freezing is mainly composed of the segregation frost heave caused by the icing of migration water. The in-situ phase change frost heave caused by the icing of water existing in the soil accounts for a small proportion of total frost heave. The methods to control frost heave mainly include restraining the migration of water in the soil, reducing the frozen-heave factor of the soil and limiting the thickness of the frozen ring.

2. The results of surface displacement obtained from three control measures show that combination of the pre-grouting and heating pipes has the best effect on frost heave control. Pre-grouting is critical to frost heave control by reducing the permeability and frozen-heave factor of the soils and inhibiting the migration of moisture. The heating pipes are an effective supplement to frost heave control by limiting the development rate of frozen soils and the thickness of frozen ring.

REFERENCES

Bronfenbrener, L. & Bronfenbrener, R. 2010. Modeling frost heave in freezing soils. *Cold Regions Science & Technology* 61(1):43.

Harlan, R.L. 1973. Analysis of coupled heat-fluid transport in partially frozen soil. *Water Resources Research* 9(5): 1314.

He, P. et al. 2001. The progress of study on heat and mass transfer in freezing soils. *Journal of Glaciology and Geocryology* 23(1): 92.

Hu, X.D. 2009. Laboratory research on properties of frost heave and thaw settlement of cement-improved Shanghai's grey-yellow sand. *Journal of China Coal Society* 34(3): 334.

Luo, Y.B. et al. 2017. Stability analysis of super-large-section tunnel in loess ground considering water infiltration caused by irrigation. *Environmental Earth Sciences* 76(22): 763.

Shi, P.X. et al. 2017. Experimental and analytical study of jacking load during microtunneling gongbei tunnel pipe roof. *Journal of Geotechnical & Geoenvironmental Engineering* 144(1): 05017006.

Zhang, M.Q. et al. 2006. Evaluation technique of grouting effect and ITS application to engineering. *Chinese Journal of Rock Mechanics and Engineering* 25(z2): 3909.

Zhang, M.Y. et al. 2017. Effect of temperature gradients on the frost heave of a saturated silty clay with a water supply. *Journal of Cold Regions Engineering* 31(4): 04017011.

Zhang, Q.S. et al. 2015. Model test of grouting strengthening mechanism for fault gouge of tunnel. *Chinese Journal of Rock Mechanics and Engineering* 34(5): 924.

Zhou, J.S. et al. 2006. Experimental research on controlling frost heave of artificial frozen soil with intermission freezing method. *Journal of China University of Mining & Technology* 35(6): 708.

Author Index